Germ Cell Tumours

Germ Cell Tumours

Edited by

C.K. Anderson
W.G. Jones
A. Milford Ward

Taylor & Francis
London

Alan R. Liss
New York

First published 1981 by Taylor & Francis Ltd, London and
Alan R. Liss, Inc., 150 Fifth Avenue, New York, NY 10011

ISBN 0-8451-3001-3

Library of Congress Catalog Card Number 81-83947

Printed and bound in Great Britain by Taylor & Francis (Printers) Ltd,
Rankine Road, Basingstoke, Hampshire RG24 0PR.

CONTENTS PAGE

PREFACE

This book contains the proceedings of a Conference on the subject of Germ Cell Tumours held at the Devonshire Hall of Residence, the University of Leeds, between 24th and 26th March, 1981. The Conference was held under the auspices of Leeds University and the Yorkshire Regional Cancer Organisation. Members of our respective departments (in particular Dr C K Anderson and Dr W G Jones) were heavily involved in the organisation of the meeting as well as presenting papers. As far as we know this is the first occasion such a conference has been held in the United Kingdom.

The primary objective of the Conference was to bring together interested individuals from a wide variety of clinical, para-clinical and laboratory disciplines to promote the exchange of information and ideas and to foster future co-operation wherever possible. This objective, we believe, was fulfilled and much valuable new information was presented which is contained in this publication. One clear message to emerge from the presentations was the considerable progress that has been made during the past few years in understanding the behaviour and treatment of these malignancies, (as illustrated in practically every section of this publication). As a result, the outlook for patients with these relatively rare tumours has improved dramatically and has swung from the dejected pessimism of former years to the bright optimism of the present. A further objective of the Conference was to publish the proceedings at the earliest opportunity and both editors and publishers are to be congratulated on this achievement.

We must also congratulate members of the Testicular Tumour Sub-Group of the Yorkshire Urological Cancer Research Group (a group working within the Yorkshire Regional Cancer Organisation) for conceiving and organising this Conference, which was obviously a most successful scientific venture. The hard work and efforts of all contributors and the excellence of their presentations are also greatly appreciated.

C C Bird,
Professor of Pathology

C A F Joslin,
Professor of Radiotherapy

The University of Leeds
July 1981

FOREWORD

Significant advances have been made in the field of germ cell tumours in the last few years. As a result, the Testicular Tumour Sub-Group of the Yorkshire Urological Cancer Research Group (YUCRG), a working body of the Yorkshire Regional Cancer Organisation (YRCO), decided that the time was ripe to hold a conference, bringing together workers from the many fields of scientific endeavour involved, to promote a useful interchange of knowledge and ideas. Hopefully, a result of the Conference will be further co-operation between individuals and groups with the ultimate aim of improving the management of patients.

The Conference was held in a very friendly atmosphere, in the pleasant surroundings of Devonshire Hall. The Organising Committee (whose constitution appears on pages xv and xvi) acknowledge the assistance of the Offices of the Registrar and Bursar of the University of Leeds for allowing the conference to be held in such a setting. The ministrations of the Domestic Bursar of Devonshire Hall and his staff are gratefully acknowledged since they contributed considerably to the smooth running and success of the Conference. The Organising Committee has to thank its many friends and colleagues in the YUCRG, the MRC Testicular Cancer Working Party and the EORTC Urological Group for helpful advice and for their contributions to the Conference.

The organisation of the Conference was undertaken by the Secretariat of the Yorkshire Regional Cancer Organisation. They worked extremely hard on the preparation and running of the meeting and subsequently to help produce this volume.

Everyone involved with the Conference would like to thank those organisations detailed on page xvii, without whose financial support and encouragement the Conference could not have taken place. Thanks are due to the local representatives of the various pharmaceutical companies involved for their help during the Conference and also to the Leeds Medical Sciences Club for the loan of the lecture timer (with a fearsome buzzer, for those who were too loquacious) which proved invaluable in keeping lectures to time, allowing a rather full programme to run to schedule.

One of the main aims of the Conference was to produce a permanent record of the proceedings as speedily as possible after the close of the meeting, and the editors wish to thank the publishers, Messrs. Taylor and Francis and in particular, Dr J Cheney, Senior Scientific Editor, for all the help and encouragement given to enable fulfilment of this aim. Thanks are also due to Mr Peter Woodward of Lundbeck Ltd., for his help and advice regarding the publication.

The editors also wish to record their grateful thanks to Mr John Scullion, Senior Medical Photographer and Mr Lyndon Cochrane, Graphic Designer, The Department of Medical Illustration, YRCO, not only for producing the majority of the figures in this book, but also for the preparation of many of the slides shown at the Conference. Thanks are due to the various secretaries and typists at the YRCO, The Royal Hallamshire Hospital Sheffield, and Cookridge Hospital, Leeds, who were involved in the preparation of the manuscript, but in particular to Vivienne Broadbent for organising and undertaking the work of typing the manuscript onto a'word processor. The editors also wish to acknowledge the various contributions made to the conference and the publication by Mr Peter Corbett. The editors wholeheartedly thank each of the contributors to this volume, from those who presented valuable new information in formal presentations, to those who contributed to the discussion sessions, and also to those who conducted such sessions (detailed on the section pages) in particular Professor John Blandy for his mastery of this art.

Finally we should, indeed must, acknowledge the enormous contribution made, not only to the Conference, but also the the production of this publication, by Miss Elaine Walker, Project Secretary at the YRCO. Her hard work and involvement in every aspect of this venture is to be admired and we are greatly indebted to her for this exceptional commitment.

C K Anderson
W G Jones
A Milford Ward

Leeds and Sheffield
July 1981

Yorkshire Regional Cancer Organisation

and the

University of Leeds

Germ Cell Tumour Conference

Devonshire Hall

Leeds, 24-26 March 1981

CONFERENCE CHAIRMAN

Dr C K Anderson,
Senior Lecturer & Honorary Consultant in
Urological Pathology, Department of Pathology,
University of Leeds, LS2 9JT

CONFERENCE SECRETARY

Dr W G Jones,
Lecturer & Honorary Consultant in Radiotherapy,
University Department of Radiotherapy,
Cookridge Hospital,
Leeds LS16 6QB

YRCO PROJECT SECRETARY

Miss E Walker,
Yorkshire Regional Cancer Organisation,
Cookridge Hospital,
Leeds LS16 6QB

ORGANISING COMMITTEE

Mr I Appleyard,
Consultant Urological Surgeon,
Airedale General Hospital,
Keighley.

Dr J A Child,
Consultant Haematologist,
Leeds General Infirmary,
Leeds.

Mr P J Corbett,
Lecturer,
University Department of Radiotherapy,
Cookridge Hospital,
Leeds.

Dr A Milford Ward,
Director, Supraregional Protein Reference Unit,
Royal Hallamshire Hospital,
Sheffield.

Mr B Richards,
Consultant Urological Surgeon,
York District Hospital,
York.

Mr M R G Robinson,
Consultant Urological Surgeon,
Pontefract General Infirmary,
Pontefract.

ACKNOWLEDGEMENTS

The organisers of the Germ Cell Tumour Conference, Leeds, wish to place on record their appreciation of the support given by the following:

BRISTOL-MYERS CO. LTD. (MEAD JOHNSON).
CYANAMID OF G.B. LTD. (LEDERLE LABORATORIES).
ELI LILLY & CO. LTD.
FARMITALIA CARLO ERBA, (MONTEDISON GROUP).
LUNDBECK LTD.
UNIVERSITY OF LEEDS.
UNIVERSITY OF LEEDS DEPARTMENT OF RADIOTHERAPY.
YORKSHIRE CANCER RESEARCH CAMPAIGN.
YORKSHIRE REGIONAL CANCER ORGANISATION.

GERM CELL TUMOUR CONFERENCE, LEEDS, 24th - 26th MARCH 1981

LIST OF PARTICIPANTS

Mr R S Adib,
Wakefield

Professor A Akdas,
Ankara, Turkey

Mrs R Allen,
Sheffield

Dr P L Amlot,
London

Dr J Amuesi,
Leeds

Dr C K Anderson,
Leeds

Mr I Appleyard,
Keighley

Dr D Ash,
Leeds

Dr D G B Ashley,
Swansea

Dr A Barrett,
Sutton

Dr I D Barrett,
Leeds

Dr R Basting,
Donau, Germany

Mrs G Bates,
Sheffield

Dr R H J Begent,
London

Dr J O W Beilby
London

Mr L L Beynon,
Edinburgh

Dr B Bergman,
Sheffield

Dr C Biedermann,
Basel Switzerland

Dr L A Birchall,
Coventry

Professor C C Bird,
Leeds

Miss J Blackburn,
Birmingham

Professor J P Blandy,
London

Dr A R Bradwell,
Birmingham

Dr L A Brown,
Leeds

Dr L J R Brown,
Leeds

Dr P K Buamah,
Newcastle upon Tyne

Dr M Burrow,
Leeds

Dr M Carr,
Leeds

Dr S C Cartwright,
Leeds

Dr J A Child,
Leeds

Mr P B Clarke,
Leeds

Dr P H Cole,
Northampton

Dr C D Collins,
London

Dr F Contreras,
Madrid, Spain

Mr P J Corbett,
Leeds

Dr A J Coup,
Barnsley

Dr D Daponte,
Leeds

Dr M J Evans,
Cambridge

Dr R G B Evans,
Newcastle upon Tyne

Dr S Fairweather,
Birmingham

Dr S Fletcher,
Edinburgh

Dr P Flogren,
Lund, Sweden

Professor H Fox,
Manchester

Dr H G Frank,
Leeds

Dr C F Graham,
Oxford

Miss A Grail,
Sheffield

Dr J A Green,
Southampton

Mr I Grigor,
Leeds

Dr K M Grigor,
Edinburgh

Dr R Hadden,
Exeter

Mr P A Hamilton-Stewart,
Bradford

Mr B W Hancock,
Sheffield

Mr W H Hendry,
London

Dr E Heyderman,
London

Dr B Hogan,
London

Dr M Holmes,
Sheffield

Dr H F Hope-Stone,
London

Mr A L Houghton,
Leeds

Dr G K Jacobsen,
Herlev, Denmark

Dr K W James,
Cardiff

Dr W G Jones,
Leeds

Dr H Joos,
Salzburg, Austria

Professor C A Joslin,
Leeds

Dr M Karadem,
Leeds

Dr S B Kaye,
Glasgow

Dr H S Kellett,
Leeds

Dr A H Laing,
Oxford

Dr A R Lyons,
County Antrim

Dr R A J McIlhinney,
Sutton

Mr C Maddox,
Leeds

Dr J R Mann,
Birmingham

Mr J L Marcus,
Coventry

Dr J R W Masters,
London

Dr A Milford Ward,
Sheffield

Dr V Moshakis,
Sutton

Dr T Muller,
Leeds

Dr A J Munro,
Edinburgh

Dr E S Newlands,
London

Dr B Norgaard-Pedersen,
Sonderborg, Denmark

Dr R D T Oliver,
London

Dr A Olsson,
Lund, Sweden

Dr R Ozols,
Bethesda, USA

Dr N S Panesar,
Leeds

Dr C Parkinson,
London

Dr R G Pearcey,
Leeds

Professor M J Peckham,
Sutton

Dr G Pizzocaro,
Milan, Italy

Dr K Pollard,
Leeds

Dr H Prichard,
Merseyside

Dr J Pritchard,
London

Dr G Read,
Manchester

Dr F Reiss,
Strasbourg, France

Mr B Richards,
York

Miss E Robertson,
Cambridge

Mr M R G Robinson,
Pontefract

Mr R I Rothwell,
Leeds

Dr E D Rubery,
Cambridge

Dr R Sandland,
London

Mr I V Scott,
Derby

Dr F Searle,
London

Dr R Silvestrini,
Milan, Italy

Mr P H Smith,
Leeds

Mr P J B Smith,
Bristol

Dr P J Spaander,
Leiden, Holland

Dr G Stoter,
Amsterdam, Holland

Dr U Studer,
Bern, Switzerland

Dr C A Sunderland,
Oxford

Dr C J Tyrrell,
Plymouth

Professor L M van Putten,
Rijswijk, Holland

Dr F E von Eyben,
Malmo, Sweden

Professor L Wahlqvist,
Umea, Sweden

Dr J A H Waterhouse,
Birmingham

Dr T E Wheldon,
London

Dr C Williams,
Southampton

Mr R E Williams,
Leeds

Dr E Wiltshaw,
London

Dr H Yu,
Leeds

INTRODUCTION

Chairman of Session
Dr C.K. Anderson

GERM CELL TUMOURS - THE CURRENT STATE OF THE ART AND PROBLEMS IN CLINICAL MANAGEMENT

W.G. Jones
Lecturer in Radiotherapy and
Honorary Consultant Radiotherapist
University Department of Radiotherapy
Cookridge Hospital, Leeds LS16 6QB

Germ Cell Tumours are very rare. They account for only a fraction of 1% of the total incidence of cancer in the population. Germ Cell Tumours exhibit a great variety of phenomena which excite interest in a very wide variety of specialised scientific and clinical disciplines, (Figure 1).

Each of these broadly defined specialities finds different facets of these diseases to concentrate on. Obviously the main interests of the clinical and pathology groups are directed towards the proper, hopefully early, diagnosis of the patient, thorough investigation, application of the correct therapy and then monitoring of the effects of therapy

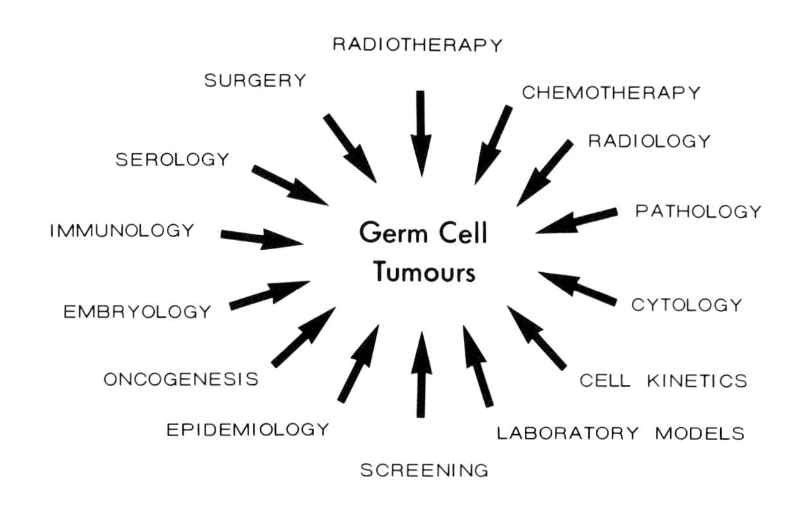

FIGURE 1 - Interests in Germ Cell Tumours

and follow-up. The interests of the laboratory scientists are perhaps somewhat different since these tumours provide them with a host of fascinating phenomena worthy of intensive investigation. Germ cell tumours themselves are used as laboratory models to study oncogenesis, embryology etc. and because of their somewhat primitive nature they provide the protein chemist, the immunologist and serologist with excellent material. The main aim of this Conference is to promote the exchange of information between these different disciplines, highlighting not only progress but also problems. It is hoped that this will lead to a better understanding, not only of the diseases themselves, but also of the specific problems encountered by groups of researchers in other disciplines. As an end result of this, perhaps even better clinical management of patients will result.

Improvements in the results of treatment seen over the last decade or so have been due to many factors. Perhaps the most important has been the development of effective cytotoxic chemotherapy (7,24) and the development of ways to deal with the side effects of these very toxic combinations of drugs. It can never be stressed too often that these regimes have potentially lethal effects. Improved machinery and techniques have been developed in radiotherapy and similar advances have been seen on the surgical front. Perhaps one important advance has been the centralisation of patients to centres large enough to carry out sophisticated evaluation of the extent of disease, to gain expertise in patient and treatment management and possibly to co-operate in clinical trials. The ability to detect and accurately define the extent of metastatic disease using diagnostic imaging techniques such as Computerised Tomographic Scanning, ultrasonography and, perhaps in the future the use of radio-immunolocalisation, will have a great impact on improving clinical management. Serology will be referred to later, and will be dealt with at some length. A better appreciation of the biology and behaviour of these tumours by clinicians is bound to have a beneficial impact on clinical management.

Germ Cell Tumours belong to a group of malignant diseases in which the multidisciplinary approach is well established with involvement of surgeon, radiotherapist and chemotherapist. However, it has been the development over the last 6-8 years of effective, but highly toxic, chemotherapy that has been of the greatest importance in not only improving the outlook for individual patients but perhaps acting as the catalyst for stimulation of scientific endeavour on many fronts. The development of this improved chemotherapy has been well described by Williams, 1977 (26). However, we must not be complacent, and realise that it takes time for the news of improved results and techniques to be disseminated. Hill, 1978 (13) showed that over one-third of testicular swellings were initially misdiagnosed. Even today we see occasionally patients who have had their testicular swellings explored through a trans-scrotal approach which opens up potential alternative pathways of lymphatic dissemination. This is obviously a matter of educating surgeons. Germ Cell Tumours occur most commonly in the testis, less commonly in the ovary and less commonly still in extragonadal sites (8) including the anterior mediastinum, the retroperitoneum, particularly in the renal area, the pre-sacral region, the pineal and occasionally the parotid (23), (Figure 2).

The pathogenesis of extragonadal germ cell tumours is thought to be due to a failure of some primordial germ cells to migrate completely to the gonad region or to the dislocation of totipotential cells during embryogenesis. At present patients who present with extragonadal tumours have a worse prognosis than those patients with testicular lesions

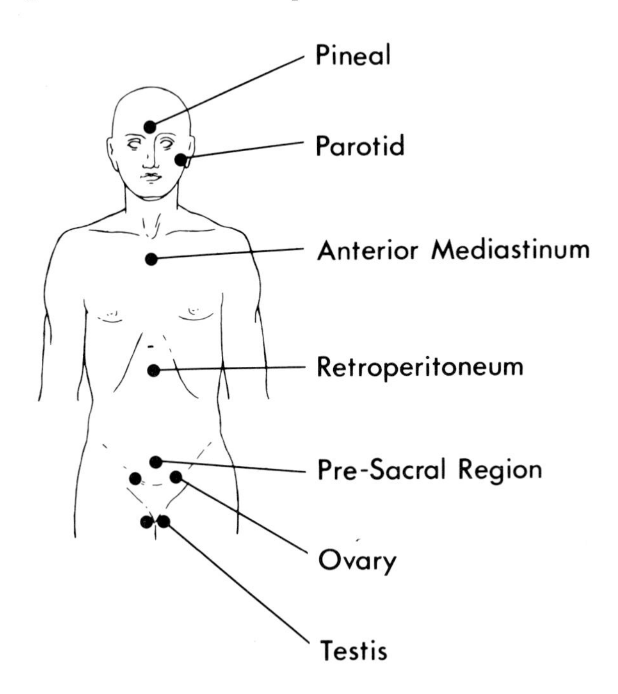

FIGURE 2 - Germ Cell Tumours: Site of Occurrence

(8,23). This may be due to a delay in making the diagnosis at an earlier stage, later clinical presentation for example in a mediastinal tumour, and is probably a function of the bulk of the tumour and its amenability to radical surgery.

There is a weight of evidence, particularly from researchers in Denmark, of an increased incidence of testicular tumours in maldescended testes (16). Whether tumour initiation is a direct effect of the maldescent, perhaps increased temperature, or whether some other agent exists which not only causes the maldescent but also predisposes to neoplasia is an interesting quandary.

When malignant change takes place the morphology, pathology and behaviour of the tumour and its possible metastases depend upon the line of differentiation or lines of differentiation in mixed tumours, whether this is towards tissue of embryonic or extra-embryonic structure. Thus in mixed tumours it is possible that the metastases may have different cellular compositions from each other and the primary tumour and it is thus possible that different deposits will behave differently even in the same metastatic site, such as lung, in the same patient. They may show a greater or lesser aggressive behaviour compared with each other. It is interesting to speculate what event initiates this malignant change.

It is probably pertinent to mention here a few of the peculiar phenomena that germ cell tumours exhibit clinically. As with certain other tumours spontaneous regression of primary or metastatic lesions may occur rarely with germ cell tumours. Franklin, 1977, reported six cases of spontaneous regression in patients with testicular tumours occuring between 1961 and 1974 (10). A case of spontaneous regression of metastatic testicular carcinoma in a patient with bilateral sequential testicular tumour has recently been reported from San Diego (19). This phenomenon must excite the interest of immunologists and other therapists.

Maturation to benign disease, first reported by Merrin et al, 1975 (18), now begs the following questions: Is this a natural phenomenon, with malignant tissue maturing or differentiating into benign tumour? Is this phenomenon hastened by chemotherapy? Perhaps, most importantly, is de-differentiation possible after this change to the mature state? If so, this will have a major impact since approximately one third of patients with residual masses after chemotherapy have "benign" teratoma. What is the mechanism involved in this differentiation? Perhaps the chemotherapy destroys the malignant cells leaving intact the mature, benign teratomatous elements, or perhaps the kinetics of the tumour are altered causing the totipotent malignant germ cells to form benign mature teratoma.

Another peculiarity recently reported is that of acceleration of disease after debulking or cytoreductive surgery. Lange et al (17) recently reported 8 cases of sudden dramatic exacerbation of disease after such surgery. Whether this is due to more aggressive cells being left behind, an alteration in cellular kinetics or an alteration in the immune response caused by the surgery or anaesthesia is open to question.

Malignant germ cell tumours commonly produce specific oncofoetal proteins referred to as marker substances which are not only detectable in serum but are accurately measurable and give a good correlation with the body's tumour burden (15). Techniques of immunocytochemistry may be a useful tool to the histopathologist aiding diagnosis and detection of different elements within the tumour (15,12). The commonly produced markers are alphafetoprotein and the beta sub-unit of human chorionic gonadotrophin (AFP and beta HCG). The amounts of these substances detectable in the serum depends on the type of tumour or mixture of tumour types. These markers are very helpful in assessing response to clinical therapy and in detecting relapse. That is, of course, if they were present at the time of initial diagnosis and were, therefore, being produced by the primary tumour. The need for pre-operative specimens should be stressed. Not all germ cell tumours are marker producing, which makes the search for new markers worthwhile. Recent reports from the Charing Cross group positively correlate the prognosis of patients to the highest level of serum tumour markers seen (11). Germ Cell Tumours may exert other endocrine effects on their host as described by Fox & Reeve, 1979, (9). The production of beta HCG by a small percentage, in the order of 5%, of seminomas is a puzzling phenomenon. The raising of specific antisera to AFP and beta HCG is leading to a very exciting era in nuclear medicine. Radiolabelling of these antisera causes localisation of isotope to those areas of the subject carrying tumour cells bearing these oncofoetal proteins on their cell surfaces, or in the cytoplasm, when the antiserum is injected (4). Radioimmunolocalisation has great potential in the field of improving diagnostic accuracy and may, perhaps, have therapeutic implications in the future. The "biology" of these tumours is facinating and laboratory models will hopefully help unravel some of our present problems.

Recent surveys in Scandanavia and the UK suggest that in

these geographical areas there has been a real increase in the incidence of testicular germ cell tumours over the last few years (5,6,13). Data from the retrospective survey of the information in the Yorkshire Cancer Registry appears later in this volume. This increasing incidence prompts further thought about dysgenesis and oncogenesis. The epidemiology of these tumours may provide clues to their causation (22). I just wonder that, with the usual latent period, whether radioactive fall-out from nuclear explosions and the advent of tight fitting underclothes increasing the intrascrotal temperature are in any way contributing factors. Certainly as we have seen maldescent is a causative factor. A leading article in the British Medical Journal, 1980 (1), highlighted the need for early detection of testicular tumours and pointed to possible ways in which this could be done, such as the use of self examination programmes, screening serology, and diagnostic ultrasound for suspicious masses. Dr Skakkebaek in Copenhagen last year suggested needle biopsy, particularly in infertile men, and has described a disturbingly high frequency of atypical germ cells akin to the carcinoma in situ seen at other sites (2,3). Certainly self examination and the use of diagnostic ultrasound seem very promising. However we must also educate doctors to recognise that early detection is essential and that any testicular swelling may be tumour, and remember that in the UK a large proportion of such swellings are dealt with by general surgeons. For extragonadal germ cell tumours earlier diagnosis is perhaps a little more difficult, but certainly we are seeing this condition being included more frequently in the list of differential diagnoses and serological estimations are being performed more frequently.

Unfortunately pathological classifications have been developed from conflicting philosophies in the past. In the United Kingdom and Europe the classification of Collins & Pugh and its derivatives are used. In North America the Armed Forces Institute of Pathology classification is used. This difference in philosophy and nomenclature has confounded useful interchange of results of treatment etc. However, there has been progress towards a common understanding on both sides of the Atlantic and there is now a great degree of equivalence between systems as shown in Table 1.

An even worse situation exists with anatomical staging, seemingly each institution or group devising its own system.

Table 2 shows the Royal Marsden system (21) not only because it is accepted by the Medical Research Council's working group on testicular tumours, but mainly because it illustrates the realisation that bulky disease, and disease

COMPARISON OF NOMENCLATURE OF CERTAIN TESTICULAR TUMOURS

("Germ Cell" Tumours)

Pugh (1976)	Mostofi and Price (1973)	World Health Organisation (1977)
Seminoma Typical Spermatocytic	Seminoma Typical Spermatocytic Anaplastic	Seminoma Typical Spermatocytic Anaplastic
Teratoma Differentiated (TD)	Teratoma Mature Immature	Teratoma Mature Immature
Malignant Teratoma Intermediate (MTI)	Embryonal Carcinoma with Teratoma (Teratocarcinoma)	Embryonal Carcinoma with Teratoma (Teratocarcinoma)
Malignant Teratoma Undifferentiated (MTU)	Embryonal Carcinoma Adult Polyembryoma	Embryonal Carcinoma
Malignant Teratoma Trophoblastic (MTT)	Choriocarcinoma	Choriocarcinoma
Yolk-Sac Tumour (YST)	Embryonal Carcinoma Infantile (Juvenile)	Yolk-Sac Tumour
Combined Tumour (CT) Seminoma and Teratoma	Embryonal Carcinoma with Teratoma, with or without Other Elements	Tumours of more than one Histological Type Other Combinations (The components should be specified)

TABLE 1 - Comparison of Nomenclature of Certain Testicular (germ cell) Tumours

TESTICULAR TUMOURS

Staging classification

(Peckham, et al., Royal Marsden, 1979)

STAGE	
I	Radiographically no evidence of metastases
II	Radiographically metastases confined to abdominal nodes.
	A. Max. diameter of metastases < 2 cm
	B. Max. diameter of metastases 2-5 cm
	C. Max. diameter of metastases > 5 cm
III	Involvement of supra and infradiaphragmatic lymph nodes. No extralymphatic metastases.
	Abdominal status:
	A, B, C as for Stage II
IV	Extralymphatic metastases. Suffixes:
	O - Nodes radiographically negative, A, B, C as for Stage II
	Lung status:
	L_1 <3 metastases
	L_2 multiple < 2 cm max. diameter
	L_3 multiple > 2 cm diameter
	Liver status:
	H_+ = liver involvement

TABLE 2 - Staging Classification: Testicular Tumours

at certain sites, for example, the liver, makes for a worse prognosis, and is thus taken into account in this way.

The use of modern diagnostic imaging with ultrasound or with computerised tomographic scanning is improving the accuracy of staging (14,25,27). The development of not only effective first line chemotherapy (24) but now also effective salvage chemotherapy with combinations including the agent VP16-213 has been the major step forwards (28). The use of markers, improved accuracy of staging and an improved understanding of the conditions, particularly the gathering of information from clinical trials, together with improved therapies, and more logical use of the order in which they are used, and improved treatment strategies, must all contribute to better patient management, (Figure 3).

In the face of possible cures in these patients it is imperative that we study the acute, medium and long term effects of the therapies. We must be aware of potential hazards when using such toxic treatments, for example, the increased risk of oncogenesis (20). Since these diseases commonly strike at relatively young individuals we must strive to achieve as normal a life as possible for those whom we expect to adopt a normal life expectancy after treatment. Hopefully this conference will achieve its aims by examining a number of issues raised here.

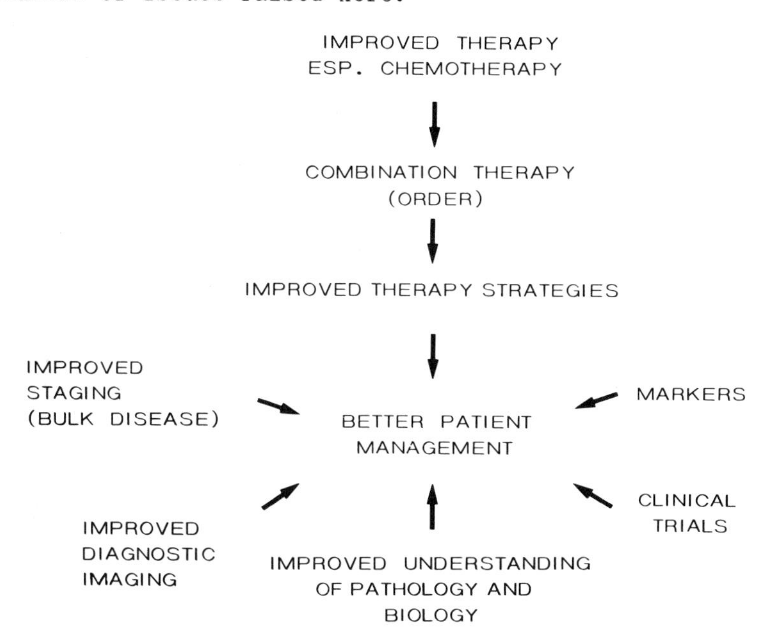

FIGURE 3 - Improved Management of Germ Cell Tumours

REFERENCES

1. Anon., 1980. Earlier diagnosis of testicular tumours. British Medical Journal, i, 961.

2. Anon., 1980. Testicular biopsy for early detection of testicular tumour. British Medical Journal, i, 426-427.

3. Anon., 1980. Early testicular cancer. Lancet, ii, 1175.

4. Begent, R.H.J., Searle, F., Stanway, G., Jewkes, R.F., Jones, B.E., Vernon, P., and Bagshawe, K.D., 1980. Radioimmunolocalisation of tumours by external scintigraphy after administration of ^{131}I to human chorionic gonadotrophin: preliminary communication. Journal of the Royal Society of Medicine, 73, 624-630.

5. Clemmesen, J., 1969. Statistical studies in the aetiology of malignant neoplasms. No 3 - Testis Cancer. Basic Tables, Denmark, 1958-62, Acta Pathologica et Microbiologica Scandanavia, 209, suppl. p xv.

6. Corbett, P.J., and Jones, W.G., 1979. Improved management of testicular tumours. British Medical Journal, i, 1143-1144.

7. Einhorn, L.H., and Donohue, J.P., 1977. Cisdiammine-dichloroplatinum, vinblastine, and bleomycin combination chemotherapy in disseminated testicular cancer. Annals of Internal Medicine, 87, 293-298.

8. Feun, L.G., Samson, M.K., and Stephens, R.L., 1980. Vinblastine (VLB), Bleomycin (BLEO), cis-Diammine-dichloroplatinum (DDP) in disseminated extragonadal germ cell tumours (a Southwest Oncology Group study). Cancer, 45, 2543-2549.

9. Fox, H. and Reeve, N.L., 1979. Endocrine effects of testicular neoplasms. Investigative and Cell Pathology, 2, 63-73.

10. Franklin, C.I.V., 1977. Spontaneous regression of metastases from testicular tumours - a report of six cases from one centre. Clinical Radiology, 28, 499-502.

11. Germa-Lluch, J.R., Begent, R.H.J. and Bagshaw, K.D., 1980. Tumour marker levels and prognosis in malignant teratoma of the testis. British Journal of Cancer, 42, 850-855.

12. Heyderman, E., 1980. The role of immunocytochemistry in tumour pathology: a review. Journal of Royal Society of Medicine, 73, 655-658.

13. Hill, J.T., 1978. Misdiagnosis of testicular tumours. Journal of the Royal Society of Medicine, 77, 737-740.

14. Husband, J.E., Peckham, M.J. and MacDonald, J.S., 1980. The role of abdominal computed tomography in the management of testicular tumours. Computerised Tomography, 4, 1-16.

15. Javadpour, N., 1980. The role of biologic tumour markers in testicular cancer. Cancer, 45, 1755-1761.

16. Krabbe, S., Skakkebaek, N.E., Berthelsen, J.G., Eyben, F.V., Volsted, P., Mauritzen, K., Eldrup, J., and Nielsen, A.H., 1979. High incidence of undetected neoplasia in maldescended testes. Lancet, i, 99-100.

17. Lange, P.H., Hekmat, K., Bosl, G., Kennedy, B.J., and Fraley, E., 1980. Accelerated growth of testicular cancer after cytoreductive surgery. Cancer, 45, 1498-1506.

18. Merrin, C., Baumgartner, G. and Wijsman, Z., 1975. Benign transformation of testicular carcinoma by chemotherapy. Lancet, i, 43-44.

19. Much, J.R., Greco, C.M., and Green, M.R., 1980. Spontaneous regression of metastatic testicular carcinoma in a patient with bilateral sequential testicular tumour. Cancer, 45, 2908-2912.

20. Nefzger, M.D., and Mostofi, F.K., 1972. Survival after surgery for germinal malignancies of the testis - part II - effects of surgery and radiation therapy. Cancer, 30, 1233-1240.

21. Peckham, M.J., McElwain, T.J., Barrett, A., and Hendry, W.F., 1979. Combined management of teratoma of the testis. Lancet, ii, 267-269.

22. Schottenfeld, D., Warshauer, M.E., Sherlock, S., Zauber, Leder, M., and Payne, R., 1980. The epidemiology of testicular cancer in young adults, American Journal of Epidemiology. 112, 232-246.

23. Sinniah, D., Prathap, K., and Somasundram, K., 1980. Teratoma in infancy and childhood, a ten year review at the University Hospital, Kuala Lumpar. Cancer, 46, 630-632.

24. Stoter, G., Sleijfer, D.T., Vendrik, C.P.J., Schraffordt Koops, H., Struyvenberg, A., Van Oosterom, A.T., Brouwers, T.M., and Pinedo, H.M., 1979. Combination chemotherapy with cis-diamminedichloroplatinum, vinblastine and bleomycin in advanced testicular non seminoma. Lancet, i, 941-945.

25. Tyrrell, C.J., Cosgrove, D.O., McCready, V.R., and Peckham, M.J., 1977. The role of ultrasound in the assessment and treatment of abdominal metastases from testicular tumours. Clinical Radiology, 28, 475-481.

26. Williams, C.J., 1977. Current dilemmas in the management of non-seminatous germ cell tumours of the testis. Cancer Treatment Reviews, 4, 275-297.

27. Williams, R.D., Feinberg, S.B., Knight, L.C., and Fraley, E.E., 1980. Abdominal staging of testicular tumours using ultrasonography and computed tomography. Journal of Urology, 123, 872-875.

28. Williams, S.D., Einhorn, L.H., Greco, F.A., Oldham, R., and Fletcher, R., 1980. VP16-213 salvage therapy for refractory germinal neoplasms. Cancer, 46, 2154-2158.

EMBRYOGENESIS

Chairman of Session
Dr C.K. Anderson

INITIATION OF MOUSE TERATOMAS

C.F. Graham
Department of Zoology, University of Oxford,
South Parks Road, Oxford

In a peculiar sense the analysis of mouse teratomas in the 1970's has been one of an academic success story in the sense that a large number of new facts have been discovered, a large number of biologists have been employed and a large number of grants have been successfully awarded. To set against this success story we must set a failure story which is the failure of the biologists, who have so far exploited the mouse system, to extend their work into the field of human biology. Most of us are now trying to isolate stem cells from humans but we are all short of primary human testicular and ovarian teratomas and there are now a number of workers who would travel many miles to pick up material direct from the operating theatre. If that is not possible, we would invite you to freeze it down so that we can collect it subsequently. So, one of the explanations for our inabiliby to say definite things about the cell biology of the human teratoma is simply that we do not have the stem cells of such tumours in culture. This is in contrast to the mouse system where, following the pioneering work of Martin Evans and Gail Martin, it has been shown that it is possible to clone stem cells of the mouse teratoma and that these cells as individuals can give rise to a multitude of cell types and can differentiate into the full range of cell types that are found in adult mice. Further, during this differentiation step such cloned cells can be shown to prove that the individual malignant cells of the teratoma can differentiate and become non-malignant in the sense that they are no longer transplantable and all such work has depended upon the ability to manipulate cells in culture.

The second thing I want to say about the difference between mouse and human material is that in the mouse it has been possible completely to circumvent the problem of studying rare, spontaneous tumours and to a large extent the biologists have worked with tumours obtained by inducing embryos to produce teratocarcinomas by taking them out of the uterus and putting them into extra-uterine sites. So what I would like to do in the rest of this presentation is first to describe what little is known about the origin of the spontaneous tumours in mice and I will be talking about testicular and ovarian tumours and then go on and talk about work that has been done on a much more widely used system, that of inducing tumours by transplanting embryos to extra-uterine sites. Right at the beginning can I say that whilst the biologists working on mice always refer to the stem cells of the mouse tumour as embryonal carcinoma we have no confidence at all that that name reflects an identicality with the embryonal carcinoma cells in human tumours. It is simply a usage which is applied to the mouse by analogy with the human, but that analogy is probably very weak and subsequently I will try and call them prinicpally the stem cell of the mouse teratocarcinoma to avoid difficulties.

First of all I will discuss the spontaneous testicular tumours of mice. What I want to do is to show that from genetics we can say that it is highly probable that the spontaneous testicular tumours, the spontaneous ovarian tumours, the embryo induced tumours and the gonad induced tumours may well have completely different pathologies involved in their origin. Now the principle way in which this has been shown has been by studying the genetics of these spontaneous tumours and first of all I am going to talk about the testicular ones. Now in mice, as in humans, testicular tumours are relatively rare. In fact they are very much rarer in mice in the sense that, except in one strain, only three cases of such tumours have ever been reported. The strain in question is strain 129J, and this strain is exceptional in the sense that as Leroy Stephens found, it produces a 1% incidence of such tumours. Leroy Stephens also noticed that if 129J had placed on its background a gene producing infertility in the homozygous state then that gene increased the frequency of spontaneous tumours to 5% and then a further mutation or event in the sub-line of 129J giving rise to 129J Sv ter increases the incidence to 30%. So there is a strong genetic element in the formation of spontaneous testicular tumours. Now it is of some interest that Leroy Stephens went on to show that spontaneous testicular tumours were in no way due to the uterus in which the embryo developed and I mention this fact because you may know a recent paper in the Lancet which shows that amongst the siblings of people who develop testicular teratomas, there is quite a high incidence of foetal defects which might suggest that in the human, teratomas develop in patients of mothers who have uteri which are poor at supporting the growth and development of normal foetuses. There is absolutely no evidence that the genetic affect on teratoma incidence in mice is mediated by such a defective uterus, or defective pregnancy, in fact quite the reverse. What Leroy Stephens showed by transplanting ovaries from the 129J strain to mice of other strains which did not have spontaneous tumours was that it was the ovary with the germ cells which carried the defect, or gene, which led to this high incidence of tumours. Now simply because he had developed a strain with this high rate of tumour incidence he was able to trace back the embryology of the formation of growth inside the testis to very early stages of development. He found that the origin of small patches of cells which resemble the early forms of the stem cells of mouse teratocarcinoma were developing in the testis by about the 16th day of gestation.

The other fact I want to draw to your attention to about the male tumours is that the tumours which develop are XY which means there is no evidence that the cells which give rise to the tumour have been through any form of meiotic division.

Now passing on to the ovarian spontaneous tumours; again they are very rare in mice , indeed they are much rarer than in humans, and once again it has been the development of a particular strain with ovarian teratoma which has led to an analysis of the growth and differentiation of these cells. If I can give you an example of how rare they are in other strains of mice; an American pathologist of the 1920's called Slee dissected 20,000 female mice from a whole range of different strains and only found one possible case. In strain LT the situation is completely different; here about 100% of females have inside their ovaries eggs which are undergoing stages of embryogenesis and so they can be thought of as carcinogenetic embryos. If you look at the chromosomes of these tumours what is found is that some are haploid and some are diploid. It is also clear that by doing the appropriate crosses that the majority of the tumours which develop are post-meiotic; in other words there has been segregration and re-arrangement of the genes in the female which gave rise to that tumour. And so the situation is similar to the situation in the human ovarian teratomas where again there is evidence of meiotic and post-meiotic cells in the tumours which develop. So, if I can summarise the spontaneous tumours of the mouse, the situation is different from that of the human in the sense that they are very rare, it is also different from the human in the sense that a very strong genetic element has been proved, particularly by the increased incidence of these tumours in the male 129J strain with various genetic alterations.

I would now like to move to the induction of tumours by embryo transplant and again I would make the point that the pathological origin of these different types of mouse tumour can be very different and that we should be cautious about grouping the pathologies of different gonadal tumours into some prime event which always happens in each form of the tumour. First of all the embryo-derived tumours; what is done here is to take the embryo out of the uterus and place it in an extra-uterine site, most commonly beneath the testicular capsule or kidney capsule. In general, if such an experiment is made in an early stage of development then 50% of the embryos will give rise to growing teratomas which are subsequently transplantable. I would draw your attention to two exceptions to this rule, the first exception involves a strain called C57 black, and what one would say about that strain also applies to the second exception which is a strain called AKR. It has been shown that in both these strains if you take an embryo of that strain and transplant it to the kidney or testis of a mouse of the same strain then teratoma formation is extremely rare.

In contrast if you take the embryo from the strain and place it in a F1 hybrid between that strain and another strain of mouse, then the incidence of teratoma formation corresponds to that of most other strains. This shows a direct host involvement in the formation of teratomas and that is quite distinct from the absence of host involvement apparent in the spontaneous formation of these tumours. Lastly, another method for making and initiating these tumours is to transplant the gonads of very young mice. Mouse embryos, usually pre-meiotic, can give rise to teratomas when the gonad is transplanted to another site, most usually to the testis. In that situation it so happens that for the particular strains of mice of which the embryo can form a teratoma, the gonad cannot. So the conclusion from this survey would be that there may be at least four different origins of teratomas in the mouse and I think that this has now been sufficiently emphasised but has not been thought about in quite such detail with human material.

So, if I can draw a summary of conclusions so far it would be that the stem cells of teratoma are probably derived from totipotent cells which are found either in the early testis of the male foetus, or in the ovary of the female by way of egg carcinogenesis, or by embryo transplantation or by transplantation of the gonads which do themselves contain totipotent cells.

Following blastocyst formation on the 7th day of mouse

development a structure called an egg-cylinder appears which contains within it some cellular complexities, but the only important point to notice that it is only one part of this structure that, when transplanted, can give rise to teratomas as demonstrated by Leroy Stephens. It is also only a part, as shown by Richard Gardener, which can give rise to a whole mouse, while the other parts of this complicated egg cylinder are concerned with forming the extra-embryonic membrane which is discarded at birth. So the relationship with totipotency is relatively clear. This shows another fact - the success of forming teratocarcinomas by transplanting embryos at different stages of development showing that as you go from the 1, 2, 4 to 8 blastocyst to egg cylinder stages so the frequency with which you obtain teratomas is increased and at the 12th, 13th and 14th day of development tumours can also be formed by transplantation of the gonad, or gonadal ridges, from the foetus into extra-uterine sites. The stem cells which you obtain from any of these various routes show large numbers of common properties with both primordial germ cells and with the cells of the early embryo. I want to mention one particular experiment done by Colin Stewart in which he takes two early mouse embryos at the 16 cell stage and puts a group of stem cells between them and watches what happens. What happens is that these stem cells are slowly engulfed by the two surrounding embryos - it is called a Micky Mouse experiment, the ears curling round the face of the stem cells - to form a single composite structure. When that composite structure is transplanted back into a uterus you obtain a chimeric mouse displaying both the white cells of the normal embryo and the dark cells obtained by transplantation of the teratoma. This experiment confirmed a large body of work by Richard Gardener and his colleagues, and by Beatrice Mintz and her colleagues in the United States, showing that the stem cells of these mouse tumours can apparently integrate into the mouse embryos and form perfectly normal structures. However, in a complicated mix of cells of this kind it is very difficult to know what the individual cells are doing and consequently many workers have concentrated on the behaviour of the stem cells in culture to try and find out what the relationship is between the malignant properties of these cells and their differentiation and loss of malignancy. Brigid Hogan will be later on describing the methods which are used to induce the stem cells to differentiate.

Finally, I would like briefly to address myself to the question of what it is about stem cells and about their differentiated derivatives that makes them either capable of forming tumours or renders them non-malignant.

Probably part of the answer lies in growth requirements of the stem cell. The stem cell is an extremely undemanding cell in the sense that it can grow in culture with salt solutions, amino acids and with the addition of very few ingredients; only insulin, transferrin and laminin are required to support the growth of that cell. That distinguishes it from most fibroblasts and epithelial cells, for instance, which require large amounts of either foetal calf serum or substitutes for foetal calf serum which have growth factors which you add to the cells in culture to make them grow. Indeed it has been shown that when differentiated cells are formed from the stem cells then the non-malignant differentiated cells require specific growth factors and these growth factors can either be insulin, epidermal growth factor or dexamethasone and as these cells become addicted to these factors for growth so they also acquire the receptors to pick them up. This is in contrast to the stem cell of the tumour which does not require these special serum factors to grow. Another property of the stem cells has been shown recently by Denise Barlow in that their growth is completely insensitive to interferon, despite the fact that interferon is received by the cells and the cells show a partial response to it. Interferon may have some antiviral action, nevertheless it does not change the growth rate of the stem cell. So the stem cell appears to be a cell which is extremely undemanding, a cell which can grow without added growth factors and its growth is not regulated in the obvious ways.

Lastly, I would say that the work of Peter Stern, in particular, has recently shown that the stem cell is particularly sensitive to killing by unprimed lymphocytes, the so called natural killer cells of the immune system, and also that the stem cells, lacking the major histo-compatibility antigens, are in no way susceptible to attack by the more normal arms of the immune system. So if I can summarise, what I have tried to say is first of all there is strong evidence that the occurrence of spontaneous testicular tumours in the male has a genetic basis. From the embryo transplants it is clear that there can also be a strong environmental influence on the initiation of tumours. The cells appear to escape and grow uncontrollably because they are emancipated cells and not dependent upon growth factors and are not susceptible to the more normal forms of immunological surveillance.

REFERENCES

1. Graham, C.F., 1977, Concept in Mammalian Embryogenesis. Ed. M.I. Sherman, MIT Press, Cambridge, Massachusetts. pp. 315-394.

2. Damjonov, I., Solter, D., and Skreb, N., 1979, Pathology of Tumours in Laboratory Animals, Vol. 2., Ed. Turusov, V.S., Published by International Agency for Research on Cancer, Lyon, pp. 655-669.

ARE TERATOCARCINOMAS FORMED FROM NORMAL CELLS?

M.J. Evans
Department of Genetics, University of Cambridge,
Downing Street, Cambridge.

Teratocarcinomas are fascinating tumours because their growing stem cell population is, typically, pluripotential. This potentiality for differentiation is expressed in the tumour which is characterised by the presence within it of many diverse types of tissue each type of which is ultimately derived from the tumour stem cell lineage.

There are two important aspects of the differentiative capacity of the tumour stem cells. On the one hand the differentiated tissues are themselves primarily non-tumourigenic and so it is possible to study the loss of cell malignancy, and on the other their differentiative capacity has proved to be explicable by their very close similarity to early embryo cells. Thus, studies of the properties and differentiation of teratocarcinoma stem cells parallel those on early embryos. The outstanding advantage of teratocarcinoma stem cells is that they may be grown in tissue culture as pure clonally-derived populations in large amounts, making them readily available for study and experimental manipulation. Their disadvantage in comparison with normal embryos is that they themselves are not entirely normal. They have been derived from tumours, often after a long passage history, and frequently also have a long passage history in vitro. Numerous different lines exist each with distinct properties and most with karyotypic abnormalities. Normal cells do not form tumours and conversely tumour cells are not normal. This concept lies at the heart of much of the study of tumour cell biology. Malignant teratocarcinoma stem cells spontaneously differentiate into benign cell types and normal embryos, or primordial germ cells, and are able to initiate teratocarcinoma formation at a relatively high frequency.

Is it reasonable to regard this process as a malignant transformation and cellular differentiation as a spontaneous reversion from malignancy? One alternative possibility is that the teratocarcinoma stem cell is essentially a cell showing a completely normal embryonic phenotype; a cell which is both pluripotential and has a high rate of proliferation. Alterations of the differentiative capacity, growth rate and karyotype often found in cells from tumours or cells with a long passage history (in vivo or in vitro) may very possibly be secondary modifications.

Experimentally, teratocarcinomas may be induced to form in mice from ectopically implanted early embryos or primordal germinal ridges. One criterion providing support for the idea that the teratocarcinoma stem cells arise from normal embryos or primordial germ cells would be the ready reversibility of the process and indeed embryonal carcinoma cells from in vivo embryoid bodies, or embryonal carcinoma cells in tissue culture, are able to recolonise an early embryo. Differentiation of embryonal carcinoma cells in vitro closely parallels early post-implantation stages of embryonic development, (1).

Embryonal carcinoma cell lines in tissue culture have been initiated from solid teratocarcinomas and from embryoid bodies but in either case any "malignant transformation" involved will already have taken place. Until now, however, it has not been possible to establish pluripotential cells into tissue culture directly from an embryo. This paper concerns such direct establishment of pluripotential cells in tissue culture from embryos. These cells have all the properties of mouse embryonal carcinoma cells and form teratocarcinomas upon re-injection into a histocompatible mouse, (2).

The inter-relationships between teratocarcinoma stem cells and early embryos (Figure 1) suggest that these stem cells may be directly homologous to a particular embryonic cell type normally present in the embryo at an early stage of development.

This possibility has been discussed at greater length elsewhere, (3). In essence comparison of cell surface properites by the use of a monoclonal antiserum (M1.22.25) to a Forssman antigenic determinant, (4,5) and comparisons of total cell protein synthesis by 2D gel electrophoresis, (4,6) indicated that embryonal carcinoma cells were most similar to the embryonic epiblast prior to proamniotic cavity formation at an early post-implantation stage of embryonic development. Cells of the inner cell mass of blastocysts and cells of the embryonic ectoderm of egg cylinder stages are not equivalent and indeed do not grow directly into cultures of pluripotential cells in culture.

The early post-implantation embryos of 5 day p.c. are not

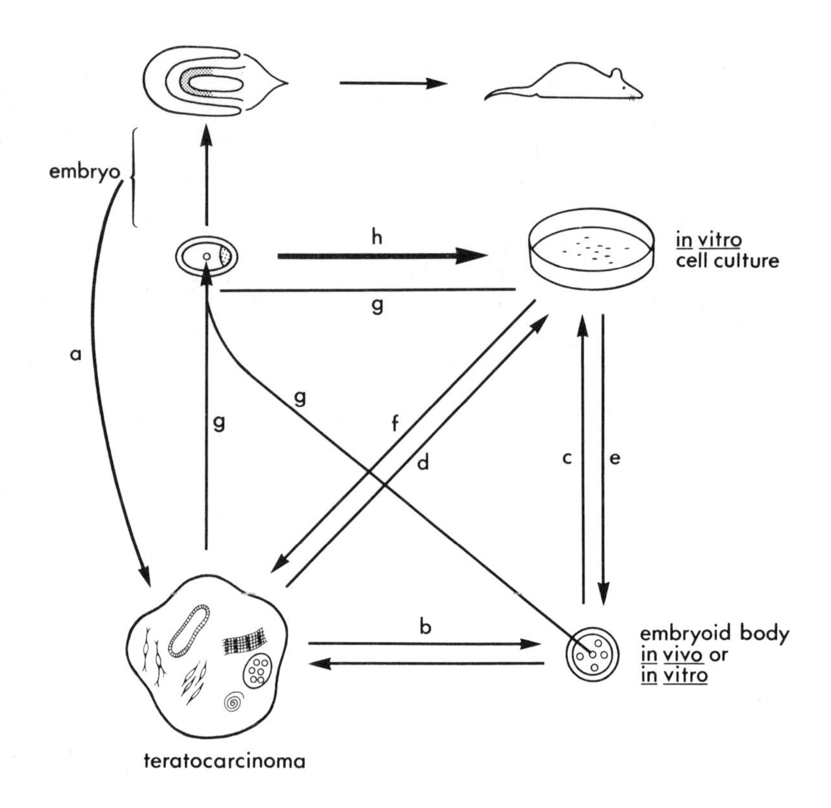

FIGURE 1 - Interrelationships of cell lines, teratocarcinomas and embryoid bodies with normal mouse embryos. The arrows indicate routes of cell transfer:
a) formation of teratocarcinoma by implantation of embryos ectopically;
b) formation of embryoid bodies from teratocarcinoma and *vice versa*;
c) derivation of cell culture from embryoid bodies;
d) cell culture directly from solid tumours;
e) differentiation of embryoid bodies from culture;
f) formation of solid tumours on reinjection of cells from culture;
g) transfer of embryonal carcinoma cells either from cell culture or from the core of an embryoid body or from a solid tumour back to a blastocyst. All these procedures may result in chimerism of the resulting mouse;
h) direct culture of pluripotential cells from an embryo. Reproduced from Evans & Kaufman (Nature, in press).

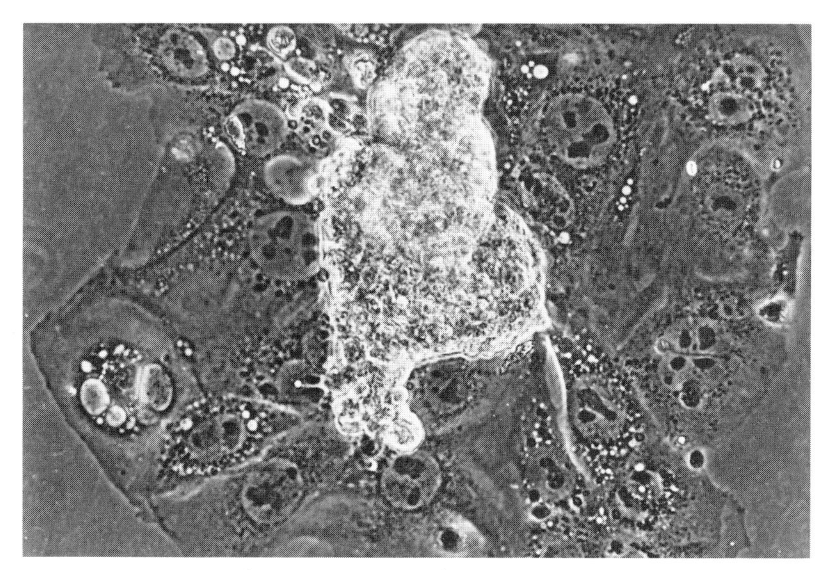

FIGURE 2 - Delayed blastocyst after 4 days in culture. Pluripotential cells may be isolated by disaggregation and passage culture of the central mass of cells which have arisen from the blastocyst inner-cell mass.

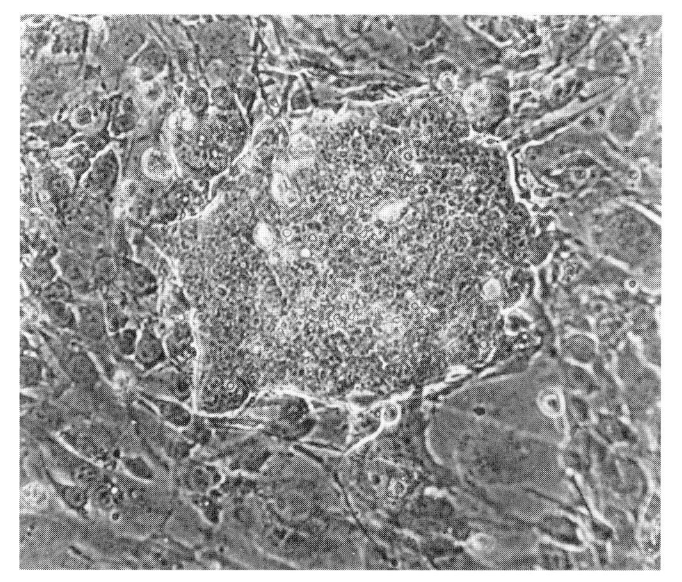

FIGURE 3 - A nest of primary pluripotential embryonic cells growing on a fibroblast feeder layer.

easy to dissect and to date attempts at direct growth of cells in vitro from the dissects have not been successful. We chose therefore, to arrive at this stage of cellular development via in vitro culture of intact blastocysts. Moreover, in order to increase the cell number in the explanted blastocysts we chose to use embryos which had undergone an implantational delay in vivo.

Mice were ovariectomised on the afternoon of day two and a half of pregnancy and injected subcutaneously with 1mg of Depo-Provera (Upjohn). The embryos progress to the blastocyst stage and enter a state of siapause just prior to implantation. Embryos in implantational delay hatch from the zona but remain free-floating to the uterine lumen. The cell number increases to at least twice that of non-delayed blastocysts and the primary endoderm is sometimes formed but no further development takes place until, under the control of hormonal stimuli, implantation occurs. The delayed blastocysts were recovered after 4 to 6 days. They were cultured either separately or in a small groups in small drops of tissue culture medium under paraffin oil on tissue culture plastic dishes for 4 days (Figure 2).

Disaggregation of such explanta by trypsin treatment and passage on inactivated feeder layers in the manner previously described for teratocarcinoma cells (7) gives rise to cultures containing cell colonies of a number of morphologies among which are recognisable giant trophoblast cells, endodermal cells and cell colonies which are very similar in appearance to embryonal carcinoma cells (Figure 3). These latter colonies were picked out, passaged and grown up into mass cultures.

Where these cells were derived from embryos of 129 mice the cell lines were re-injected subcutaneously into adult male mice. In all cases tested so far injection of 10^6 or more cells have given rise to a typical teratocarcinoma (Figure 4). To date, fifteen lines of distinct origin have been isolated including lines from outbred mouse stocks.

These cell lines appear at present to be typical mouse embryonal carcinoma cells. Their only difference being that (probably by virtue only of their short passage history) they posses a karyotype unchanged from that of the embryo from which they were isolated. This means that euploid pluripotential cells in tissue culture are now readily available, and will probably provide good vehicles for the introduction into the mouse germ line of genomes manipulated in vitro. It is also possible to isolate specifically abnormal genotype from mouse stocks. We have already isolated a line karyotypically marked by a Robertsonian translocation, (unpublished results Bradley and Evans).

This straightforward isolation of malignant teratocarcinoma cells from normal embryos strongly re-confirms the results obtained by transplantation of embryos to ectopic sites in vivo. That is that there is no "malignant transformation" involved, embryonal carcinoma cells are primarily cells showing a particular normal embryonic cell phenotype. (They may acquire secondary changes, such an alteration of karyotype, during their tumour passage.) Viewed in this light their non-malignant behaviour with their differentiation is not surprising nor is their ability successfully to colonise an embryo to give rise to a normal chimeric mouse.

Embryonal carcinoma cells provide the strongest example for the model of many tumours being the abnormal progression of an essentially normal cell phenotype. It would seem to be extremely important to discover whether human teratocarcinomas fit the now increasingly well understood mouse model and indeed whether similar models are applicable to other solid tumours.

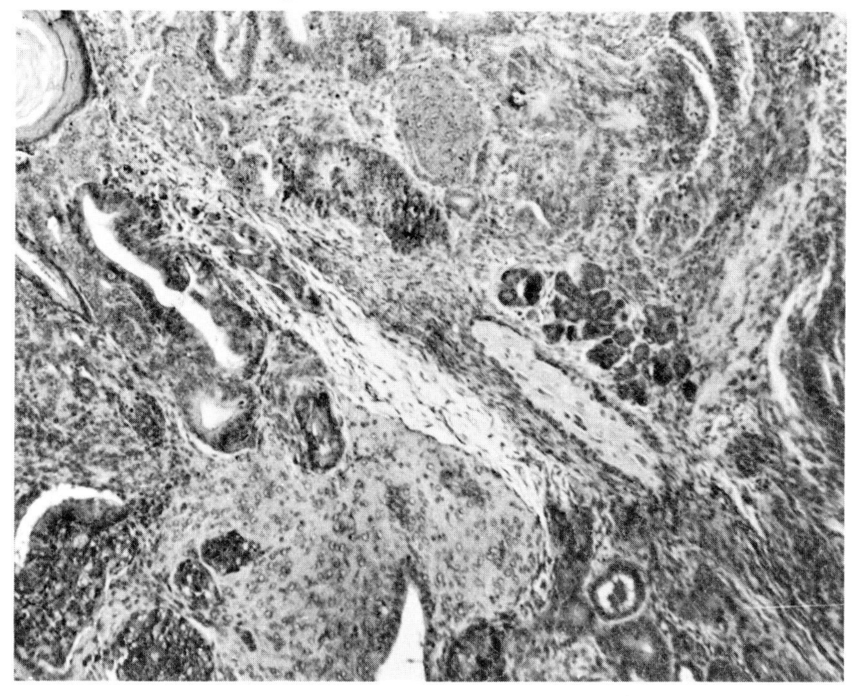

FIGURE 4 - Cross section of a teratocarcinoma formed by subcutaneous innoculation of one million embryo cells into a syngeneic mouse.

Acknowledgement

I gratefully acknowledge the support for this work provided over many years by the Cancer Research Campaign.

REFERENCES

Reference 1 includes comprehensive reviews available:

1.1 Damjanov, I., and Solter, D., 1974. Experimental teratoma. Current Topics in Pathology, 59, 69-130.

1.2 Hogan, B.L.M., 1977. Teratocarcinoma cells as a model for mammalian development. In: J. Paul (Ed) Biochemistry of Cell Differentiation II, Vol. 15, University Park Press, Baltimore, pp. 333-376.

1.3 Martin, G.R., 1975. Teratocarcinomas as a model system for the study of embryogenesis and neoplasia: Review, Cell,5, 209-243.

1.4 Martin, G.R., 1978. Advantages and limitations of teratocarcinoma stem cells as models of development. In: M.H. Johnson (Ed) Development in Mammals 3. North-Holland Publishing Co., Amsterdam, New York, Oxford, pp. 225-265.

1.5 Pierce, G.B., 1967. Teratocarcinoma: Model for a development concept of cancer. Currrent Topics in Developmental Biology, 2, 223-246.

1.6 Stevens, L.C., 1967. The biology of teratomas. Sdv. Morph. 6, 1-32.

2. Evans, M.J., and Kaufman, M.H., 1981. Establishment in culture of pluripotential cells from mouse embryos. Nature, (In Press.)

3. Evans, M.J., 1981. Teratocarcinoma formation from ectopic implantations of mouse embryos - a discussion of the origin of embryonal carcinoma cells and of the possibility of their direct isolation with tissue culture from embryos. Journal of Reproduction and Fertility, 2, 625-631.

4. Evans, M.J., Lovell-Badge, R.H., Stern, P.L., and Stinnakre, M.G., 1979. In: Cell Lineage, Stem Cells and Cell Determination. INSERM Symposium No. 10. (Ed) N. Le Louarin. Elsevier/North-Holland and Biomedical Press. pp. 115-129.

5. Stinnakre, M.G., Evans, M.J., Willison, K.R., and Stern, P.L., 1981. Expression of Forssman antigen in the post-implantation mouse embryos. Journal of Embryology and Exerimental Morphology,61, 117-131.

6. Lovell-Badge, R.H., and Evans, M.J., 1980. Changes in protein synthesis during differentiation of embryonal carcinoma cells, and a comparison with embryo cells. Journal of Embryology and Experimental Morphology, 59, 187-206.

7. Martin, G.R., and Evans, M.J., 1975. Differentiation of clonal lines of Teratocarcinoma Cells: Formation of Embryoid Bodies In vitro. Proceedings of the National Academy of Sciences of the United States of America, 72, 1441-1445.

DISCUSSION

Dr. Norgaard-Pedersen

I have one comment to make about Dr. Jones's lecture. We agree with most of his points, but it is our experience that you may have negative markers before operation but during the course of disease you may see an increase in these markers or during relapse or recurrence or progression of disease. So you may have negative markers before operation but during the course of disease or progression you can see them appear.

Dr. Jones

That is very interesting because it is in contradiction to our experience that if markers have been positive before the operation and then go negative then we can pick up relapses later. If they are negative before the operation then it is unusual for them to go positive later. Perhaps I could seek the advice of Dr Milford Ward on that, I think you would agree?

Dr. Milford Ward

I would indeed and one or two of our cases would confirm the Danish experience, but think marker negative before surgery cannot be completely ignored.

Dr. Norgaard-Pedersen

Out of 1,000 testicular cancer patients I think it is about 10 or 15 where we have observed this phenomenon.

Dr. Jones

I think that is very important and we may well have to alter our surveillance techniques to include patients who were marker negative before operation.

Dr. Norgaard-Pedersen

We should still follow the markers although they were negative pre-operativly.

Dr. Graham

I wonder if I could raise a general point about the current use of tumour markers in following the human teratoma as follows: that in the mass, the expression of characters such as the production of alpha-fetoprotein (AFP) would be associated with non-malignant, non-transplantable cells, while in the human situation it is routine to follow teratoma with Beta-HCG and AFP and further there is good evidence this is a good way of following the progression of the tumour. To us this appears bizarre because in our view the human stem cell should not produce either AFP or HCG. It should be a much more primitive cell developmentally. So either these tumour markers have practical importance in the human because it so happens that a rapidly growing stem cell spontaneously throws off HCG and AFP - producing cells which are themselves non-malignant, or it is that the human tumour is different from the mouse in the sense that each human teratoma contains within it choriocarcinoma and yolk-sac tumour. What I really want to draw your attention to is that maybe the fact that these markers have practical importance in the human is telling us that the tumours of mice and human are completely different and I wondered if someone would like to comment on that.

Dr. Evans

I would like to comment more broadly than that. I do not believe that I have seem anything that convinces me that the cell biology of the human tumour is in any way different from what we know about the cell biology of the mouse tumour and in that case I think that what is happening in the human tumour is that it is extremely difficult to follow because every individual has a separate origin, every type of tumour has gone through its own separate progression and I think, therefore, it is effectively "a mess". I just think that it would be "a mess" if we had not had reproducible mouse lines, we would be in exactly the same position with the mouse. For instance, when I started out with a tumour that Leroy Stevens had given me, which he said was a transplantable tumour and indeed it was, I wanted to know what the stem cells looked like so that I could recognise them histologically. I waded through many boxes of slides without actually finding any cells that looked like embryonal carcinoma and indeed in those early tumours there are very few cells there. However, they were transplantable and from those tumours we isolated cell lines. I now know very well what embryonal carcinoma cells look like but we only find them in tumours which differentiate rather poorly and I think that might be the problem in the human case.

Dr. Rubery

We had one patient in Cambridge who had a very small primary which was one centimetre in diameter which was all in situ seminoma and well differentiated teratoma. Within a month of removal of that primary the abdomen was full of tissue which was what looked like malignant yolk-sac tumour and the serum AFP level was extremely high. I suppose you could still postulate a stem cell giving rise to all these malignant-looking AFP cells but it gives the impression that AFP-producing tissue can be very malignant, certainly from a clinical point of view. I think we had one other patient with a similar sort of thing. We had several patients with differentiated primaries and malignant teratomatous metastases which contained a lot of AFP which would suggest to me that there must be something very malignant going on in these metastases.

Dr. Ashley

I think we have plenty of precedent in human pathology for cells which are part of undoubted malignant tumours, while at the same time being functioning cells. We find virilising tumours, we find feminising tumours, we find islet cell tumours and many types of malignant disease in which the tumour cell is actively producing something which is released into the circulation. I think in the case of the human tumour we are thinking of something which has undergone a considerable number of generations of development. There are many millions of cells in a human tumour. I think in the human we are dealing with something which is possibly at a more advanced stage in its own evolution than you are in the experimental tumours in mice.

Dr. Hogan

It is wrong to think that all mouse tumours when they differentiate give rise to non-malignant differentiated cells. I think when you transplant them over many generations they can give rise to differentiated cell types which are malignant as well. You can get yolk-sac carcinomas, you can get neuroblastoma-like tumour cell lines from there and also you can then culture these differentiated tumour lines, so one does not want to give the impression that every mouse teratocarcinoma stem cell, when it differentiates, gives rise to a non-malignant cell type.

Dr. Graham

Could I quickly agree and say of course there are multiple lines of differentiation. It is the case that teratomas contain within them choriocarcinoma, yolk-sac tissue and perhaps mesenchyme elements as well. Then it is a quite subtle kind of tumour because there is either a stem cell which has a mistake in it which gives rise to other stem cells which have mistakes in them or there are multiple independent events occuring along different lines of differentiation. So I think there is a real problem and it is dodging the issue a little bit to say that they are just a bit old compared to the mass and there is a point of biological interest here.

Dr. Rubery

I think they are likely to be a bit old compared to the mass because you do not passage most human tumours and some of the last ones have been passaged for 15 years or so now and growing very fast all the time.

Professor Fox

I would just like to ask the biologists why they always seem to produce malignant germ cell tumours, while the most common germ cell tumour by far is the benign mature teratoma of the ovary, which is a common tumour and yet all the talk has been about teratocarcinoma and nothing about the biology of the mature benign germ cell tumour. I should also like to mention the process of maturation. Every mature teratoma must sometime in its life have been an immature teratoma and therefore there must be a natural process of maturation.

Professor van Putten

I have a question concerning the histo-compatibility antigens. Do these tumours also have species-specific antigens or is there cross-reactivity from mice to rats?

Dr. Graham

The embryonal carcinoma of mice does not express H2 and, as far as I know, the rat material has not been looked at sufficiently to know about its cross-reactivity with mouse cells. Could I quickly answer Professor Fox's point, which is that the so-called benign tumours do not grow rapidly. No-one has seriously attempted to grow cells from them in culture and so the 50% of embryos which do not take and do form benign tumours are not studied. We like fast and nasty things.

HISTOPATHOLOGY

Chairman of Session
Dr C.K. Anderson

HISTOLOGICAL FEATURES OF PROGNOSTIC SIGNIFICANCE IN GONADAL GERM CELL TUMOURS

J.O.W. Beilby
The Bland Sutton Institute of Pathology,
The Middlesex Hospital Medical School,
London W1P 7PP

Within the last five years our series of solid ovarian teratomas and testicular germ cell tumours diagnosed as teratoma have been studied, (1, 2, 3, 7, 8). The neoplasms were correlated with their clinical behaviour to select morphological features of possible prognostic significance. Although in the ovary the benign behaviour of the common cystic teratoma (the dermoid cyst) is established, the solid teratoma frequently arouses alarm until proved to be composed solely of mature somatic structures. Difficulty with this group of neoplasms has arisen when immature somatic tissues are found; these are often regarded as malignant because the application of conventional methods of histological grading are unreliable. We consider that ovarian neoplasms labelled immature teratoma can be expected to follow a benign course unless the tumour has ruptured before, or at the time of, operation causing dissemination of immature somatic tissues, often neuroectodermal in type, which can implant on the peritoneum and give rise to adhesions, ascites, intestinal obstruction, deep vein thrombosis and possibly pulmonary embolism. Widespread vascular or lymphatic spread in such cases is unusual and symptomatic surgery, therefore, plays an important role in management. By contrast, tumours with extraembryonic elements, better called mixed germ cell tumours, tend to metastasise widely and rapidly by virtue of the behaviour of their yolk sac and/or choriocarcinomatous components. Neoplasms composed solely of one or both of these extraembryonic elements can be relied upon to behave in a viciously malignant manner.

On the basis of the findings in ovarian germ cell tumours a series of testicular tumours previously diagnosed as "teratoma" was assessed, using the same broad principles described above, and assigned to one of three groups. Those composed exclusively of somatic tissues either mature or immature were the only tumours referred to as teratoma. The second group was exclusively extraembryonic, either yolk sac or choriocarcinoma and the third group, which incorporated somatic and extraembryonic tissues and occasionally seminoma, was called mixed germ cell tumour. The patients with teratoma showed a low mortality whereas yolk sac tumours proved highly malignant. In mixed germ cell tumours the malignant nature of the yolk sac component was maintained even when combined with somatic elements, but when seminoma was also present the survival rate was significantly improved. Yolk sac tumours occur more frequently in adult testicular neoplasms than hitherto suspected and in mixed germ cell tumours can be expected to dictate behaviour except when combined with seminoma. This modified approach to the classification of testicular germ cell tumours has largely arisen from comparative ovarian and testicular studies and from the widespread use of tumour marker proteins in particular AFP (alpha-fetoprotein) and HCG which have revealed inconsistences in current methods of classification.

The variable patterns of yolk sac tumour have induced terminological confusion in most classifications and the prognostic importance of this extraembryonic component has not always been fully appreciated. Mostofi and Price (4) and also the World Health Organisation (5) classifications of testicular tumours include yolk sac tumours as a sub group of embryonal carcinoma; their illustrations show a morphologically similar pattern in both adults and children. Paradoxically, however, only to the infant tumour is the name yolk sac given. In 1976 Pugh (9) described yolk sac tumour, which previously had been called orchioblastoma, as a neoplasm confined to infants and excluded yolk sac tumours from his classification in adults. An inherent part of both American and British classifications is the implication that the undifferentiated areas in malignant teratoma intermediate (MTI) and malignant teratoma undifferentiated (MTU) mature into somatic tissue associated with a better prognosis. Although in principle one accepts the existence of primitive tissue with multipotent capacity, many focal undifferentiated areas represent morphological variations of yolk sac tumour with a very high metastatic potential. Yolk sac elements have been reported in several series of adult testicular tumours, but the incidence recorded has been extremely variable. Talerman, (10) for example, in 1975 reported a figure approximating 40% and his series indicated that yolk sac tumour was associated with a poor prognosis, a sentiment with which we would agree.

In 1976, in a retrospective study of malignant teratoma Neville et al (6) detected "yolk sac tumour foci" in approximately two thirds of their cases. They suggested that the morphology of yolk sac tumour might be much more diverse than previously appreciated. The reported discrepancy in incidence may well be due to the adoption of different morphological criteria for diagnosis. As a result of comparative studies of ovarian and testicular germ cell tumours unconventional morphological patterns such as acinopapillary and solid nodular are considered to be variations of yolk sac differentiation. Although some of these may be regarded as controversial, subsequent immuno-peroxidase reactions for tumour markers in particular AFP, alpha-antitrypsin and transferrin have shown such areas to be positive. Alpha-fetoprotein has been reported in the cells of "embryonal carcinoma in addition to yolk sac tumour." However, the morphological overlap between some neoplasms referred to as "embryonal carcinoma" and yolk sac tumour is such as to induce little surprise over any common biological manifestations. There is sufficient evidence to suggest that yolk sac tumour of the adult testis is a specific neoplastic entity and must therefore be incorporated into any classification of germ cell tumours. However, it would be preferable to define the relationship between yolk sac tumour and alphafetoprotein (AFP) more exactly and to assess the natural history of this neoplasm prospectively.

In gonadal tumours the grave implication of malignant trophoblastic tissue is established. This tumour tissue is the site of gonadotrophin production and a positive pregnancy test in patients with testicular, or for that matter ovarian neoplasms, has been associated with a poor prognosis. Since the advent of radioimmunoassay, elevated serum HCG levels or that of sub unit have been found where the traditional criteria for the diagnosis of trophoblastic tumour have been lacking although syncitial giant cells have been seen. This raises the question, should all such tumours be labelled trophoblastic on biochemical grounds alone or must the conventional histological criteria for the diagnosis of trophoblastic tumour be widened and the natural history reassessed?

Any tumour classification is largely a compromise based on features that are not only biologically precise but easily defined in routine diagnostic work and also of proved prognostic value. Because germ cell tumours are so frequently heterogenous sampling error has always been a problem which emphasises the value of serum markers and the importance of correlating biochemical and histological characteristics.

REFERENCES

1. Beilby, J.O.W., 1978. Germ cell and sex cord mesenchymal tumours of the gonads. In: Recent Advances in Histopathology, No. 10, Eds: Anthony P.P. and Wolf, N. Churchill Livingstone, Edinburgh, London and New York.

2. Beilby, J.O.W., and Parkinson, C., 1975. Features of prognostic significance in solid ovarian teratoma. Cancer, 36, 2147-53.

3. Beilby, J.O.W., and Todd, P.J., 1974. Yolk sac tumour of the ovary. Journal of Obstetrics and Gynaecology of Brtish Commonwealth, 81, 90-94.

4. Mostofi, F.K., and Price, E.B., 1973. Tumours of the male genital system. Armed Forces Institute of Pathology, Atlas of Tumour Pathology, 2nd Series, Fascicle 8, Washington D.C. pp. 7-84.

5. Mostofi, F.K., and Sobin, L.H., 1977. Histological typing of testis tumours. World Health Organisation, Geneva. pp. 27-31.

6. Neville, A.M., Grigor, K., and Heyderman, E., 1978. Biological markers of human neoplasia. In: Recent Advances in Histopathology, No. 10, Eds: Anthony, P.P. and Wolf, N. Churchill Livingstone, Edinburgh, London and New York. pp. 23-44.

7. Parkinson, C., and Beilby, J.O.W., 1977. Features of prognostic significance in testicular germ cell tumours. Journal of Clinical Pathology, 30, 113-119.

8. Parkinson, C., and Beilby, J.O.W., 1980. Testicular germ cell tumours: should current classification be revised? Investigative Cell Pathology, 3, 135-140.

9. Pugh, R.C.B., 1976. Testicular tumours - introduction. Pathology of the Testis, Ed: Pugh, R.C.B. Blackwell, Oxford, London, Edinburgh and Melbourne. pp. 139-159.

10. Talerman, A., 1975. The incidence of yolk sac tumour (endodermal sinus tumour) elements in germ cell tumours of the testis in adults. Cancer, 36, 211-215.

IMMUNOCYTOCHEMISTRY

Eadie Heyderman
Department of Morbid Anatomy,
St. Thomas's Hospital Medical School,
London SE1 7EH

The value of the monitoring of tumour markers in the management of testicular tumours has already been mentioned. I am going to talk about the value of localising these tumour markers in tissue sections of the original orchidectomy specimen using the immunoperoxidase technique, which is similar to fluorescent techniques except that instead of using fluorescein as a label for the antibody, a very stable enzyme, horseradish peroxidase is employed (1). The advantage of this technique is that the reaction product is a brown polymer which is insoluble and the sections can be counterstained like an ordinary histological section. Because peroxidase is present in many tissues, including red blood cells, white cells, peroxisomes of the liver and many other tissues, it is necessary to inhibit the non-specific peroxidase activity. These days most of the anti-sera used are affinity-purified or monoclonal antibodies. The control used is to absorb the antiserum with the antigen against which its directed and this shows that only in such a way can all the activities be abolished. It is well known that testicular tumours can make a wide variety of materials, there are all the placental proteins such as human chorionic gonadotrophin (HCG), human placental pregnancy-specific glycoprotein, alpha-fetoprotein (AFP) and other yolk sac products which may include transferrin, ferritin, possibly fibronectin, carcinoembryonic antigen, the epithelial membrane antigens, possibly HLA substances, and blood group substances. So far, attention has been directed mainly at the first three groups, but there are some tumours which are marker silent in that they do not produce either AFP or HCG and there is an urgent need for markers in these patients. It is obviously too late to detect metastases when they are detectable clinically and the probably reason for the great progress in the treatment of testicular tumours is that they can be picked up so easily using biochemical markers, so that in the next few years it is hoped that new markers for these tumours that do not produce the conventional markers will be discovered.

In choriocarcinoma of the uterus, there is both syncytiotrophoblast and cytotrophoblast. Very often HCG appears at a vesicular stage within the syncytiotrophoblast while the cytotrophoblast is devoid of the material. A skin metastasis of a testicular choriocarcinoma can be used to demonstrate two techniques used together. Twenty minutes before the metastasis was removed a very small dose of labelled thymidine was injected into the biopsy site; the skin nodule was removed, fixed in the usual way and sections were cut and stained for HCG while autoradiography was also performed. In a very bizarre-looking tumour, there are the syncytiotrophoblast cells with giant cells, and multinucleated cells which are strongly positive for HCG. The thymidine labelling on the other hand is in the mononuclear cells which have little or no HCG. In the normal placenta the cytotrophoblast is the dividing, generative layer and the syncytiotrophoblast is the end state layer that makes the placental proteins. We are carrying out similar labelling studies with other tumours because we can use these techniques for looking at the homogeneity of tumours. We know from cell kinetic studies that tumours divide in different groups, that they are not homogenous populations and we know from immunoperoxidase labelling studies that we have variability in the staining. what we intend to do is to compare techniques in this sort of way. When the same tumour is transplanted into an immune-deprived mouse and grows there as a xenograft, it will develop as large cells which have masses of HCG and smaller cells which do not. Very few of the cells actually form syncytial giant cells, just individual cells. Choriocarcinoma cell lines grown on a cover slips demonstrate that the cells containing HCG are giant cells, although they are mononuclear, while the smaller cells, which are supposedly the equivalent of the cytotrophoblast, do not contain HCG. Now potentially this is of great value clinically being able to stain whole cells.

The Charing Cross Group have shown that when there are very high levels of HCG, there are raised levels within the CSF, but because of the blood brain barrier the CSF levels are about a 30th of the level in the circulation. Clearly it is important to know whether a patient with meningeal symptoms has actually got metastatic tumour within the central nervous system or not because it may be necessary to decide whether to give intrathecal chemotherapy. In such a case, the CSF may contain bizarre cells and it is possible to stain such cells by immunocytochemical methods, so that this is one of the ways in which we can use the technique to look at body fluids as well as looking at tissue sections.

A lady of 76 actually presented with a lump in the breast but when she was examined she had an enormous abdominal mass and it was decided to deal with the abdominal tumour first. She had a huge ovarian mass, which was removed and she had a needle biopsy of her breast performed at the same time. The ovarian tumour turned out to be mainly adenocarcinoma with a small area of Brenner tumour and the needle biopsy of her breast showed anaplastic carcinoma. However, while she was still in the ward recovering from her operation the tumour in her breast grew rapidly until it looked as if it was about to ulcerate and so she had a mastectomy and the histology showed a choriocarcinoma. Primary choriocarcinoma of the breast is extremely rare so the obvious thing to do was to go back to the original ovarian tumour and look for evidence of primary choriocarcinoma there. Much of the tumour was necrotic, but there were few areas with non-specific giant cells which were not diagnostic of choriocarcinoma. On staining the sections for HCG the giant cells have abundant HCG, so here is a woman of 76 with an ovarian choriocarcinoma combined with adenocarcinoma and Brenner tumour that metastasised widely as choriocarcinoma. When she died she had deposits of choriocarcinoma everywhere and her circulating level before death was a quarter of a million units of HCG. So here we are using the stain in the diagnosis of choriocarcinoma which would otherwise have been extremely difficult in this very necrotic tumour.

Now Dr. Beilby has already mentioned the giant cells that one sees not uncommonly in testicular tumours. According to the dogma that Dr. Beilby was talking about in order to make the diagnosis of choriocarcinoma of the testis, one should have syncytiotrophoblasts and cytotrophoblasts together in a villous structure. We have all seen giant cells that look like syncytiotrophoblast; they look just like the individual normal placental site reaction trophoblastic cells, but it has always been taught that they are unimportant. They look like syncytiotrophoblast however; they are eosinophilic, they are vacuolated, but they do not have the right combined morphology, although they do stain positively for HCG. These HCG positive cells are also found in seminoma and in dysgerminoma of the ovary. In patients with dysgerminoma and raised levels of HCG the clinicians may be worried that the pathologist had missed a focus of choriocarcinoma. Now one can never exclude sampling error unless you cut serial sections of every bit of tumour, but what we can certainly say is that there are giant cells which stain strongly for HCG so the tumour is a dysgerminoma with cells that make HCG.

We have altogether now looked at 251 testicular tumours; of those that we would classify as trophoblastic malignant teratoma, all of them had HCG, which is what you would expect. Of the group which would either be called malignant teratoma intermediate or undifferentiated with or without seminoma, half of them had HCG positive cells, and in our group 17% of seminomas had HCG positive cells. So from this we can say that if you have a man who has high levels of HCG but in whom the histology reveals no evidence of choriocarcinoma it may be because they have these individual trophoblastic giant cells without conventional choriocarcinoma. Neither Dr. Parkinson nor I have found any prognostic significance in the testis in the presence of these giant cells. I suspect it is because the clinicians treat them so well that they all get better. I wonder if perhaps we went back 50 years before any really effective treatment was available we would discover whether there would have been any prognostic significance then, but at the moment there is not. In yolk sac tumours it is common to find HCG positive cells in very close apposition with the other extra embryonic elements. About 80% of the tumours that have HCG also have human placental lactogen. Now HPL has not turned out to be particularly useful tumour marker in the circulation, but it may be useful looking at the main tumour if one is trying to tie this up with gynecomastia.

Now I do not want you to think that every giant cell has HCG in it; we have looked at over 50 anaplastic carcinomas of various sorts and have not found HCG in any of these. There are also foreign-body type giant cells in testicular tumours and when we come to stain these for HCG they are, of course, negative; so this is a highly specific stain and in spite of the publications by McManus and his group we have not found HCG is made by every single tumour cell. It seems to be that at least the Beta sub-unit of HCG is made by a particular morphological type which is recognisable in an H & E section.

I am going to discuss briefly the localisation of AFP, since Dr. Beilby has already mentioned that the demonstration of AFP is fickle and difficult. Very often in yolk sac tumours in the adult testis one sees a rather microcystic pattern with a quite different lacey pattern of staining rather than the staining seen in syncytial giant cells with HCG and occasionally one sees positive staining in differentiated elements. I wish to refer to the article published recently in the Journal of Clinical Pathology about a seminoma (2). This seminoma was put into immune-deprived mice by Dr. Raghavan. In the primary tumour there are a few very small, slightly cystic areas but the patient had raised serum levels of AFP. It was decided to stain the tumour for AFP even though it was morphologically a seminoma. Small clumps of cells stained for AFP and at higher power you could see that these cells, which still look morphologically like seminoma, contained AFP. The transplanted mouse had high levels of AFP in its serum and there was a high level of AFP in the tumour. It is an opinion shared by many of us that, what we are looking at in seminomas and teratomas is a spectrum of tumours, that they are not as separate as one would perhaps assume, from the dichotomy of germ cell tumours from seminomas. However, I think I ought to repeat again that there are very few tumours which are easy to stain for AFP and in the majority of cases one gets results that are very difficult to interpret. It is very important, I think, that in the future one should be able to identify the sort of elements in a tumour that can produce AFP but I do not think we are yet in a state of being able to do so.

Now it has been shown that CEA is not a good marker for testicular tumours and I think the reason is that only a certain number of testicular tumours make it but the tumours which have gut-like differentiation, which form glandular elements, can localise CEA within them. It may be that if one chose the tumours in which one can demonstrate CEA in the original material those tumours would be the ones worth monitoring. What then are we using these techniques for? We are already using the immunoperoxidase localisation of tumour markers in diagnosis. If we have a testicular tumour to look at and want to know if there are areas of trophoblastic differentiation, we do not spend hours looking at each section, but stain them all for HCG and it is very easy indeed to make the diagnosis.

I think it is going to be impossible for us to monitor every marker in every patient with every sort of tumour. Obviously it is much more sensible to stain the primary tumour for a variety of markers and then to monitor the most useful ones. Professor Fox has apparently shown in the case of the colon and Dr. Walker in the case of the breast that the presence of the alpha sub-unit of HCG imparts the worst prognosis. Neither Dr. Parkinson nor I have shown that it makes any difference in the testis; between us we now have nearly 400 tumours and it does not seem to worsen prognosis. We can look at aspects of cell differentiation, both in tissue culture and xenografts, we can also combine these studies with cell labelling studies and finally we can classify our tumours, not only purely morphologically but also functionally. So we can say this patient has a malignant teratoma of the testis of intermediate kind, it has differentiation into a variety of elements and it also contains, let us say, HCG and XYZ and these are the markers worth following and I think this is the way forward for us as Pathologists. I think in the past we have made great progress, purely looking at the morphology, but I think now what we have to do is look at function as well.

REFERENCES

1. Heyderman, Eadie, 1979. Immunoperoxidase technique in histopathology: applications, methods and control. Journal of Clinical Pathology, 32, 971-978.

2. Raghavan, D., Heyderman, Eadie, Monaghan, P., Gibbs, J., Ruoslahti, E., Peckham, M.J., and Neville, A.M., 1981. Hypothesis: when is a seminoma not a seminoma? Journal of Clinical Pathology, 34, 123-128.

CURRENT PROBLEMS IN CLASSIFICATION OF TESTICULAR GERM CELL TUMOURS

Constance Parkinson
Institute of Urology,
Department of Pathology,
St. Paul's Hospital,
24 Endell Street,
London WC2

The original aim of tumour classification based on description was to allow and encourage documentation and communication of findings, thus, epidemiological and aetiological information which contributes to health service planning and patient management is accumulated. This is exemplified by the work of Cancer Registries and World Health Organisation (W.H.O.), for which authorities it is of paramount importance that classifications have universal application. It soon became apparent that certain neoplasms followed a distinctive natural course and responded in a predictable manner to a specific form of therapy. This in turn facilitated the development of classifications with prognostic value which are generally selected in routine diagnostic practice. The reproducibility of such schemes and the speed, cost and safety measure involved in their application are the limiting factors. In contrast, when considering a research or prototype classification the aims are legion and the speed and cost of application are not always relevant. If research findings do prove to have descriptive or prognostic value, before advocating their acceptance by tumour registries and diagnostic laboratories, it is important to consider the restrictions in these environments mentioned above.

Tumour classifications can be based on any aspect which facilitates division into groups. Such features include macroscopic appearance, histogenesis, morphological, immunological and biochemical characteristics and the extent of spread. Although classifications of testicular tumours commonly utilise histogenetic and morphological features (14, 22), the application of immunocytochemical and serological findings have also been discussed (11). The extent of the tumour may be defined, both histologically on examination of the orchidectomy specimen and clinically, to give a pathological and clinical stage. The latter results give the most valuable prognostic information, the histopathologists' role being limited to the examination of lymphadenectomy specimens and tumour masses persisting after therapy (6).

The development of the two major classifications of testicular germ cell tumours will be outlined (Figure 1) and the problems which may arise in their use by cancer registries, their application to diagnostic histopathology and their possible integration with research findings will be discussed.

The Evolution of Morphological Classifications

In 1946 the Armed Forces Institute of Pathology (A.F.I.P.) published their classification of testicular tumours in which four structural patterns were described - seminoma (S), embryonal carcinoma (E.C.), teratoma (T) and teratocarcinoma (T.C.) (4). Some of the problems discussed are still unresolved. Friedman and Moore were aware of the difficulty involved in the diagnosis of anaplastic areas in seminoma

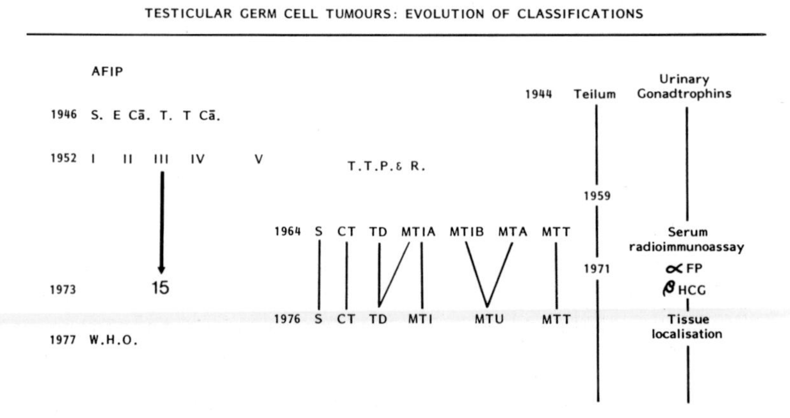

FIGURE 1

because of their resemblance to embryonal carcinoma, but they did not include a separate anaplastic seminoma group. Chorioepithelioma (choriocarcinoma, Ch.C.) was incorporated as a sub-group of embryonal carcinoma because of the morphological similarity betwen cytotrophoblast and embryonal carcinoma and the frequent association of syncytiotrophoblastic elements and embryonal carcinoma. Friedman and Moore warned against calling any apprently differentiated teratoma "benign", because metastases sometimes occur, and preferred the term "adult teratoma". Teratocarcinoma was the name proposed for those tumours containing a mixture of teratoma and any other malignant element including seminoma, embryonal carcinoma and chorioepithelioma. Follow-up was limited in 1946 but as the series was extended sufficient information was available by 1952 to allow Dixon and Moore (3) to define five prognostic tumour groups:

I	S
II	E.C. ± S
III	T ± S
IV	T + E.C./Ch.C. ± S
V	Ch.C. ± E.C. ± S

This classification was based on the observations that seminoma as a pure tumour was associated with the best prognosis, but did not change the highly malignant course of embryonal carcinoma. Conversely teratoma did ameliorate the poor prognosis associated with embryonal and chorio-carcinoma.

In the most recent classification published by the A.F.I.P. there is a return to the descriptive approach based on the four morphological patterns of seminoma, embryonal carcinoma, teratoma and chorio-carcinoma (14). As these neoplasms occur in pure form, or in combination, there are 15 possible groups in the Mostofi and Price classification. The W.H.O. classification (15) only differs from that published by the A.F.I.P. in occasional sub-groups and will not be discussed separately.

The British Testicular Tumour Panel convened in 1958 and in 1964 published their monograph in which seven separate groups were described in terms of morphology and behaviour (2). Their views on seminoma were similar to those expressed previously in the A.F.I.P. publication, but teratomas were divided into six groups based on three prognostic findings. Firstly, the more differentiated the teratomatous element the less malignant the tumour, reflected in the apparent decrease in malignant potential from malignant teratoma anaplastic (MTA) to malignant teratoma intermediate B (MTIB) to malignant teratoma intermediate A (MTIA) to teratoma differentiated (TD). Secondly, trophoblast was associated with a high mortality regardless of whatever other elements were present in the tumour, therefore malignant teratoma trophoblastic (MTT) formed a separate group. Thirdly, patients whose tumours contained seminoma in addition to teratoma had better survival rates than those with teratoma alone resulting in the combined tumour group (CT). As a result of a larger series (23) immature teratomas were found to resemble TD in behaviour so were transferred to this group from the MTIA class. Likewise, MTIB neoplasms had a similar prognosis to MTA; therefore these two groups were combined in the malignant teratoma undifferentiated (MTU) class and the MTIA nomenclature was simplified to malignant teratoma intermediate (MTI).

Classification and Cancer Registries

Information on the coding of tumours by registries in England and Wales was made available by the Office of Population, Censuses and Surveys (O.P.C.S.) and further details were obtained from the South Thames Registry. The morphological classification used by registries in England and Wales is that incorporated in the International Classification of Disease for Oncology (I.C.D.-O.) (10); the one exception is the Birmingham Registry which uses Systematised Nomenclature of Pathology (S.N.O.P.) (26), but their data is converted to I.C.D.-O. when sent to the O.P.C.S. In Scotland from 1980 onwards cancer registry cases have been coded according to the 9th revision of I.C.D., which is identical with the morphological axis of I.C.D.-O.

All the population based cancer registries outside Britain, listed in "The Directory of On-Going Research in Cancer Eipdemiology" (16), were asked what coding system they used for testicular tumours and the reason for their selection. Replies were received from 34 registries in 18 different countries. I.C.D.-O. was the classification used in 27 instances. Four registries in America and Canada preferred the A.F.I.P. or W.H.O. classifications and Yugoslavia also used the latter. In Norway the Manual of Tumour Nomenclature and Coding (13) was favoured and in Jamaica, where the incidence of testicular tumours is low, a morphological code was not applied.

It may appear paradoxical that neither of the histological classifications routinely used by histopathologists are those commonly applied to morphological coding in registries. However, despite its name, the aim of the I.C.D.-O. was to produce an internationally acceptable coding nomenclature, not a classification. Therefore, it could be argued that the concurrent diagnostic use of two classifications has induced the numerous morphological groups relevant to testicular neoplasms in the I.C.D.-O. The sections relating to germ cell tumours (906 - 910) combined the nomenclature of the Mostofi and Price (14) and Pugh (22) classifications, thereby producing 22 groups with 23 sub-groups. Code numbers frequently overlap, thus, for example, the same tumour could be designated 9081/3 Teratocarcinoma or 9083/3 Malignant Teratoma, intermediate type. In addition, although some allowance is made for coding of mixed germ cell tumours in 9081/3, (Teratocarcinoma - mixed embryonal carcinoma and teratoma) and 9101/3 (choriocarcinoma combined with teratoma), there is no provision for coding seminoma as an element of a mixed neoplasm. These are some of the problems which must enhance the difficulties and errors in the transcription of a pathological report into the I.C.D.-O, especially at registries where a pathologist is not available to interpret the original morphological opinion.

Despite these disadvantages it is apparent from the replies received from tumour registries that the popularity of I.C.D.-O. reflects the desire for uniformity and comparability of results essential for the functioning of a registry. In addition, countries which are members of the W.H.O. are committed to the use of the I.C.D.-O. It follows that changes in the classification of testicular germ cell neoplasms used by registries would be most effectively achieved by the W.H.O. but could only follow the acceptance of a uniform system by diagnostic histopathologists.

Classifications in Diagnostic Use

Some of the problems arising from the use of two classifications have been outlined above. In this section the differences, similarities and the extent to which these classifications are interchangeable will be examined. Differences are apparent in three aspects: histogenesis and nomenclature, histological observations and interpretation and the associations found between morphology and prognosis.

In contrast to Mostofi and Price (14) the T.T.P. & R. were reluctant to extrapolate the experimental work indicating the germ cell origin of teratoma from animals to man (23). In view of the demonstration of antigens shared by germ cells and 'embryonal carcinoma' in both animals and man (5, 8, 9, 25), and the transition from in situ malignant change in spermatogonia to teratoma (24) the germ cell concept is no longer disputed. The nomenclature debate is best read in the text of its original proponents (14, 22), but its practical relevance to the interchangeability of the classifications is seen below.

Histological differences concern the interpretation of anaplastic seminoma and the definition of choriocarcinoma (trophoblastic tumour). In the Mostofi and Price classification the diagnosis of anaplastic seminoma was based on finding more than three mitotic figures per high power field and its association with a higher mortality than typical seminoma. The T.T.P. & R. did not define anaplastic or aggressive seminoma as a separate group but observed that the difference in three-year survival for the three grades of "differentiation" and mitotic rate were significant ($p < 0.02$ and $p < 0.05$ respectively) (28). More recently Percarpio et al (21) found that anaplastic seminoma and typical seminoma had a similar response to therapy. Thus, the clinical relevance of the anaplastic seminoma is still in dispute. The T.T.P. & R. diagnosed trophoblast only when cyto and syncytio-trophoblast arranged in a papillary pattern were seen, whereas Mostofi and Price preferred the term choriocarcinoma and described it as cyto and syncytio-trophoblast often with an appearance reminiscent of early villi. Despite the slight descriptive differences the illustrations shown in the relevant texts are very similar and in practice differences in diagnosis would not be frequent.

From a clinicopathological viewpoint the most important differences between the American and British systems relate to the prognostic significance of morphological groups and their interchangeability. The recent A.F.I.P. publication advocates a descriptive classification but expresses survival rates according to the prognostic groups defined by Dixon and Moore (3). Thus, although certain nomenclature may be easily translated (e.g. teratoma to TD, teratoma plus embryonal carcinoma to MTI, embryonal carcinoma to MTU), the percentage of patients surviving is given for teratoma ± seminoma, teratoma + embryonal carcinoma/choriocarcinoma ± seminoma and embryonal carcinoma ± seminoma. Conversely, for followers of the Mostofi and Price classification, the C.T. and M.T.T. groups of the T.T.P. & R. cannot be translated directly to the A.F.I.P. nomenclature in the absence of a detailed histological report, as C.T. indicates seminoma + one or more germ cell element and M.T.T. includes choriocarcinoma ± one or more germ cell tumour elements.

Therefore, when publishing the results of therapy for international comparison it is essential to express the pathological diagnosis according to both classifications. In the future it would be valuable to return to a descriptive morphological diagnosis for a trial period as the prognostic groups in both current classifications developed as a result of tumour response to forms of therapy that have changed.

The Integration of Diagnostic Classification and Research Findings

The validity of current diagnostic classifications has been questioned by the recent observations on extraembryonic elements discussed by Dr. Beilby, (20), and by Dr. Heyderman's findings on tumour markers. It is interesting and salutory to see that these controversies are not entirely new (Figure 1). In 1944 Teilum (29) published his findings on homologous tumours of ovary and testis; in 1959 (30) he described extra-embryonic tumours of yolk sac origin in both adult gonads and subsequently was involved in the early papers on the utilisation of alpha-fetoprotein (AFP) as a tissue and serum marker for this neoplasm (18, 29, 30, 31).

In contrast, both the American and British systems classify yolk sac tumour as a neoplasm of infancy although they illustrate or refer to yolk sac like areas in adult tumours. Recent morphological series have reiterated the fact that yolk sac tumour is found in the adult testis, although its incidence has varied possibly as a result of differing morphological criteria (17, 19, 27, 32).

It was hoped that the morphological differences would be resolved by the localisation of alpha-fetoprotein by an immunoperoxidase technique. This has proved successful in short series (1, 31), but in routine use results have been inconsistent.

Even if only the lowest rate of occurrence of yolk sac tumour were to be accepted (27), it is apparent that this tumour must be included in descriptive classifications of germ cell tumours in the adult, but its place in a prognostic classification is undecided. Wurster et al (32) found no difference in the 5 year survival in patients whose tumours contained this element whereas other workers (19, 27) attributed a poor prognosis to its presence. Moreover, the stage at orchidectomy, known to have a profound effect on survival, was not available in any of these series. Therefore, as yet, there is insufficient information on which to incorporate yolk sac tumour into a prognostic classification.

There has always been a discrepancy in the finding of urinary gonadotrophins, or elevated levels of serum HCG, and the detection of its apparent morphological equivalent - trophoblastic tumour or choriocarcinoma. Friedman and Moore (4) were aware of this and explained it by their observation that many embryonal carcinomas showed some trophoblastic differentiation. As strict morphological criteria for trophoblastic tumour (choriocarcinoma) upheld by distinctive behaviour developed in both classifications, the gap between morphology and biochemical findings widened. This was emphasised by the many publications that followed the widespread use of serum radio immuno-assay for βHCG. It is now established that serum HCG in patients with testicular germ cell tumours is most commonly associated with isolated syncytial giant cells (7). As these syncytia closely resemble syncytio-trophoblast and the highly malignant nature of classical trophoblastic tumour is well recognised, their relevance in prognostic classifications is frequently discussed. It was tentatively suggested that isolated HCG positive giant cells might adversely affect prognosis in teratomas (7). The presence of such cells was investigated in 89 early stage MTI tumours from the Testicular Tumour Panel files but no statistically significant difference in survival was demonstrated between 58 patients where such cells were present and the 39 where they were not detected. The relevance of HCG positive giant cells in seminoma to both descriptive and prognostic classifications is thus in dispute.

In summary both AFP and HCG localised in testicular tumour tissue enhance a descriptive classification but in contrast to their use as serum markers in predicting therapeutic response, recurrence and stage, have not been shown to have prognostic value. The scientific problems involved in the application of immuno-histochemistry in a diagnostic laboratory are a cause for concern (12) and in this setting the limiting factors of time and cost must be considered.

Further information on the pathology of testicular germ cell tumours can only be derived from series in which stage and treatment are defined and, as these neoplasms are uncommon, collaboration is essential. In such a study histopathologists must be prepared to use and compare the different morphological classifications available and to be less rigid in their prognostic groupings as therapy changes. It is essential that any such pathology panel should include an immunocytochemist, a biochemist and an experimentalist so that there is constant interchange with these disciplines.

ACKNOWLEDGEMENTS

I would like to thank the members of the Cancer Registries who replied to my queries and especially Messrs. Skeets and Thompkins of the South Thames Cancer Registry who answered numerous questions and supplied many valuable references.

REFERENCES

1. Beilby, J.O.W., Horne, C.H.W., Milne, G.D., and Parkinson, C., 1979. Alpha-fetoprotein, alpha-1-antitrypsin and transferrin in gonadal yolk sac tumours. Journal of Clinical Pathology, 32, 455-461.

2. Collins, D.H., and Pugh, R.C.B., 1964. Pathology of testicular tumours. British Journal of Urology, Supplement to Volume 36.

3. Dixon, F.J., and Moore, R.A., 1952. Tumours of the male sex organs. A.F.I.P. Atlas of Tumour Pathology, VIII. 31b and 32, Washington D.C.

4. Friedman, N.B., and Moore, R.A., 1946. Tumours of the testis. Military Surgeon, 99, 573-593.

5. Gachelin, G., Fellows, M., Guenet, J.L., and Jacob, F., 1976. Developmental expression of an early embryonic antigen common to mouse spermatozoa and cleavage embryos and to human spermatozoa: its expression during spermatogenesis. Developmental Biology, 50, 310-320.

6. Hendry, W.F., Barrett, A., McEwain, T.J., Wallace, D.M., and Peckham, M.J., 1980. The role of surgey in the combined management of metastases from malignant teratoma of the testis. British Journal of Urology, 52, 38-44.

7. Heyderman, E., 1978. Multiple tissue markers in human malignant testicular tumours. Scandanavian Journal of Immunology, 8, Supplement 8, 119-126.

8. Hogan, B., Fellows, M., Avner, P., and Jacob, F., 1977. Isolation of a human teratoma cell line which expresses F9 antigen. Nature, 270, 515-518.

9. Holden, S., Bernard, O., Artzt, K., Whitmore, W.T., and Bennett, D., 1977. Human and mouse embryonal carcinoma cells in culture share an embryonic antigen (F9). Nature, 270, 518-520.

10. International Classification of Diseases for Oncology. 1976. World Health Organisation, Geneva, Switzerland.

11. Kurman, R.J., Scardino, P.T., McIntire, K.R., Waldmann, T.A., and Javadpour, N., 1977. Cellular localisation of alpha-fetoprotein and human chorionic gonadotrophin in germ cell tumours of the testis using an indirect immunoperoxidase technique. Cancer, 40, 2136-2151.

12. Leathem, A., Watts, G., and Atkins, N., 1981. Controlling antisera in immuno-histochemistry. Journal of Clinical Pathology, 34, 226.

13. Manual of Tumour Nomenclature and Coding, 1968. American Cancer Society.

14. Mostofi, F.K., and Price, E.B., 1973. Tumours of the male genital system. Armed Forces Institue of pathology. Atlas of Tumour Pathology, 2nd Series, Fascicle 8, Washington D.C.

15. Mostofi, F.K., and Sobin, L.H., 1977. Histological typing of testis tumours. World Health Organisation, Geneva.

16. Muir, C.S., Wagner, G., and Davis, W., 1978. Directory of on-going research in cancer epidemiology. International Agency for Research on Cancer, Scientific Publication 26.

17. Neville, A.M., Grigor, K., and Heyderman, E., 1978. Biological markers of human neoplasia. In: Anthony, P.P., Woolf, N., (Eds.). Recent Advances in Histopathology. Churchill Livingston. Edinburgh, London and New York. pp. 23-44.

18. Norgaard-Pedersen, B., Albrechtsen, R., and Teilum, G., 1975. Serum alpha-fetoprotein as a marker for endodermal sinus tumour (yolk-sac tumour) or a vitelline component of 'teratocarcinoma'. Acta Pathologica et Microbiologica Scandinavica, 83A, 573-589.

19. Parkinson, C., and Beilby, J.O.W., 1977. Features of prognostic significance in testicular germ-cell tumours. Journal of Clinical Pathology, 30, 113-119.

20. Parkinson, C., and Beilby, J.O.W., 1980. Testicular germ-cell tumours: should current classification be revised? Investigative Cell Pathology, 3, 135-140.

21. Percarpio, B., Clements, J.C., McLeod, D.G., Sorgen, S.D., and Cardinale, F.S., 1979. Anaplastic seminoma. An analysis of 77 patients. Cancer, 43, 2510-2513.

22. Pugh, R.C.B., 1976. Testicular tumours - introduction. In: Pugh, R.C.B., (Ed.). Pathology of the Testis. Blackwell. Oxford, London, Edinburgh and Melbourne. pp. 139-159.

23. Pugh, R.C.B., and Cameron, K.M., 1976. Teratoma. In: Pugh, R.C.B, (Ed.). Pathology of the Testis. Blackwell. Oxford, London, Edinburgh and Melbourne. pp. 199-244.

24. Skakkebaek, N.E., 1978. Carcinoma in situ of the testis: frequency and relationship to invasive germ-cell tumours in infertile men. Histopathology, 2, 157-170.

25. Solter, D., and Knowles, B.B., 1978. Monoclonal antibody defining a stage-specific mouse embryonic antigen (SSEA-1). Proceedings of the National Academy of Sciences, U.S.A., 5, 5565-5569.

26. Systematised Nomenclature of Pathology, 1965. College of American Pathologists. Chicago, Illinois.

27. Talerman, A., 1975. The incidence of yolk-sac tumours of the testis in adults. Cancer, 36, 211-215.

28. Thackray, A.C., and Crane, W.A.J., 1976. Seminoma. In: Pugh, R.C.B., (Ed.). Pathology of the Testis. Blackwell. Oxford, London, Edinburgh and Melbourne. pp. 164-198.

29. Teilum, G., 1944. Homologous tumours in the ovary and testis, contribution to classification of the gonadal tumours. Acta Obstetricia et Gynecologica Scandinavica, 24, 480-503.

30. Teilum, G., 1959. Endodermal sinus tumour of ovary and testis. Comparative morphogenesis of the so-called mesonephroma ovarii (Schiller) and extra-embryonic (yolk-sac-allantoic) structures of the rat's placenta. Cancer, 12, 1092-1105.

31. Teilum, G., Albrechtsen, R., and Norgaard-Pedersen, B., 1974. Immunofluorescent localisation of alpha-fetoprotein synthesis in endodermal sinus tumour. Acta Pathologica et Microbiologica Scandinavica, 82A, 586-588.

32. Wurster, K., Hedinger, C. and Maienberg, O., 1972. Orchioblastomatige Herde in Hodenteratomen von Erwachsenen zur Frage der Eigenstandigkeit des Orchioblastoms. Virchows Archives (Pathol. Anat.), 357, 231-242.

ALPHA-FETOPROTEIN (AFP) IN TUMOUR TISSUE AND SERUM FROM PATIENTS WITH GERM CELL TUMOURS OF THE TESTIS

G. Krag Jacobsen, M. Jacobsen and P. Praetorius Clausen
Department of Pathology, Herlev Hospital,
2730 Herlev,
Denmark

47 consecutively operated germ cell tumours of the testis (20 seminomas and 27 non-seminomas) were examined with indirect immunoperoxidase technique for the presence of alpha-fetoprotein (AFP) in the various tumour components. All patients had pre-operative serum values of AFP determined. 22 of the 27 non-seminomas were positively stained for AFP. AFP was demonstrated in 14/14 yolk sac tumour (YST) components, in 11/19 embryonal carcinoma (EC) components and in 9/17 teratoma (T) components.

12 patients had elevated serum AFP. Two of these had positively stained pure EC, one had positively stained T only, while a positive YST-component was present in the tumour tissue from the others. 10 patients had normal serum AFP values despite various positively stained tumour components.

AFP was not found in tumour tissue or serum from patients with seminomas.

Our investigation indicates that tissue staining for AFP may minimise difficulties of diagnosing YST-components. In addition, AFP-demonstration in EC supports the theories of histogenic relationship between EC and YST. Furthermore, the tissue examination indicates which of the patients should be followed with serum measurements. Finally, the immunotechniques leave the way open for investigations of the prognostic significance of AFP-containing tumour tissue with and without elevated serum values which has not yet been determined.

DISCUSSION

Dr. Grigor

First of all the last paper, that of Dr. Jacobsen et al, that showed AFP and various different types of teratoma and I think this shows that really staining for AFP is not a specific marker for yolk-sac tumour. I have mentioned this before and I think this is a demonstration of that. Now Drs. Beilby, Heyderman and Parkinson will be glad to hear that I agree with just about everything they said although sometimes I do not do so in private. When the article by Parkinson and Beilby appeared in 1977 it was thought that this cleared up everything. I think that it just made for a lot of confusion. I think the classification that Dr. Beilby and Dr. Parkinson suggested seems to me to be simplifying the non-seminomas tumours into either well differentiated teratomas or choriocarcinomas and all the rest get lumped together as yolk-sac tumour. Although you did not actually imply this I think that this was the impression that was given by that paper. I think yolk-sac should be a positive diagnosis rather than a negative diagnosis and once you have accepted choriocarcinomatous elements and yolk-sac elements and differentiated elements there is still an awful lot of undifferentiated areas of tumour which must be considered. I think as far as a prognostic classification is concerned we should really have some kind of unity and I suggest that the Pugh classification is the best basis for classifying tumours and perhaps modifications should be fitted into that classification.

Dr. Pizzocaro

In every serious case we see that histology and the serum markers dictate prognosis by themselves, but when I consider that a patient has stage 1 disease and when I compare this case with a patient who has stage 2 disease as a separate entity I was unable to find any difference in prognosis in either stages judged by histology and cell markers alone. So I ask, are histology and serum markers a reliable prediction of the course of the disease and not only of the presentation of the disease?

Dr. Beilby

I take this point entirely but I think it is semantic. What

Dr. Parkinson and everybody else was trying to point out is that the reports that we give on these complex, mixed tumours must give some indication as to how the tumours are likely to behave. If it has a seminoma with it then it is possible the effect of this is going to improve prognosis. If it has trophoblast with it, then trophoblast does not have an intrinsic blood supply, it parasatizes its host and therefore it spreads very rapidly by the bloodstream. If it has yolk-sac tumour, or you can call it embryonal carcinoma if you must, then one has got to accept the existence of something with multipotent qualities. Many of the tumours called embryonal carcinoma, with the connotations that that name has, are in fact committed to a yolk-sac differentiation and would be expected to produce AFP. Embryonal carcinoma has already been said by Kerman-Norris and others to produce AFP; this is not particularly surprising if it has already differentiated in a yolk-sac direction. Yolk-sac will spread by the lymph nodes and by the bloodstream and has a prognostic connotation, if it is combined with somatic elements the natural history of the disease would be that of the yolk-sac tumour, not of the somatic component.

Dr. Sandland

I would like to support what Dr. Heyderman said about the spectrum of testicular tumours in which we see seminoma as part of the same entity as the others. I am speaking as a clinician and one not infrequently sees patients who start off with a seminoma but one may find the pattern of spread is that of a trophoblastic teratoma. One may see this associated, of course, with elevated serum HCG concentrations and occasionally late in the course of the disease in that type of patient one sees AFP appear in the serum as well. It is very difficult to fit this picture together with a classification which insists on separating these tumours up into neat compartments, or into a pattern of behaviour that is set at the outset. I would also like to say that in assessing prognosis I think we also have to appreciate that the rapid changes in treatment have occurred in the last few years make it very difficult to assess prognosis in terms of histology alone, we must consider therapy as well. Trophoblastic teratoma is in fact an extremely drug-sensitive tumour and theoretically one would expect to have a good prognosis. In our hands it does, providing that the tumour bulk is small in the first place and it is only when the tumour bulk is very large, as shown by extremely high levels of HCG at the outset, that one can predict a poor prognosis for these patients.

Dr. Anderson

In reply to Dr. Grigor, I was greatly impressed by the paper of Parkinson and Beilby when it appeared in 1977 and I remain of the same opinion - I am unrepentant. It is difficult to remember now the intellectual confusion of those days. Advances in this field have enabled us to see solutions to problems that appeared difficult then. I should hate to think that in 4 years time we shall still be at the same intellectual base we are at today; so we do make progress and incorporate advances in attitudes and practices into our everyday approach to clinical problems, almost without realising it. In 1977 this paper offered us a view of the vitally important Danish work translated to British practice. You could then, if you so wished, attempt to use information thus gained to suggest a much modified classification of testicular tumours. On the other hand you could take the conventional classification of the British Testicular Tumour Panel and Registry and usefully expand it to include additional categories of tumour whose significance was now becoming apparent. The realisation dawned on us that it was vitally important for pathologists to identify areas of yolk-sac tumour in a positive way. I fully accept that many workers in several countries all contributed to this change of attitude, but for me this paper was of great importance and I know it was to others also.

THE CONCEPTUAL BASIS OF THERAPY

Chairman of Session
Dr W.G. Jones

NON-SEMINOMATOUS GERM CELL TUMOURS OF THE TESTIS: TREATMENT OPTIONS

M J Peckham
Institute of Cancer Research
and
The Royal Marsden Hospital
London & Surrey

In the age group 25-34 the commonest form of male malignancy is now testicular cancer. Testicular tumours are increasing in incidence, an increase which appears to extend back to the early years of the century. It is estimated that in England and Wales the chance of a 15 year old male developing a testicular tumour before the age of 50 is approximately 1 in 500 (2). The aetiology of testicular cancer is unknown. There is no evidence that known predisposing factors such as testicular maldescent have shown a concomitant increase in incidence. Approximately one third of testicular tumour patients have subnormal sperm counts prior to therapy and approximately one half of this group are azoospermic, as reported by Barrett et al later in this volume. It has also been shown that, in a minority of infertile males, large aneuploid cells, designated as carcinoma-in-situ cells, are present in the seminiferous tubules, in some instances scattered focally throughout the testis (11). About 50% of men with these changes progress to invasive testicular malignancy within 5 years. These observations suggest that there may be a pre-existing abnormality of germ cell tissue in a proportion of men developing tumours. On the other hand, in the majority of patients there is no obvious associated abnormality and many have fathered normal children prior to detection of the tumour. The immediate prospects for modifying tumour incidence by identifying and eliminating causative factors are bleak. Similarly it is difficult to envisage the development of effective methods for early detection although prompt referral and treatment for the clinically obvious tumour is essential.

The clinical problem

In the Danish National Study approximately 55% of non-seminoma patients presented with Stage I disease and the

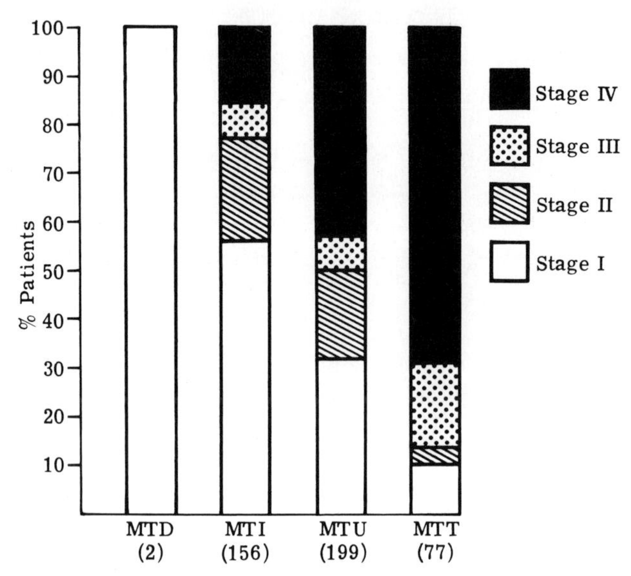

Teratoma Testis, Royal Marsden Hospital 1962 - 1979
Histology and stage distribution

FIGURE 1

TABLE 1

Histology of primary testicular teratomas in relation to local extent of disease (data from Pugh & Cameron, 1976) (9)

Extent of primary tumour	Percentage distribution by histology MTI (123)*	MTU (102)	MTT (11)
P_1 (confined to testis or rete)	67	32	55
P_2 involvement epididymis and/or lower cord	18	33	36
P_3 involvement of upper cord	15	34	9

*Number of patients

remainder with metastases (Schultz, H., personal communication).

The relationship between stage and histology in the Royal Marsden series is shown in Figure 1. Teratocarcinoma (malignant teratoma intermediate (MTI)) is predominantly associated with early stage disease whereas embryonal carcinoma (malignant teratoma undifferentiated (MTU)) and trophoblastic teratoma (MTT) tend to occur with advanced stage presentations. In keeping with the higher tendency of MTU and MTT to metastasise is the more locally aggressive behaviour of these subtypes. As shown in Table 1 spermatic cord involvement is less common with MTI than is the case with MTU and MTT.

Treatment options

Treatment options in the management of testicular non-seminoma are summarised in Table 2. Essential changes from previous practice derive from the development of effective chemotherapy. The traditional treatment methods for early stage disease; radiotherapy and radical node dissection, have been challenged. It has been argued that radical node dissection in addition to its therapeutic role may be justified by its contribution as a staging procedure. However, the increasing precision of clinical staging methods and the effectiveness of chemotherapy for small volume disease largely invalidate this claim. Excellent treatment results can be obtained in Stage I disease by radiotherapy and deferred chemotherapy (8), radical node dissection and deferred chemotherapy (14) and radical node dissection and adjuvant chemotherapy (12). As discussed below preliminary data suggests that equally good results may be achieved with orchidectomy and deferred chemotherapy without the proportion of patients requiring chemotherapy being increased.

Tumour volume dependency of chemotherapy response and implications for treatment selection.

As shown in Figure 2 the size of metastases exerts an important influence upon chemotherapy response. The implications of these observations for treatment selection is as follows:

(a) It is realistic to carry out an 'orchidectomy-only' study in Stage I patients since the 15-20% of patients who relapse can be detected with a small tumour load

TABLE 2
Possible treatment options in the management of testicular non-seminoma

Treatment Option	Application	Rationale
Orchidectomy	Stage I. Proximal cord tumour free. Markers negative or normalising rapidly after orchidectomy	Predicted cure rate of 80% with orchidectomy. Chemotherapy effective for early relapse
Orchidectomy and radiotherapy	Employed in Stage I prior to 'orchidectomy-only' study	25% of lymphogram negative patients have occult node metastases. Deferred chemotherapy effective.
Orchidectomy and radical node dissection	Stage I & II. United States and some European centres	Accurate staging, deferred or adjuvant chemotherapy effective.
Orchidectomy and chemotherapy	a) Stage II,III & IV. Patients with small volume disease b) Stage IV. Patients with bulky generalised disease	High probability of cure Disease too extensive for local treatment methods
Orchidectomy, chemotherapy & surgery	Selected Stage II,III & IV patients with initially bulky metastases	Chemotherapy less effective in bulky disease.
Orchidectomy, chemotherapy & radiotherapy	"	" Avoids surgery
Orchidectomy chemotherapy, radiotherapy & surgery	" residual masses > 4 cm. excised	15% of patients with residual masses have +ve histology
Chemotherapy	Occult or small primary tumours in patients presenting with extensive metastases from which tissue diagnosis is made.	Chemotherapy effectively controls the primary tumour

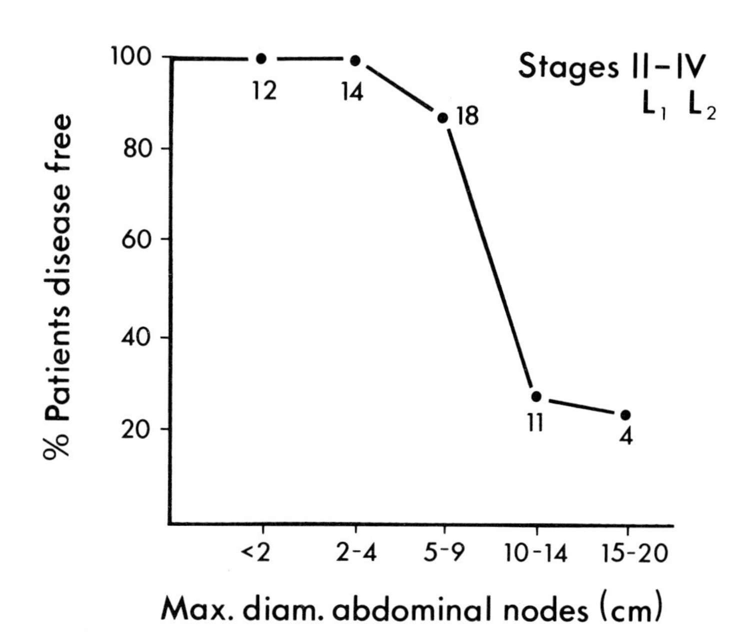

FIGURE 2 - Advanced non-seminoma testis: Influence of size of metastases on treatment outcome (The Royal Marsden Hospital, 1976-1980)

and treated promptly and effectively.

(b) Equally effective but less toxic chemotherapy should be sought for the highly curable group of patients with small volume metastases.

(c) A selected group of patients with intermediate size disease may avoid surgery (i) by employing radiotherapy as an adjunct to chemotherapy or (ii) by developing methods for discriminating between the presence of residual malignancy and either differentiated teratoma or fibrosis in residual masses.

(d) More active chemotherapy needs to be developed for patients with very bulky and widespread disease.

Current studies will be discussed under these headings.

<u>A study of no treatment other than orchidectomy for Stage I disease</u>

Routine lymph node irradiation was discontinued at the Royal Marsden in 1979. Since that time patients have been followed closely after orchidectomy if they fulfil the following criteria:

(i) the proximal spermatic cord is free of tumour

TABLE 3

Prognostic factors in patients with Stage I non-seminomatous germ cell tumours of the testis receiving lymph node irradiation

Factor	Of prognostic significance	Comment
Histology	Yes	Relapse MTU > MTI (p<0.05)
Peri-orchidectomy serum AFP and/or HCG levels raised	No	
Rate fall of serum marker levels after orchidectomy	Yes	*Rapid fall: relapse rate - 15% Slow/ incomplete: relapse rate - 64% p<0.
Tissue staining for HCG	No	
Involvement of spermatic cord by tumour	Yes	*Positive: relapse rate - 60% Negative: relapse rate - 18%

* Data from Raghavan et al, 1981 (10)

TABLE 4

Relapse in Stage I non-seminoma patients treated by orchidectomy alone

Histology	Total patients	Number relapsing
MTI	12	0
MTU	9	4
MTT	1	0
Seminoma (AFP positive)	1	0
MTD	1	0
Total	24	4 (16.7%)

Follow up 8-55 months (median 14.5)

(ii) serum alphafetoprotein and /or beta-human chorionic gonadotrophin levels are either normal at the time of orchidectomy or revert rapidly to normal after excision of the primary tumour.

The result of an analysis of prognostic factors in Stage I patients treated with radiotherapy (10) is summarised in Table 3 and preliminary results of the surveillance study shown in Table 4. To date only 4 or 24 (16.7%) men followed for a minimum period of 8 months have relapsed. All 4 have been successfully treated by chemotherapy with or without additional radiotherapy. The objective of the study is to define the prognostic significance of factors such as histology, vascular invasion and lymphatic permeation within the primary tumour. Figure 3 summarises the fate of 25 'Stage I' patients either managed by chemotherapy because of persistently elevated markers (1 patients) or entered into the surveillance study (24 patients).

Less toxic chemotherapy for patients with limited volume metastic disease

In an attempt to reduce the toxicity of the PVB

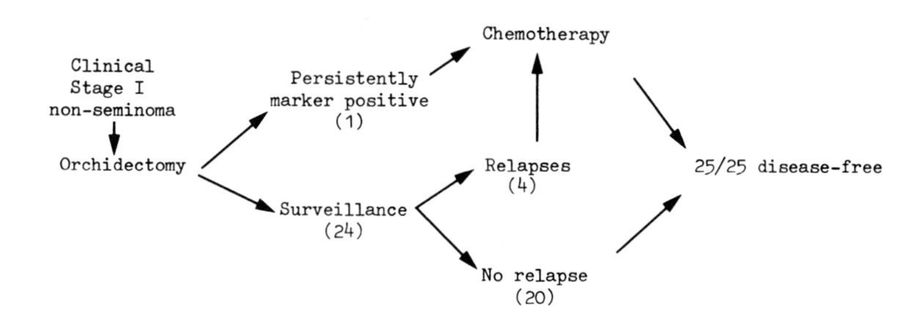

FIGURE 3 - Clinical Stage I non-seminoma patients (The Royal Marsden Hospital, 1979-1980)
(Maximum follow-up after orchidectomy 8 months)

(platinum, vinblastine, bleomycin) combination and on the basis of the performance of VP16-213 as a single agent in non-seminoma (1,3,7) vinblastine was replaced by the epipodophyllotoxin and a combination of bleomycin, epipodophyllotoxin and cis-platinum (BEP) used at the Royal Marsden Hospital since 1980 as first line treatment for patients other than those with bulky lung and abdominal disease. The latter group are entered into the Medical Research Council pilot study of bleomycin, VP16-213, vinblastine and cis-platinum (BEVIP). Initially VP16-213 in the BEP combination was used in a dose of 120 mg/m^2 days 1-5. This was modified subsequently because of myelosuppression, to 120 mg/m^2 days 1-3. It is premature to assess the performance of the regime, although preliminary experience indicates that BEP is associated with less gastrointestinal toxicity.

Radiotherapy as adjunct to chemotherapy

Previous data clearly established the efficacy of radiation in eradicating small metastases of teratoma (13). This observation together with the reduced effectiveness of chemotherapy in bulky disease led us to propose a sequence of chemotherapy followed by radiotherapy in an attempt to avoid surgery. The results of this approach have been published elsewhere (8). The feasibility of applying this form of management without enhanced toxicity has been demonstrated but in order to assess its value a prospective randomised study will be necessary. As described below recent developments in CT scanning and isotope scanning may enable benign and malignant masses remaining after chemotherapy to be distinguished by non-invasive means. If this proves possible it will allow post-chemotherapy management to be pursued on an individualised basis.

Non-invasive discrimination between benign and malignant masses remaining after chemotherapy

Despite complete normalisation of tumour markers and disappearance of extralymphatic metastases after chemotherapy, it is not uncommon for a residual mass to be detected in the retroperitoneum by CT scanning. Previously it was our policy to advocate surgery routinely because it was known that a proportion of patients harboured residual malignancy. The results of a recent analysis are summarised in Table 5. This shows that no mass less than 4 cm. in maximum transverse diameter showed histological evidence of tumour. Overall 20% of resected masses had residual

TABLE 5

Histology in relation to size of residual
masses excised after chemotherapy ± radiotherapy
for advanced non-seminomatous germ cell tumours of the testis

(data from Hendry et al, 1981) (4)

Maximum diameter of mass (cm)	Fibrosis	Differentiated teratoma	Malignant tissue
<4	8	5	
4-8	4	11	7
>8	-	4	2

malignant disease. Recently the results of a CT scan study have shown that the mean CT number may be correlated with histology. The results suggest that differentiated teratoma can be distinguished from residual malignant tissue or masses composed of fibrosis and necrosis (5). Finally, two radioiodinated monoclonal antibodies raised against human teratoma cells are being investigated for their value as tumour localising agents (6). The application of size, CT number and possibly antibody localisation should allow considerable flexibility in deciding whether post-chemotherapy surgery is necessary. It is our current policy not to recommend surgery for patients with residual masses less than 4 cm. in diameter if they have been treated with chemotherapy and radiotherapy.

More active chemotherapy in poor risk patients with disseminated bulky disease

As mentioned above, a 4 drug combination (BEVIP) is being investigated under the auspices of the Medical Research Council. Meanwhile, new drugs including a platinum analogue (CBDCA) in clinical phase I study (Dr Hillary Calvert), are being investigated in teratoma xenografts growing in immune-suppressed mice. It is hoped that this will prove to be a useful test system since it is increasingly difficult to investigate new agents in the clinic given the efficacy of current treatment methods.

Other considerations in treatment selection

In addition to stage and volume of disease, pretreatment serum marker status may be of prognostic significance although we have been unable to demonstrate this in groups of patients comparable with respect to tumour volume (8).

In addition to efficacy, toxicity is a major consideration in the design of treatment. This includes acute toxicity (which should take into account symptomatic disturbance), and more protracted damage to normal tissues. In addition to the effects of chemotherapy on lung, kidney and germinal epithelium other longer term side effects are being observed including Raynaud's phenomenon. Patients undergoing surgery for retroperitoneal disease are at risk of sympathetic nerve damage and consequent ejaculatory impotence. Finally, given the wide range of active cytotoxic

TABLE 6

Advanced non-seminomatous germ cell tumours of the testis
(The Royal Marsden Hospital, 1976 - April 1980)

Total patients	Alive	Disease-free
129	95 (73.6%)	89 (69%)

drugs available in the treatment of teratoma, drug choice should take into account the possible long term hazard of carcinogenesis.

Conclusions

As shown in Table 6, 70% of advanced stage patients are alive and disease-free (follow-up 1-5 years). Assuming that 50% of the non-seminoma patient population present with Stage I disease and that this group is uniformly curable, overall cure rates of 85% are achievable with current treatment methods. The prospects for improving these results by developing more effective chemotherapy for high risk patients with bulky disease are encouraging.

REFERENCES

1. Cavelli, F., Klepp, O., Renard, J., Rohrt, M. and Alberto, P., 1981. A phase Ii study of oral VP16-213 in non-seminomatous testicular cancer. European Journal of Cancer, 17, 245-249.

2. Davies, J.M., 1981. Testicular cancer in England and Wales: Some epidemiological aspects. Lancet, i, 928-931.

3. Fitzharris, B.M., Kaye, S.B., Saverymuttu, S., Newlands, E.S., Barrett, A., Peckham, M.J. and McElwain, T.J., 1979. VP16-213 as a single agent in advanced testicular tumours. European Journal of Cancer, 16, 1193-1197.

4. Hendry, W.F., Goldstraw, P., Barrett, A., Husband, J.E. and Peckham, M.J., 1981. Elective delayed excision of bulky para-aortic metastases from non-seminomatous germ cell tumours of the testis. British Journal of Urology, (In press).

5. Husband, J.E., Hawkes, D. and Peckham, M.J., 1981. Quantitation of therapeutic response in testicular tumours. (In preparation).

6. Moshakis, V., McIlhinney, R.A.J., Raghaven, D. and Neville, A.M., 1981. Monoclonal antibodies to detect human tumours: an experimental approach. Journal of Clinical Pathology, 34, 314-319.

7. Newlands, E.S. and Bagshawe, K.D., 1977. Epipodophyllotoxin derivative (VP16-213) in malignant teratoma and choriocarcinoma. Lancet, ii, 87.

8. Peckham, M.J., Barrett, A., McElwain, T.J., Hendry, W.F. and Raghaven, D., 1981. Non-seminoma germ cell tumours (malignant teratoma) of the testis. British Journal of Urology, 53, 162-172.

9. Pugh,, R.C.B. and Cameron, K.M., 1976. Teratoma. In : Pathology of the Testis, Ed. Pugh, R.C.B. Blackwell Scientific Publications, Oxford. pp 199-244.

10. Raghaven, D., Peckham, M.J., Heyderman, E. and Tobias, J.S., 1981. prognostic factors in clinical Stage I non-seminomatous germ cell tumours of the testis managed by orchidectomy and lymph node irradiation. (Submitted for publication).

11. Skakkebaek, N.E., Berthelsen, J.G. and Vifeldt, J., 1981. Clinical aspects of testicular carcinoma-in-situ. International Journal of Andrology, Supplement 4, 153-159.

12. Skinner, D.G. and Scardino, P.T., 1979. Relevance of biochemical tumour markers and lymphadenectomy in the management of non-seminomatous testis tumours: Current persepctive. Transactions of the American Association of Genito-Urinary Surgeons (Baltimore), 87, 293-298.

13. Tyrrell, C.J. and Peckham, M.J., 1976. The response of lymph node metastases of testicular teratoma to radiation therapy. British Journal of Urology, 48, 363-370.

14. Williams, S.D., Einhorn, L.H. and Donohue, J.P., 1980. High cure of Stage I or II testicular cancer with or without adjuvant therapy. Proceedings of the American Association for Cancer Research and American Society of Clinical Oncology, 21, 421.

THE THEORETICAL BASIS OF RESPONSE TO TREATMENT

L.M. van Putten
Radiobiological Institute TNO,
Lange Kleiweg 151,
228 GJ Rijswijk,
The Netherlands

Last year in a review of the reasons why colon tumours respond poorly to chemotherapy (12), I listed some of the possible reasons:

a. inherent cellular insensitivity to cytostatic drugs;
b. inhomogeneity in drug sensitivity among tumours of a certain type;
c. few proliferating cells in each tumour where proliferating cells respond better to drugs;
d. poor vascularisation of the tumour preventing the cytostatic drugs from reaching the tumour cells;
e. early emergence of a drug-resistant cell line after exposure to each agent;
f. surviving tumour cells proliferate rapidly during treatment and cause regrowth;
g. other factors?

Of this list points a, b and e are concerned with intrinsic cellular factors, whereas points c, d and f may be more related to properties of tumour structure, especially vascularisation.

When dealing with germ cell tumours, which are among the tumour types that respond well to chemotherapy, it would be pertinent to use the same checklist to verify what are the reasons for failure to respond to therapy and to analyse in what areas improvement is possible.

A. Inherent cellular sensitivity may gain only from the introduction of new types of drugs. This reminds us of the fact that new drug development has been in the past, and is likely to remain in the near future, the major source of improvement of treatment results.

B. Inhomogeneous drug sensitivity of tumours. Germ cell tumours respond with a relatively high percentage of remission to cytostatic treatment with combination chemotherapy. The responses to single drugs have a much lower frequency (20-40%) and this implies that many patients receive a suboptimal combination in which some of the drugs may contribute little. It is not unlikely, therefore, that the results of treatment might be improved if every patient could be treated with his individual optimal combination of cytostatic agents instead of, as is done now, with the best standard drug combination.

C. Cell kinetic factors. In general cell kinetic factors are favourable for rapidly growing tumours and it is therefore pertinent to verify whether treatment resistance occurs especially among slow growing variants. A majority of relapses after treatment is seen within two years. Delayed relapses might be a consequence of a slower growth rate so there would be an indication to watch for late-relapsing, apparently slow growing tumours in order to see whether this plays a role. Slow growing tumours might be especially less sensitive to the phase-specific drugs VLB, VCR, MTX, but it appears that the importance of cell kinetic factors is certainly smaller than the variation in intrinsic drug sensitivity and it would appear less likely that they are responsible for treatment failure.

D. Poor vascularisation might hinder drug penetration. In model tumours this is noted in large tumours that contain necrotic areas bordering on intact cells. Slow diffusion may limit cellular exposure to drugs that have a brief high peak of serum concentration followed by a rapid drop in level. It may be partially overcome by selection of agents that have a long serum half life, or by the use of a slow release form of drug. This prolonged exposure can, however, not fully compensate for a long diffusion pathway if the drug is consumed, inactivated or selectively absorbed by cells. This is illustrated by the poor penetration of MTX in multicellular spheroids even after 48 hour incubation (15).

It should be noted that poor vascularisation is in principle also important for the response to radiotherapy; poor oxygenation of tumour portions may make them radioresistant. The usual rapid shrinkage of testicular tumours will lead to rapid redistribution of blood associated with reoxygenation of hypoxic tumour areas. Similarly, the shrinkage after the first doses of a cytostatic treatment course may facilitate penetration of drugs from later doses if a similar process also leads to better blood supply to those areas of the tumour that were originally poorly vascularised.

E. Emergence of resistant tumour cell lines. This may be caused by the presence of pre-existing resistant cells, or by mutation. The probability of mutation is proportional to the number of tumour cells present so it will be less for small tumour loads. Furthermore, the hazard of development of resistance is much smaller in combination chemotherapy and if initially the tumour is sensitive to three different drugs, the likelihood of acquired resistance to all three agents is very small. This emphasises that it may be very important to adapt the combination of drugs optimal to the individual tumour sensitivity.

F. Rapid repopulation of surviving tumour cells during treatment. This is for human tumours a hypothetical reason for failure to respond to therapy. It has been identified for mouse tumours as a cause of resistance to fractionated radiotherapy when tumour doubling times are shorter than two days and average survival of control animals with a palpable tumour is less than two weeks. It seems an unlikely cause of failure in the treatment of human tumours.

G. Other factors. At present we have no hard data on other factors. For radiotherapy there is evidence for at least one experimental tumour system indicating that reduction of cell survival can explain completely the cure of the tumour (6). For chemotherapy the data are somewhat less convincing in the case of solid tumours. Resistance to adriamycin encountered in the centre of cellular spheroids can be explained by poor drug penetration or by cell kinetic factors (11). Schabel (8) has emphasised the variability of tumour volume response in Ridgway Osteogenic Sarcoma carrying mice even when identically treated (Figure 1). This is certainly different from our results in osteosarcoma C22LR of which Figure 2 is representative. Further study seems useful to identify unknown factors that may influence treatment results and we should certainly avoid the conclusion that we know all mechanisms that play a role in determining the response of tumours in mice and men. Nevertheless, the known factors are of sufficient importance to guide us in attempts at improving the results of treatment.

In summary, it appears that for germ cell tumours among the possible reasons for failure to cure by chemotherapy the most important is the possibility of a suboptimal choice of cytostatic drugs in a combination. This might be improved in

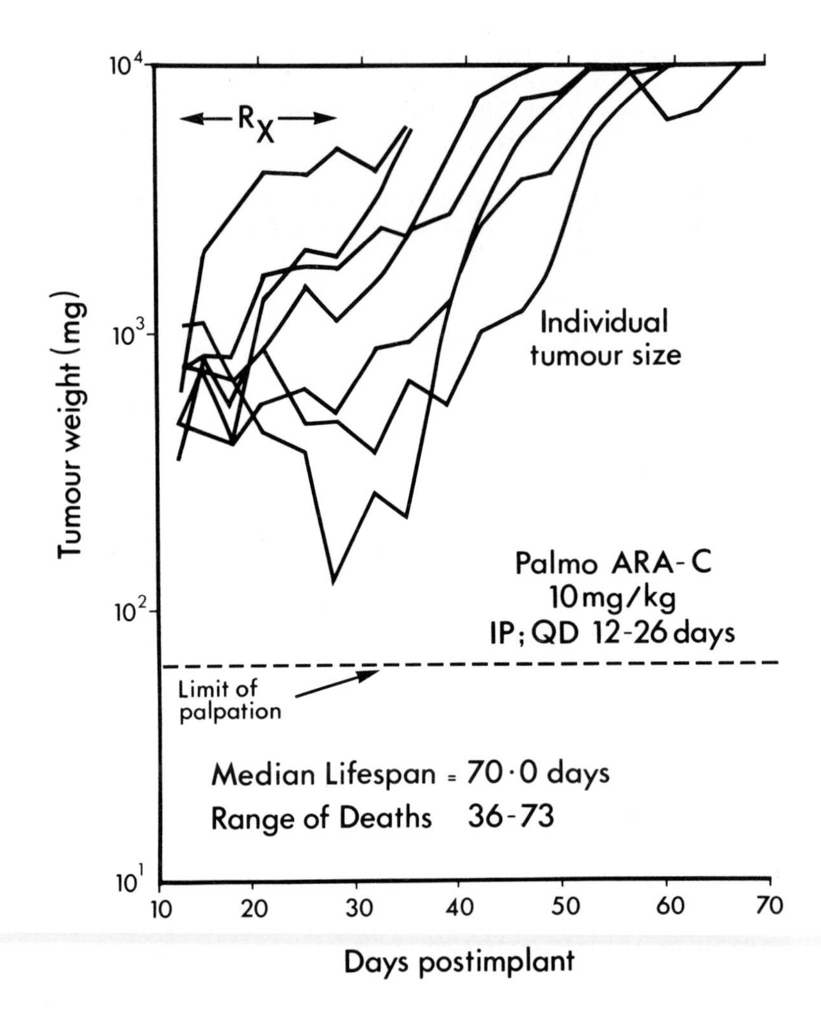

FIGURE 1 - A representative graph of the tumour volume response of Ridgway Osteogenic Sarcoma to chemotherapy. Adapted from Schabel (8).

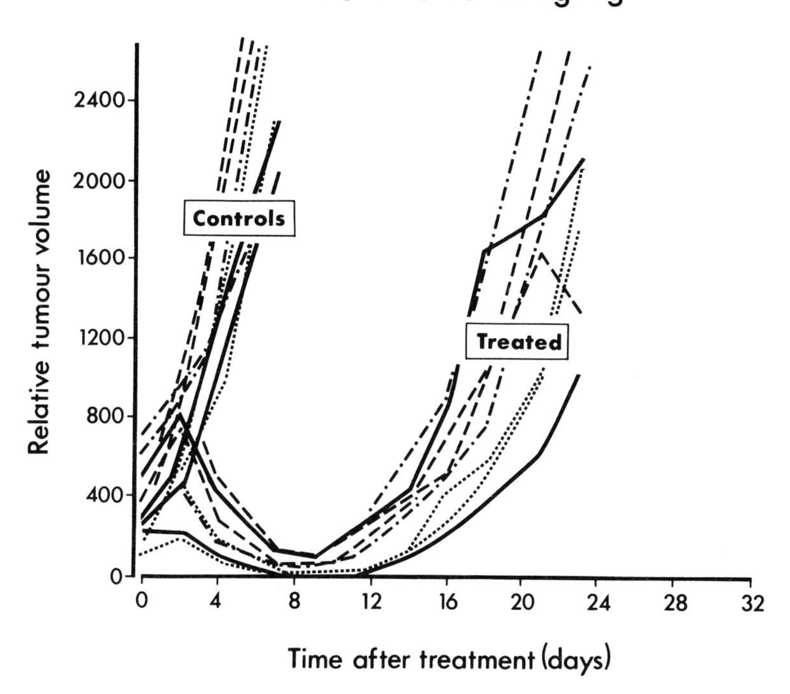

FIGURE 2 - A representative graph of the tumour volume response of osteosarcoma C22LR to chemotherapy. Note the much lower variability of response in this tumour.

TABLE 1

IMPORTANT PROPERTIES OF DRUG SENSITIVITY TEST

	Failure to give result	Results available in 10 days	"Appropriate drug exposure"	Valid endpoint	Overall relevance
Salmon (clonogenicity assay)	50%	+	--	+	?
Volm (precursor incorporation)	15%	+	--	±	?
Cytometry of drug level	?	+	--	?	?
Cytometry of "drug effect"	?	+	--	?	?
Xenograft subcutaneous	60%	--	+	+	?
Ditto under Kidney capsule	?	+	±	?	?

two ways: either by finding better drugs that are effective in an even larger fraction of tumours or, alternatively, by the selection of an individually optimal drug combination to be used for tumours on the basis of a prediction of response.

It is relevant in this respect to evaluate how far we have come to solve this problem. There are a number of different techniques available such as the clonogenic cell assay of Salmon (7) and its variants (5, 14) the tests studying whether drugs inhibit the incorporation of metabolic precursors for DNA-RNA- or protein synthesis (3, 13), tests on growth properties in tissue culture (4), flow cytofluorometric techniques measuring the incorporation of drugs into cells (9), or measuring cell kinetic disturbances caused by drugs (2) and finally drug sensitivity tests in xenografts of human tumours (1, 10) in immune-deprived mice. It is not possible to review here the publications in this area in detail but it would appear justified to indicate some of the problems involved in the application and reliability of these techniques. In the first place it is obvious that the success rate of making a prediction is limited: not all tumour samples yield cell suspensions and not all control suspensions lead to growth in vitro of a sufficient number of clones to permit a confident conclusion. A second uncertainty lies in the relevant exposure level. Probably the most sensible approach here is to search along empirical lines for an exposure intensity that proves to lead to the best correlation of test results with clinical response. In Table 1 the properties of these tests are compared in which the best results in the literature for each technique are listed. The main problems limiting the practical usefulness presently lie in the high frequencies of test failure and in the uncertainties concerning the appropriateness of drug exposure.

In summary: from this review it may be concluded that we know now probably the most important factors determing response of germ cell tumours to radio- and chemotherapy. For chemotherapy the most likely reason for failure is insufficient inherent cellular sensitivity, or the emergence of resistant lines. At this stage of our knowledge it is impossible to predict whether improvement of this situation will be reached first by the development of new cytostatic agents, or by the development of a simple and dependable drug sensitivity assay system. Both approaches should be pursued, since each of them - even if made superfluous by optimal development of the other - will be extremely useful also for treatment of other types of tumour.

REFERENCES

1. Bogden, A.E., Haskill, P.M., LePage, D.J., Kelton, D.E., Cobb, W.R., and Esber, E.J., 1979. Growth of human tumour xenografts implanted under the renal capsule of normal immunocompetent mice. Experimental Cell Biology, 4, 281.

2. Kipp, J.B.A., Barendsen, G.W., 1980. A sensitive method for the determination of vinblastine in rat blood plasma and tumour tissue using flow cytofluorometry. In: Flow Cytometry IV. (Ed.) O.D. Lacrun, T. Lindmo and E. Thorud. Universitetsforlaget, Bergen, 531.

3. Mitchell, J.S., Dendy, P.P., Dawson, M.P.A., and Wheeler, T.K., 1972. Testing anticancer drugs. Lancet, i, 955.

4. Morasca, L., Balconi, G., Erba, E., Lelieveld, P., and van Putten, L.M., 1974. Cytotoxic effect in vitro and tumour volume reduction in vivo induced by chemotherapeutic agents. European Journal of Cancer, 10, 667-671.

5. Ozols, R.F., Wilson, J.K.V., Weltz, M.D., Grotzinger, K.R., Myers, C.E., and Young, R.C., 1980. Inhibition of human ovarian cancer colony formation by adriamycin and its major metabolites. Cancer Research, 40, 4109-4112.

6. Reinhold, H.S., and de Bree, C. Tumour cure rate and cell survival of a transplantable rat rhabdomyosarcoma following X-irradiation. European Journal of Cancer, 4, 367-374.

7. Salmon, S.E., Hamburger, A.W., Soehnlein, B., Durie, B.G.M., Alberts, D.S., and Moon, T.E., 1978. Quantitation of differential sensitivity of human tumour stem cells to anticancer drugs. New England Journal of Medicine, 298, 1321-1327.

8. Schabel, F.M., Jr., 1975. Animal models as predictive systems. In: Cancer Chemotheraphy, Fundamental Concepts and Recent Advances. 19th Annual Clinical Conference on Cancer. M.D. Anderson Hospital and Tumour Institute at Houston. Year Book Medical Publishers, Chicago, pp. 323.

9. Sonneveld, P., van den Engh, G., Nooter, K., 1980. Flow cytometric determination of intracellular anthracycline levels. Annual Report REP-TNO, 1980.

10. Steel, G.G., Courtenay, V.D., Phelps, T.A., and Peckham, M.J., 1980. The therapeutic response of human tumour xenografts. In: Immunodeficient Animals for Cancer Research. (Ed.) S. Sparrow. McMillan, London, p. 179.

11. Sutherland, R.M., Eddy, H.A., Bareham, B., Reich, K., and Vanantwerp, D., 1979. Resistance to adriamycin in multicellular spheroids. International Journal of Radiation Oncology, Biology, and Physics, 5, 1225-1230.

12. van Putten, L.M., 1980. Why do colon tumours respond poorly to chemotherapeutic agents? Paper presented at the XIth International Congress of Gastroenterology, Hamburg.

13. Volm, M., Kaufmann, M., Hinderer, H., and Goerttler, K., 1970. Schnellmethode zur Sensibilitatstestung maligner Tumoren gegenuber Zytostatika. Klinische Wochenschrift, 48, 374-376.

14. van Hoff, D.D., 1980. New leads from the laboratory for treating testicular cancers. In:Therapeutic Progress in Ovarian cancer, Testicular Cancer and Sarcomas. van Oosterom, A.T. (Ed.) Nijhoff, The Hague, p. 225.

15. West, G.W., Weichselbaum, R., and Little, J.B., 1980. Limited penetration of Methotrexate into human osteosarcoma spheroids as a proposed model for solid tumour resistance to adjuvant chemotherapy. Cancer Research, 40, 3665-3668.

DISCUSSION

Dr. Silvestrini

I want to add something to what Dr. van Putten has talked about. In our Institute we are deeply engaged in performing a chemo-sensitivity test on short-term cultures. The reliability of this in-vitro system has already been assessed in many types of human tumours, such as breast cancer, non-Hodgkin's lymphomas and also testicular tumours. The results that we have obtained (recently published in the European Journal of Cancer) show that the chemo-sensitivity is a peculiar intrinsic characteristic of individual tumours. We have observed that the sensitivity to the same drug may change greatly from tumour to tumour although of the same histological type and with the same proliferative activity. More particularly in testicular tumours, we have observed that the chemo-sensitivity of the primary tumour can be different to the metastatic disease all within the same patient.

Dr. Jones

A very interesting observation which I think possibly has a clinical bearing.

Professor Peckham

Several groups have now reported that histology is not a prognostic determinate. In our series, the undifferentiated teratomas or embryonal carcinomas are doing significantly better than the intermediate teratomas.

Concerning the model systems, I think one ought to draw a distinction between a model system which is useful for screening new drugs and those for investigating the emergence of drug resistant lines from short-term predictive tests. Given that these are tumours of high chemo-sensitivity and the progress clinically on an empirical basis, and in view of the difficulty of establishing these in short-term systems, particularly the agar cloning technique, I doubt whether predictive testing is going to be useful for this group of tumours. However, I think that the use of xenograft models could be exploited to look at new drugs and may be useful for detecting drug resistance.

EPIDEMIOLOGY

Chairman of Session
Dr W.G. Jones

TESTICULAR TERATOMA IN YORKSHIRE - A RETROSPECTIVE REVIEW

P.J. Corbett, R.A. Cartwright and H. Annett
Cookridge Hospital, Leeds LS16 6QB
Yorkshire Regional Cancer Organisation
and Yorkshire Regional Health Authority

It is hard to believe today that it is less that four years since we began to treat advanced germ cell tumours of the testis with modern aggressive chemotherapeutic regimes, offering at least a proportion of such patients the chance of attaining a sustained complete remission.

We are aware that up until that time the treatment of extranodal metastatic germ cell tumours, by any modality, was essentially palliative. In order to obtain a base-line against which it would be possible to estimate the effects of our new therapeutic philosophy and technology we decided to review the records of all patients with testicular germ cell tumours which were not pure seminomas seen in the Yorkshire Region as a whole over the 15-year period from 1961-1976.

The following paper is an account of our findings relating to these cases, numbering 217 in all.

The Yorkshire Region consists of an area of 1.5 M hectares (3.7 M acres). There are large areas of mixed heavy, light and woollen industry principally around the major towns and cities. There is also a large agricultural component with some wild and rugged moorland of great natural beauty.

The Cancer Registry for the Region currently collects data from a population of about 1.7 M males. 217 cases of non-pure seminomatous germ cell tumours of the testis were registered between early 1961 and the end of 1976. The series was closed at the end of 1976. We believe that this series represents virtually all such cases occurring within our Region during this period of time, and that it reflects the true natural history of the disease within this part of the United Kingdom. It differs from many (in fact most) published series which are based on the experience of a single, and often specialised, treatment centre and which are often biased to include a high proportion of cases with metastatic disease seen as either secondary or even tertiary

TABLE 1

BTTP CLASSISICATION OF TERATOMATOUS TUMOURS OF THE TESTIS

(Pugh 1976)

Teratoma: Teratoma Differentiated (TD)
Malignant Teratoma Intermediate (MTI)
Malignant Teratoma Undifferentiated (MTU)
Malignant Teratoma Trophoblastic (MTT)

Combined Tumour (CT)

TABLE 2

CLINICAL STAGING (ROYAL MARSDEN HOSPITAL)

STAGE I	Tumour confined to testis (Lymphogram negative)
STAGE II	Positive lymphogram or other evidence of abdominal nodal disease. No supradiaphragmatic nodal disease.
STAGE III	Supradiaphragmatic nodal disease (Mediastinal and/or supraclavicular)
STAGE IV	Extralymphatic spread (Lungs, liver, brain, etc.)

referrals, some from overseas. Other reported series are based on pathological specimens from one or many centres.

The cases were identified from the Cancer Registry and, with the permission and co-operation of the Consultant Medical Staff involved, the original case notes were consulted. In a few instances where the patient did not reach or receive attention at a hospital, data from the General Practitioner and the Office of Population Censuses and Surveys were utilised, the latter class of information being automatically requested when death certification reached the Registry.

Information on presentation, treatment, clinical course and survival was extracted from the clinical notes and transferred onto a standard form, then analysed using the University of Leeds ICL 1906A Computer and the SPSS software package. In addition the log-rank program of Peto was used to provide survival data (5, 6).

Histology

This was defined using a slight modification of the classification described by Pugh and the British Testicular Tumour Panel (7).

The histological opinion, (Table 1), expressed was usually that of the local pathologist, but some specimens had been submitted to Dr. Pugh, and more recent cases to our local Testicular Tumour Panel, under the Chairmanship of Dr. C.K. Anderson.

Staging

The staging of disease at presentation was carried out retrospectively with the available information from the clinical case notes, using the 4 stage classification in its original form as instituted at the Royal Marsden Hospital (7), (Table 2).

Where a lymphangiogram had not been carried out and there was no evidence to the contrary (e.g. a positive CXR) the case was classified as "presumed Stage I". From these data the influence of various factors on the natural history and outcome of the disease were investigated and their significance will be discussed. As the series was closed at the end of December 1976 and the analysis of clinical data was commenced in June 1978 all possible cases were followed up for a minimum of 18 months and a maximum of 16 years.

TABLE 3

CRUDE INCIDENCE BY YEAR

Year	Nos/Year	Incidence/year/ 100,000 males
1961	13	0.85
1962	10	0.65
1963	18	1.18
1964	6	0.39
1965	11	0.71
1966	10	0.64
1967	12	0.77
1968	7	0.45
1969	11	0.70
1970	11	0.70
1971	15	0.96
1972	13	0.83
1973	15	0.95
1974	16	0.92
1975	22	1.27
1976	27	1.55
		Overall crude incidence 0.87
Total	217	

TABLE 4

TOTAL AGE DISTRIBUTION 1961 - 1976

Age Groups	No	%	
under 20	15	6.9	
21-30	91	41.9	65.3
31-40	55	23.4	
41-50	28	12.9	
51-60	12	5.5	
over 60	16	7.4	
Total	217		

TABLE 5

HISTOLOGICAL SUBTYPES

	No	%
MTI	26	12.1
MTU	62	28.8
MTT	4	1.9
Mixed	33	15.3
Unknown	90	41.9
Total	215	

TABLE 6

STAGE AT PRESENTATION

	No	%
Presumed I	70	35.7
Confirmed I	50	25.5
Combined I	120	61.2
II	30	15.3
III	8	4.1
IV	38	19.4
Total	196	

The major problems encountered, in common with any study based on retrospective survey of clinical notes, is the lack of positive assertions about facts known to be important. Twenty cases were lost to follow up but have been included in as far as appropriate data were valid. Some cases had left the Region and others defaulted from follow up and some were discharged. Those lost to follow up must be alive or have left the country because the death of any cancer patient in the UK is notified to the Cancer Registry. However, it is not known whether those surviving are "disease free" or "alive with tumour".

Incidence

The crude incidence rates by year of presentation, (Table 3), shows the small number of cases seen in certain years.

The overall crude incidence is 0.87 per 100,000 males per annum. There is a suggestion that, despite considerable fluctuations in the number of cases seen per year, there is a definite increase in incidence, as has also been reported from Scandanavia and from various parts of the United States (1, 2, 3). As in virtually all series reported there was a slight preponderance of right-sided lesions - of 207 cases in which laterality was recorded 52.2% of tumours occurred in the right testis, 47.3% in the left and 0.5% were bilateral.

The age distribution, (Table 4), of our series was fairly typical, approximately 65% of the cases were between 21 and 40 years of age, emphasising the social importance of this disease. When the histological sub-typing is examined, (Table 5), the 42% of cases which could only be classified as "teratoma - type unknown" is disappointing.

It perhaps represents the previous lack of specialised interest in testicular pathology. It was not felt, however, that to fully review all these slides retrospectively would be a practical or rewarding exercise. 15% of cases were classified as "mixed teratoma and seminoma" on purely morphological grounds; no seminomas were reclassified on the basis of marker results. Considering stage at presentation the proportion of cases having lymphangiograms, (Table 6), improved throughout the series, and if the "presumed" Stage I cases are included then 75% of the cases occur in the "early" stage (Stages I & II) group.

Stage III is rare with 4% of cases, and 20% of cases presented with advanced disease - lower than in many series reported from more specialised treatment centres (4). There is no evidence from our analysis to suggest that the proportion of various histological types or the age of patients at presentation changed significantly during the 15 year period. However, there is a suggestion that a larger proportion of cases are seen with "early" (Stage I & II) stage disease at presentation. This would not appear to be due merely to a shorter period between the onset of symptoms and the initiation of treatment. If anything this trend is contrary to that expected from the exploitation of more sophisticated staging techniques such as lymphangiography, CT scanning and marker studies which tend to "upstage" rather than "downstage". It may be worth bearing this suggestion in mind when considering numbers of patients potentially available for clinical trials.

Survival

Perhaps the most interesting analysis was the influence of various major factors in their relation to survival. The overall pattern for the entire series, (Figure 1), shows survival to a maximum of 16 years.

The curve does not plateau until 6 years after diagnosis when 47% of cases are left alive.

When age is related to survival, (Figure 2), the group which does worse is the small number of adolescents and juveniles (aged 11-20 years) whilst the other age groups have few differences about the 50% mark at 6 years.

Histology

This makes a considerable difference to survival. All of the few cases (four) of pure choriocarcinoma (MTT), (Figure 3), were dead by 2 years.

The mixed seminoma/teratoma cases do best, with MTI and MTU being intermediate, MTI doing better than MTU. It is perhaps minimally reassuring to find the "unspecified" group in a mean position, perhaps suggesting that they are a mixture of MTI and MTU cases.

Stage

The most dramatic influence on outcome is seen to be the effects of Stage at presentation, (Figure 4). The Stage I cases do not reach stability until 5 years after diagnosis but have an overall survival approaching 70%. The other stages appear to reach a stable outcome by about 2 years with 60% survival for Stage II and appallingly no survival

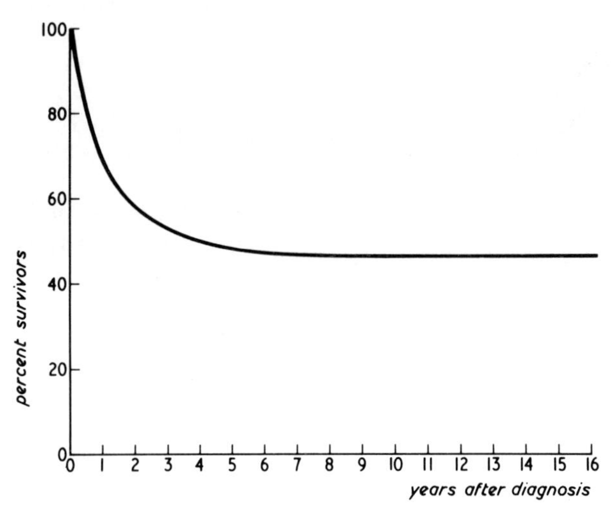

FIGURE 1

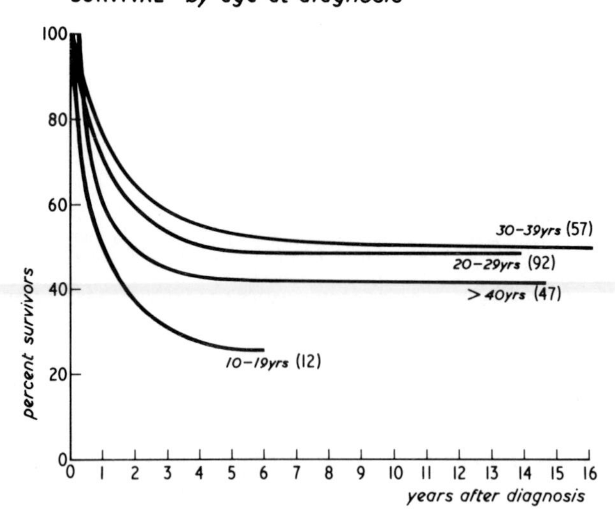

FIGURE 2

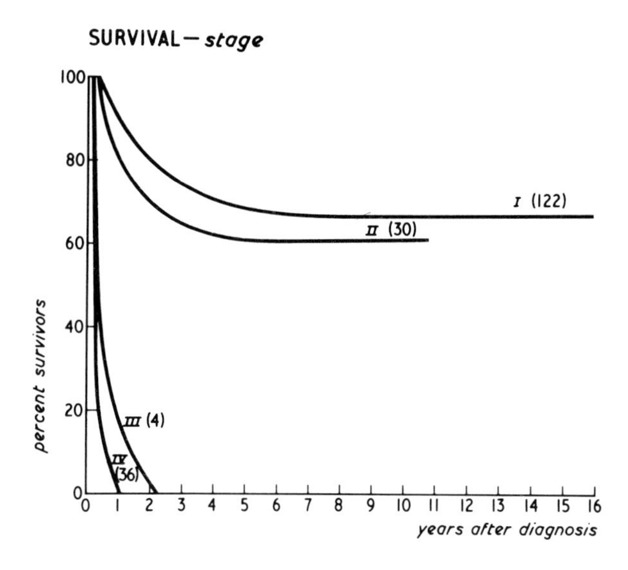

FIGURE 3

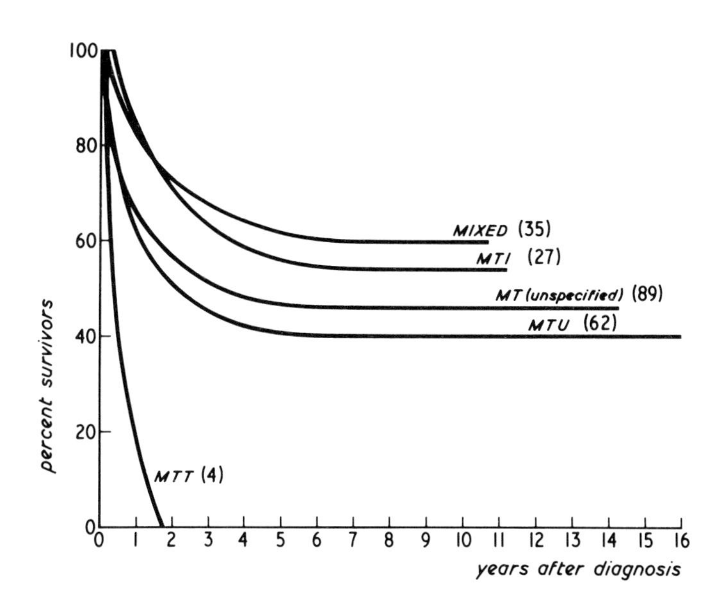

FIGURE 4

of any Stage III or IV case - all being dead by 3 years after presentation.

These data are, as a result of their origin from retrospectively reviewed clinical case notes, incomplete and unsatisfactory, coming as they did at a time when there was little specialised interest in these tumours, inadequate investigation and little hope for those presenting with, or subsequently developing, advanced disease. However, they may have some value to us in helping to identify the areas on which to concentrate our future endeavours. Their deficiencies emphasise the importance of a uniform system for diagnosis, staging and treatment and for the recording of this data.

In this region most testicular tumour slides are now submitted to our Regional Testicular Tumour Panel & Registry, and we have a uniform standard of marker assays performed by the Supraregional Assay Service at Sheffield. The concentration of investigation and treatment of these rare tumours in specialised centres is now generally accepted to be in the best interests of the individual patient and most patients from within this Region are now treated at Cookridge Hospital.

It is hoped that our next analysis will improve on the figures just reported. Since this series was closed there have been fundamental changes in classification, the recognition of the existence and significance of extraembryonic components, the introduction of markers in diagnosis and monitoring, the introduction of CT scanning and the formation of groups of biologists and clinicians with particular interest in these tumours. Along with this has evolved a radical change in treatment philosophy which has led to the current multimodal aggressive approach. All these mean that a series such as we have just reported can only have a limited value as an "historical control group". However, it may act as a baseline against which to measure any improvement that we may achieve and help us to direct our future endeavours.

Although no very definite conclusions can be reached from our data, perhaps the main points to bear in mind are that there is:

1. Possibly an increase in incidence occurring, although fortunately testicular teratomas are still rare tumours.

2. This increase may occur principally in early stage cases.

3. That in a population such as this where advanced disease was treated principally by palliative radiotherapy, single agent or "old fashioned" chemotherapy, stage was the most important factor determining survival.

4. For those patients with advanced disease survival can only improve.

REFERENCES

1. Anon, 1968. An epidemic of testicular cancer? Lancet, 2, 164-165.

2. Clemmensen, J., 1969. Statistical studies in the aetiology of malignant neoplasms - the testis, Acta Pathologica Microbiologica Scandanavia, (Suppl), 209, 15-43.

3. Krain, L.S., 1973. Testicular carcinoma in California from 1942-1969. The California Tumour Registry Experience. Oncology, 27, 45-51.

4. Peckham, M.J., Hendry, W., McElwain, T.J., and Calman, F.M.M., 1977. The multi-modality management of testicular teratomas. In: Adjuvant Therapy of Cancer. Salmon, S.E., and Jones, S.E., (Ed.) Elsevier/North Holland, Amsterdam. pp. 305-320.

5. Peto, R., Pike, M.C., Armitage, P., Breslow, N.E., Cox, D.R., Howard, S.V., Mantel, N., McPherson, K., Peto, J., and Smith, P.G., 1976. Design and analysis of randomised clinical trials requiring prolonged observation of each patient. I. Introduction and design. British Journal of Cancer, 34, 585-612.

6. Peto, R., Pike, M.C., Armitage, P., Breslow, N.E., Cox, D.R., Howard, S.V., Mantel, N., McPherson, K., Peto, J., and Smith, P.G., 1976. Design and analysis of randomised clinical trials requiring prolonged observation of each patient. II. Analysis and examples. British Journal of Cancer, 35, 1-39.

7. Pugh, R.C.B., 1976. Pathology of the Testis. Blackwell, Oxford.

EPIDEMIOLOGY OF GERM CELL TUMOURS

J.A.H. Waterhouse
Director
Birmingham and West Midlands Regional Cancer Registry,
Queen Elizabeth Medical Centre,
Birmingham B15 2TH

In fulfilment of its aim of investigating the behaviour of disease in the group rather than the individual, epidemiology has for long had to rely largely on data from mortality. It is certainly true for international comparisons, where the numbers are substantial and both the deaths and the population are usually well authenticated. Hospital series abound, but their value for epidemiological study is small because the precise description by age and sex of their catchment population is seldom known.

In the field of cancer the development within the last twenty years or so of cancer registries in many parts of the world has greatly improved the epidemiological situation. The "population based" cancer registry (as opposed to the "hospital based", which suffers from the same deficiences described above for hospital series) aims to register all cases of cancer occurring within a defined population, thus enabling the calculation of site, sex and age-specific incidence rates, comparable to mortality statistics but generally of considerably greater accuracy in diagnosis. For a number of sites, histology is more important as providing evidence of malignancy than it is to delineate sizeable sub-groups of tumours at a single site. But there are other sites where the histological breakdown is of considerable importance, and yet most cancer registries provide distributions by site of tumour, and often by histological verification, but not always by histological type.

The UICC decided to sponsor a combined histological and epidemiological study of gonadal tumours, in order to remedy this deficiency for these sites. A group of pathologists from several countries around the world met first to standardise as closely as possible their criteria and nomenclature for these sites, according to the WHO books on histological classification (3, 4). Having examined some test slides and discussed the classificatory system, and having set up as arbiters Dr. F.K. Mostofi for testicular and Dr. R.E. Scully for ovarian tumours, they returned home

TESTIS

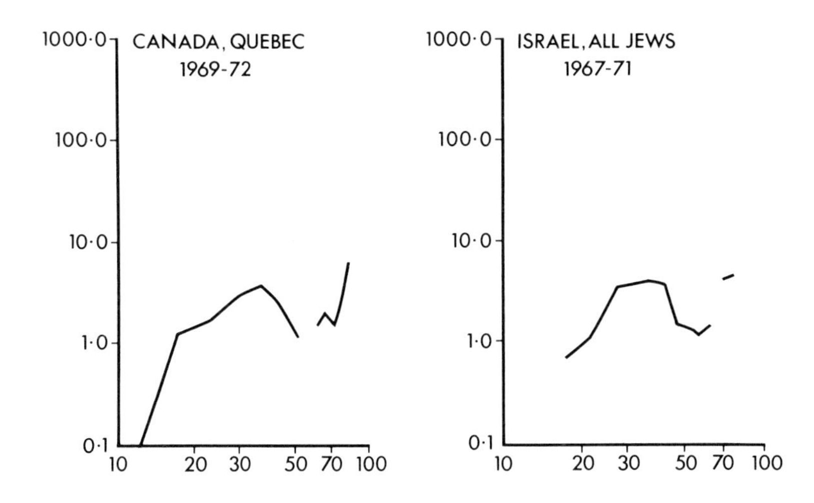

TESTIS

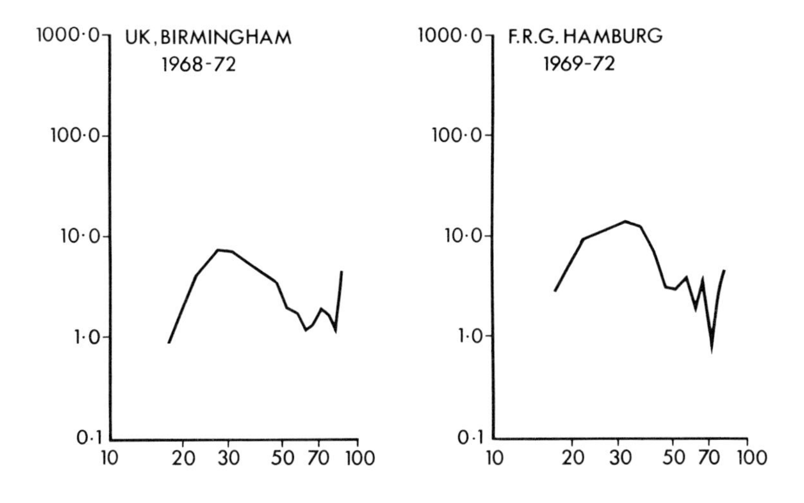

FIGURE 1a

TESTIS

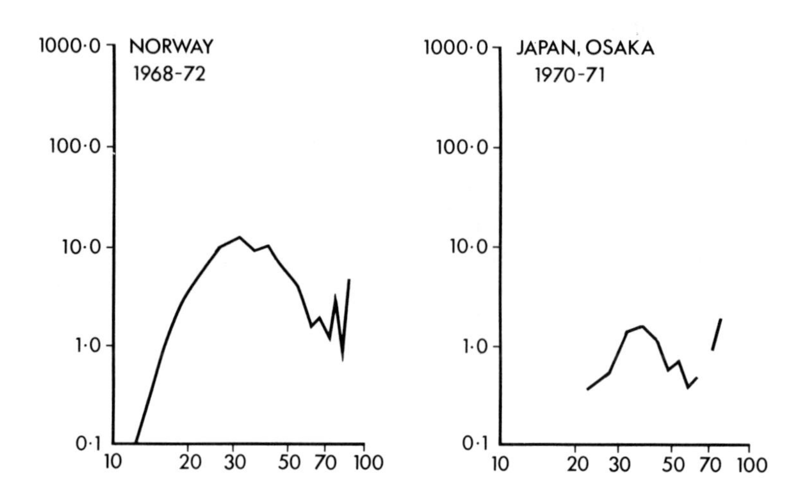

TESTIS

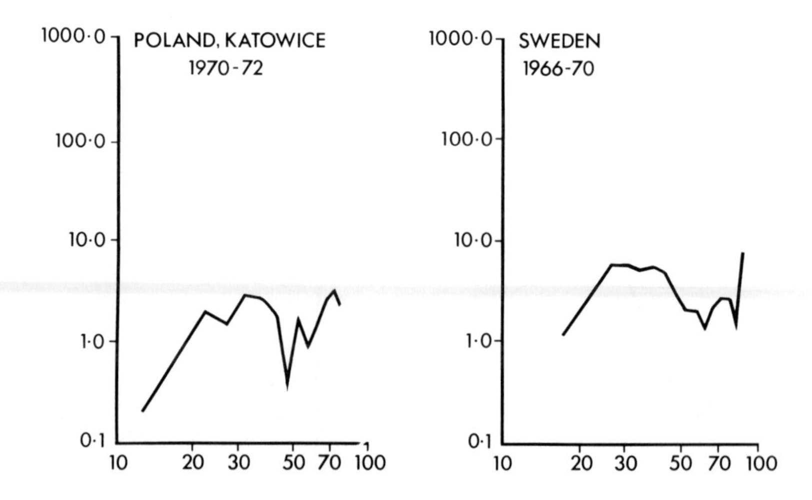

FIGURE 1b

TESTIS

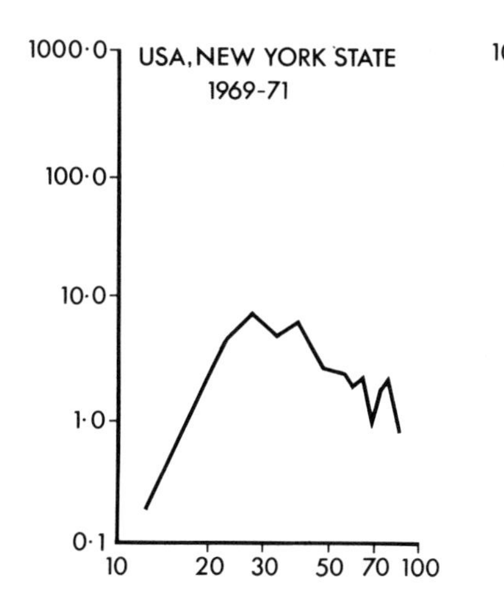

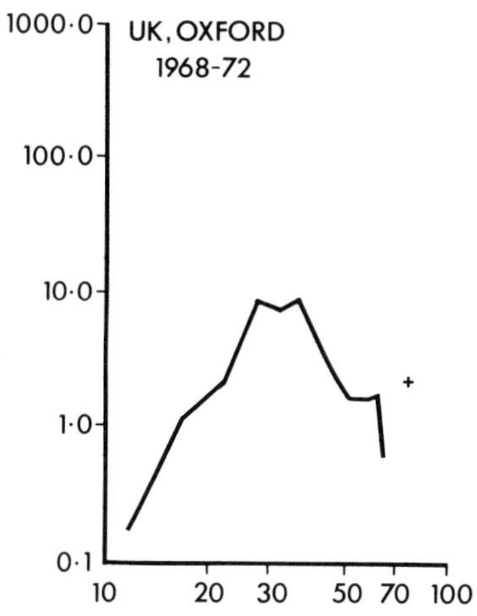

TESTIS

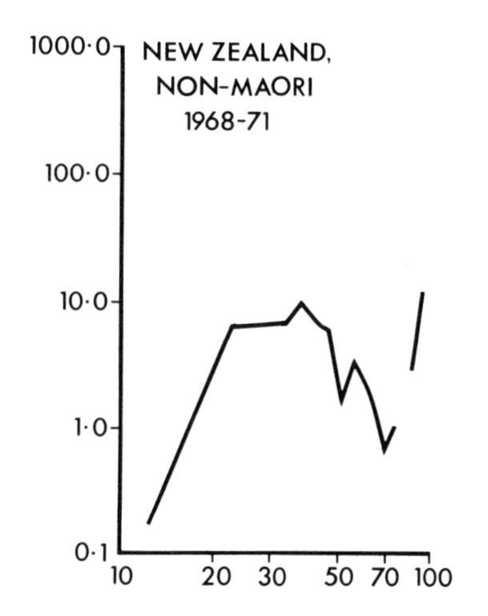

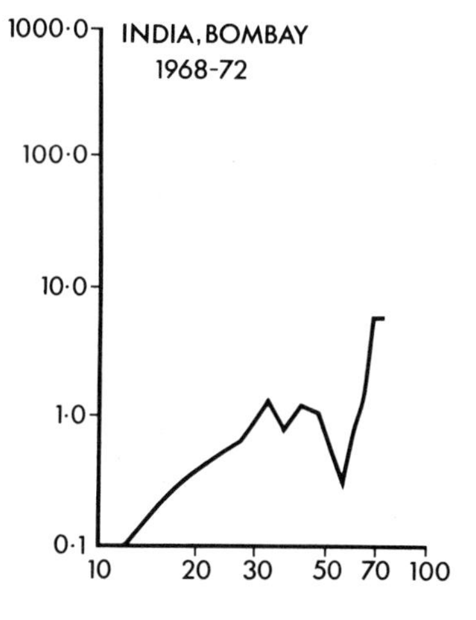

FIGURE 1c

to provide a series, each from his own area, of cases which should be consecutive, and relate to a specific time period, to link with their cancer registry data on incidence.

Testicular Tumours

Figure 1 shows the overall similarity of pattern in the age incidence curves, when plotted on double logarithmic scales, for a dozen of the cancer registries included in the third volume of Cancer Incidence in Five Continents (5). This figure, and Figure 2, are both taken from the UICC publication Cancer Risks by Site (2). Figure 2 shows that testicular tumours are increasing in frequency in most of the registries included in Volumes II and III of Cancer Incidence in Five Continents, since there are more points above the line of equality than there are below it.

For the group of countries included in the UICC Gonadal Tumour Survey, Figure 3 separates the age incidence curves for the seminomas and the other germ cell tumours. The vertical scale is logarithmic but the horizontal is not, but the figure is still closely comparable with those in Figure 1. The highest incidence for seminoma is about ten years later than the peak for other germ cell tumurs. Figures 4 and 5 subdivide, respectively for seminoma and for other germ cell tumours, the source data into groups which behave in similar ways. The first group (I) consists of the UK, New Orleans whites and Norway, and their rates are clearly above all the others. Groups II and III show very little difference and could probably be combined. The general shapes of the curves are still similar, within each histological type, showing for seminoma a sharp peak in the early thirties, declining more slowly thereafter; and for the non-seminomas a peak not so clearly marked, but rather broader, probably attributable to the heterogeneous make-up of this group. The major contributors in our own series from the West Midlands were embryonal carcinoma; embryonal carcinoma and teratoma; and a mixed group of other, these three groups being almost equal to one another in numbers of cases. Survival both of seminomas and for teratomas is comparatively good at more than 80% at five years. The mixed groups show much poorer survivals, and embryonal carcinoma has a five-year figure of 32%.

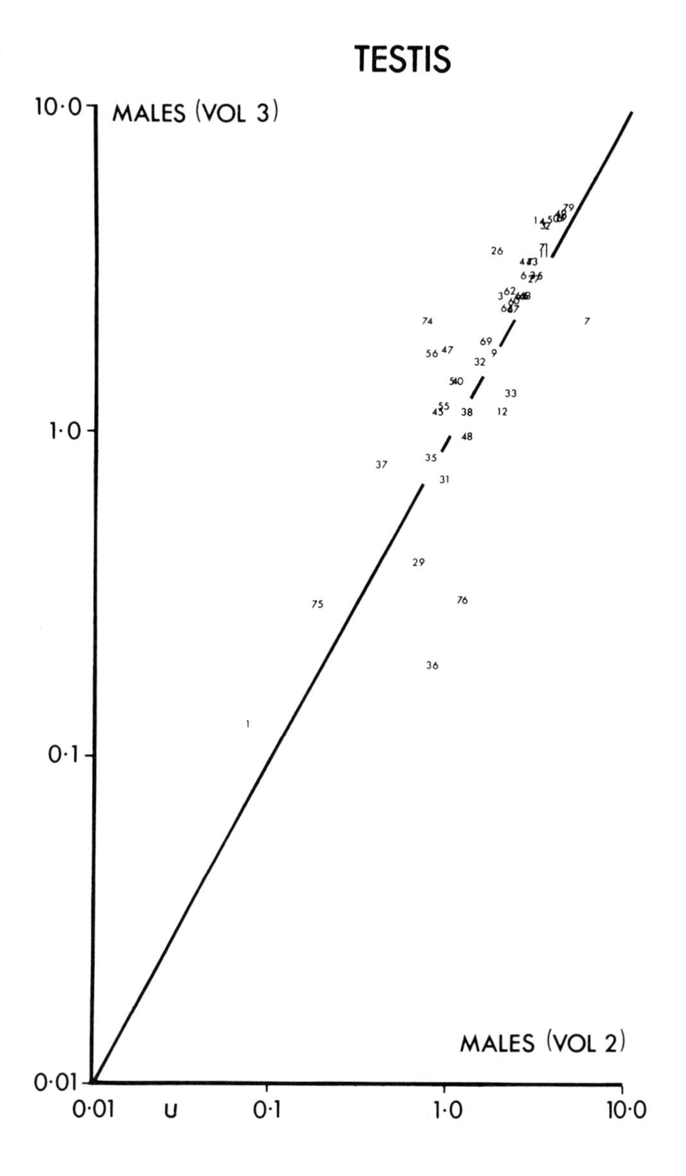

FIGURE 2

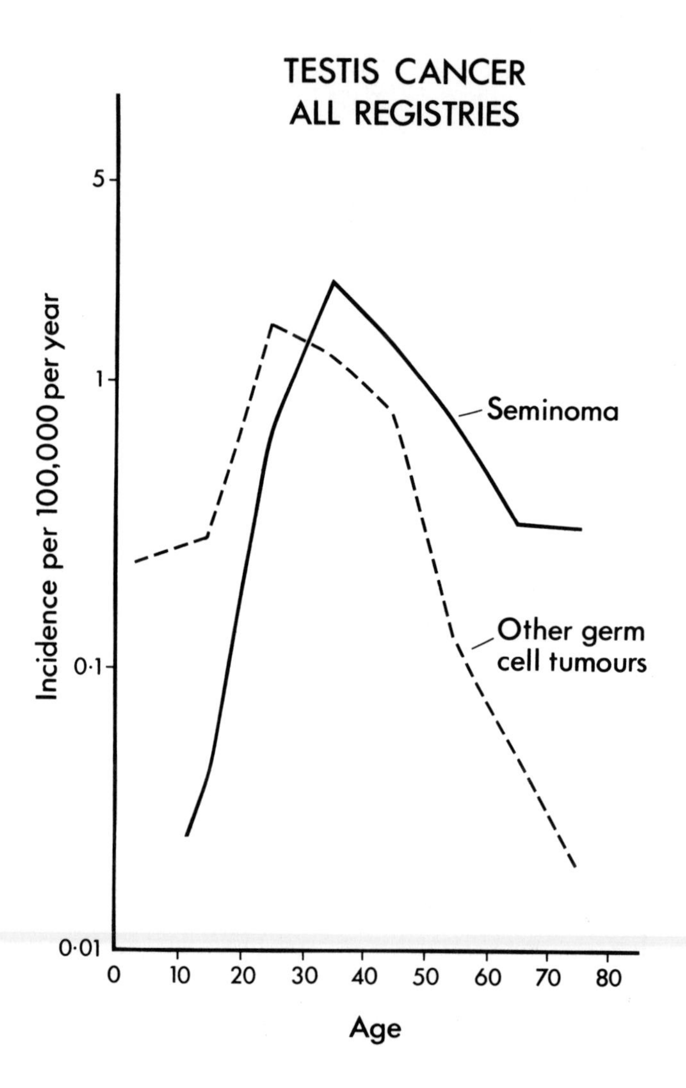
TESTIS CANCER
ALL REGISTRIES
Incidence per 100,000 per year
5
1
0·1
0·01
0
10
20
30
40
50
60
70
80
Age
Seminoma
Other germ
cell tumours

FIGURE 3

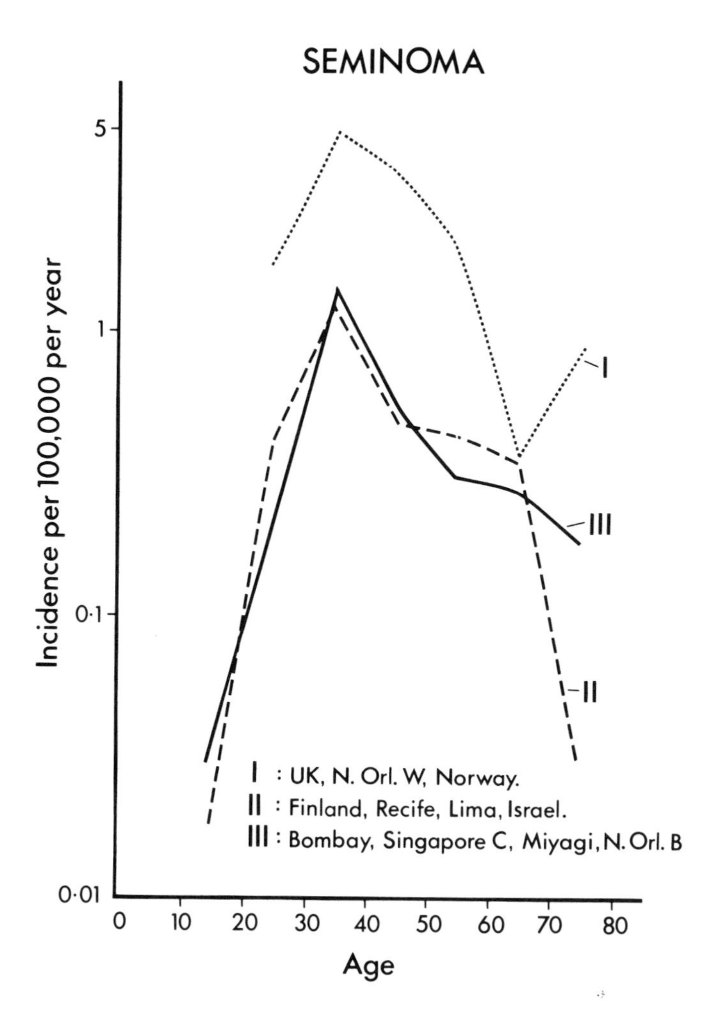
SEMINOMA
Incidence per 100,000 per year
5
1
0·1
0·01
0
10
20
30
40
50
60
70
80
Age
I
II
III
I : UK, N. Orl. W, Norway.
II : Finland, Recife, Lima, Israel.
III : Bombay, Singapore C, Miyagi, N. Orl. B

FIGURE 4

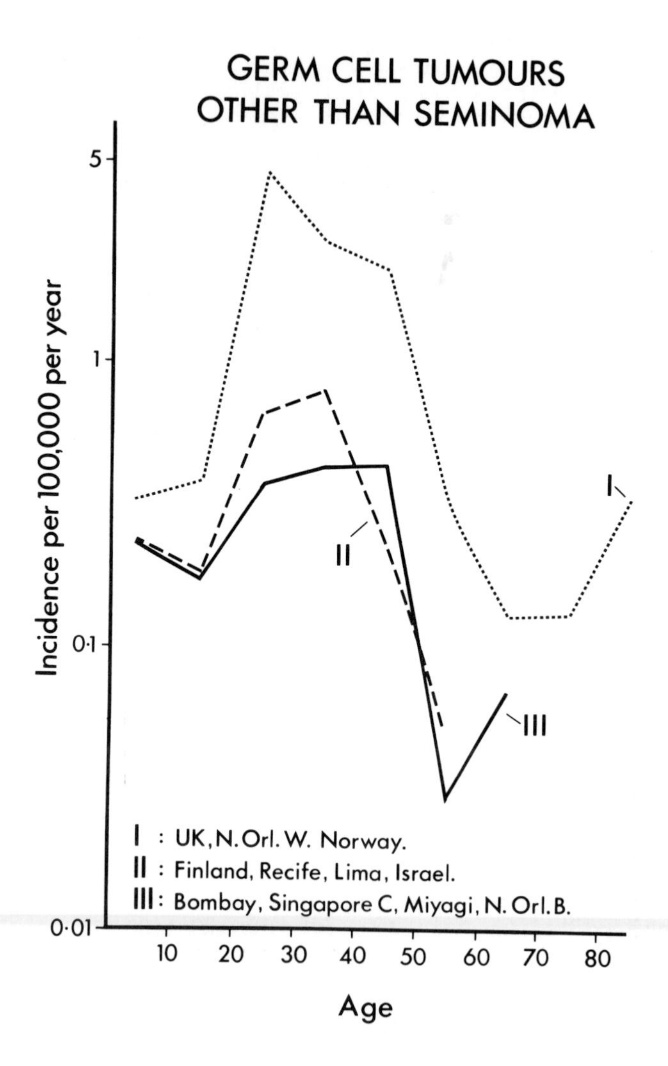
GERM CELL TUMOURS
OTHER THAN SEMINOMA
Incidence per 100,000 per year
5
1
0·1
0·01
I
II
III
I : UK, N. Orl. W. Norway.
II : Finland, Recife, Lima, Israel.
III : Bombay, Singapore C, Miyagi, N. Orl. B.
10
20
30
40
50
60
70
80
Age

FIGURE 5

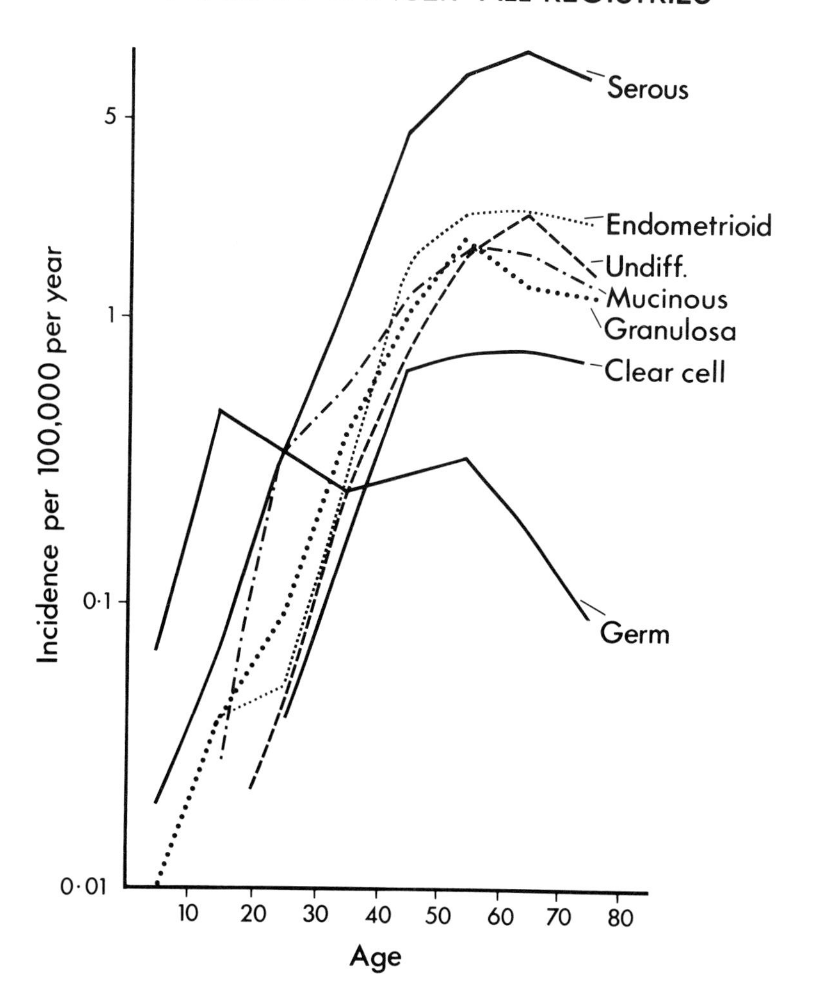
OVARIAN CANCER - ALL REGISTRIES
Incidence per 100,000 per year
5
1
0·1
0·01
10
20
30
40
50
60
70
80
Age
Serous
Endometrioid
Undiff.
Mucinous
Granulosa
Clear cell
Germ

FIGURE 6

Ovarian Tumours

Malignant tumours of the ovary are much more common than those of the testis, but while the majority of testicular tumours are germ cell tumours, it is not so for the ovary. Figure 6 shows age incidence curves for the UICC group of registries by histological type, and it demonstrates clearly the very different pattern for germ cell tumours from the remainder. These of course are those classified as malignant: in the West Midlands series of just over 500 cases of ovarian cancer there were only nine malignant germ cell tumours, four being dysgerminomas. Figure 7 groups together geographically the UICC registries to give comparative age incidence graphs for malignant germ cell tumours. Although very irregular because of small numbers there are discernible similarities in their behavior, except that the African (with New Orleans black) shows a pronounced second peak in the fifties. Survival for the malignant germ cell group of tumours is about 50% at five years.

The UICC study was based on the linkage of histological reports to cancer registry epidemiological data. For this reason it was primarily concerned with malignancy. An attempt was made, however, to assess the number of benign ovarian germ cell tumours from general hospital material available to the pathologists. When this was done, the malignant tumours amounted to less than 5% of the combined group for those of European stock, but formed a larger proportion (up to 22% in Bombay) for non-Europeans.

General

A fuller account of the UICC study will soon be published by the UICC, and I am grateful to my colleagues in that group and to the UICC for being able to draw upon some of those data. In the second volume of Cancer Incidence in Five Continents (1) there is a histological study which includes both ovary and testis. It is however a retrospective study - in contrast to the UICC's prospective one - and the classification used antedates the WHO publications. It should now become possible to organise further studies of this kind, in which histological type distributions are examined prospectively in an epidemiological way among different ethnic and geographically situated groups.

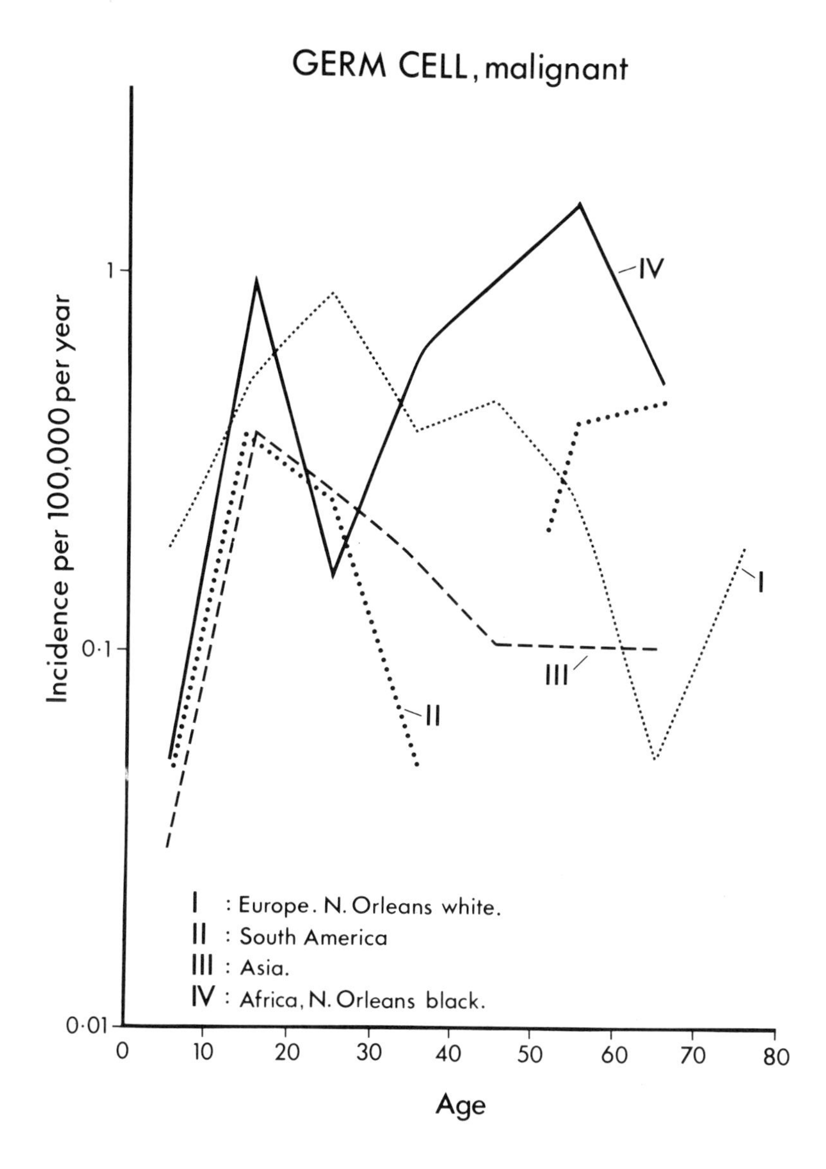
GERM CELL, malignant
Incidence per 100,000 per year
1
0·1
0·01
0
10
20
30
40
50
60
70
80
Age
IV
I
III
II
I : Europe. N. Orleans white.
II : South America
III : Asia.
IV : Africa, N. Orleans black.

FIGURE 7

REFERENCES

1. Doll, R., Muir, C.S., Waterhouse, J., 1970. Cancer Incidence in Five Continents. Volume II, UICC, Geneva.

2. Hirayama, T., Waterhouse, J., Fraumeni, J.F., 1980. Cancer Risks by Site. IARC, UICC, Geneva.

3. Mostofi, F.K., 1977. Histological Typing of Testis Tumours. World Health Organisation, Geneva.

4. Servov, S.F., Scully, R.E., 1973. Histological Typing of Ovarian Tumours. World Health Organisation, Geneva.

5. Waterhouse, J., Muir, C., Correa, P., Powell, J., 1976. Cancer Incidence in Five Continents. Volume III. IARC Scientific Publications No. 15. Lyon.

DISCUSSION

Dr. Jones

Perhaps I can suggest to Mr. Corbett that way back in the dim and distant past in embryology I was always taught that the left testicle enters the scrotum first, and if one believes what I was hinting at earlier about the temperature effects, whether you believe that or not I do not know.

Dr. Rubery

I was very interested in Mr. Corbett's review of his cases. He said that out of his 217 cases he had no cases of adults with differentiated teratomas. That was my belief in the past and over the last two years we have had three patients in Cambridge, all of whom had well differentiated teratoma in their testes but developed metastases elsewhere and I do not remember having that sort of proportion in the four years before that. I wondered if he felt that he had seen cases of well differentiated teratoma recently. Maybe the increased incidence that we are picking up is associated with a slightly different type of tumour.

Mr. Corbett

I think we have all got to watch for that and compare notes. I do not think that we have seen any differentiated teratomas recently in adults. There may have been one or two lurking in our 40% of unknowns but we felt that it was not worth going back. I do not know whether anybody else has any experience of that around the country, whether they have seen any more TDs, it may be a chance observation, maybe not. I think we will all have to compare notes in a couple of years time and see.

Dr. Rubery

You may just happen to get a cluster but it is interesting. They all happen to have massive abdominal disease as well and this makes one wonder whether this is a new phenomenon.

Mr. Corbett

Yes I know a very few cases from the States but they report them very much as isolated occurrences rather than as clusters of cases.

Dr. Anderson

There is one teratoma differentiated out of 200 in the Yorkshire Testicular Tumour Registry occurring in a young man of 27. The alarming feature is that he has an isolated tumour, approximately 2 cms in diameter in the testis, which has been very fully histologised and which appears to be an epidermoid cyst. Unfortunately, this patient also has a considerably elevated level of both AFP and Beta sub-unit of HCG so there is obviously some problem there. This case only presented recently so I cannot tell you anything about progression.

Dr. Rubery

One of our cases presented with haemorrhage in an abdominal mass. He had a small testicular swelling which was explored and all they could find was a small amount of calcified tissue there and they were so convinced that it was not a teratoma that they did not remove the testis but did a biopsy which was reported as well differentiated teratoma. They had to go back and remove the testis itself. So again it was a very small primary and the other patient also had a 1 cm tumour and again it is a very small primary which does suggest there may be some sort of consistency about these cases.

Dr. Grigor

I am sorry that Mr. Corbett decided that it was not worthwhile to look at the histology of 271 cases; it is a big task as I know having done about 300 myself. I think it is important especially in this rather unique series you have, there is very good follow up study to look at very important aspects of tumour classification and especially the yolk-sac element that you talked about at the end. Having said earlier that yolk-sac is not just a dustbin for everything that you cannot classify, I think that if you have got your definite criteria for classifying yolk-sac tumour then about two thirds of teratoma will contain it and

I would have though it would be very useful looking at the histology of these 271 cases.

Dr. Anderson

There is a problem with this series in that it involves something like 35 different regional hospitals and this presents enormous problems in retrieval of the material. If we had a research worker who could attend to this we would certainly do it, but the effort of reviewing the histology is quite beyond us because if means re-cutting all the tissue blocks from these cases and that is a formidable task.

Mr. Corbett

Can I change the subject slightly and ask a very grossly unfair question of Dr. Evans and perhaps of Dr. Waterhouse as well. Amongst the factors said to be related to the epidemiology of testicular tumours are race, nationality, religion, social class, descent, side, temperature, exposure to drugs, where people live, whether it is urban or rural, HLA typing and various other factors. In view of Dr. Evans statement that there is no suggestion of an event leading to transformation that such tumours occur in normal cells, perhaps you would like to suggest to us where we might be able to look for the possible causes in the human populations.

Dr. Evans

Yes, I would be quite prepared to speculate that there are large numbers of possible primary events that never get into any form of clinical expression. I do not know whether normal testes have been looked at in anything like enough numbers to know whether you have burnt out primaries in them, but certainly in the mouse case there is the 1% occurrence in 129 mice that Dr. Graham was talking about and most of those mice lived to a healthy old age without any sign of tumour at all, but indeed there is a 1% incidence and you find it if you do histology on the testis.

Dr. Anderson

Can I be equally unfair and suggest that one of the reasons for the relatively high incidence of spermatocytic seminoma in the United Kingdom series as against foreign series is that I think there are quite different judgement criteria applied to the entity. I think there are a number of cases that would not be acceptable as spermatocytic seminomas in the Armed Forces Institute of Pathology, while many UK pathologists who's judgement I respect would accept them and I think that is one reason that will explain the discrepancy.

Dr. Waterhouse

May I just say that maybe that is a better explanation than ours. If I had gone through all the cases and did really think that it was a bit excessive. I think there may well be something in what you said and we have applied different characteristics.

ANIMAL TUMOUR MODELS AND KINETICS

Chairman of Session
Mr P. J. Corbett

TERATOMAS IN CULTURE

Brigid L.M. Hogan
Imperial Cancer Research Fund,
Mill Hill Laboratories,
Burtonhole Lane,
London NW7 1AD

Clonal lines of embryonal carcinoma (EC) cells derived from murine teratocarcinomas (spontaneous or embryo-derived) differ widely in their ability to differentiate in vitro. Pluripotent lines can only be maintained as homogeneous populations of undifferentiated stem cells with frequent subculturing, and they start to differentiate as soon as the culture become locally dense (14), or when small aggregates of cells are transferred to bacteriological petri dishes to which they cannot adhere (12, 13). Under these conditions differentiation leads to the formation of structures known as embryoid bodies because they resemble the foetal portion of the early post-implantation mouse embryo (13). Simple embryoid bodies consist of an outer epithelial layer of endoderm cells surrounding a solid core of EC cells. In more complex or cystic embryoid bodies the inner cells differentiate into ectoderm-like tissue, surrounding a 'proamniotic cavity', and mesodermal cells, which give rise to blood islands and spontaneously contracting cardiac muscle. Further differentiation, into keratinizing epithelium, cartilage and nerve, for example, can be observed if cystic embryoid bodies are allowed to reattach and spread on a tissue culture surface. At the other end of the spectrum, nullipotent EC cells do not differentiate at all, even after prolonged culture as aggregates in suspension. Because they do not differentiate in vivo or in vitro, the classification of these cells as embryonal carcinoma is based solely on properties which they share with bona fide pluripotent EC cells, for example morphology, expression of embryonic surface antigens (such as the 'F9' (1, 14), Forssman (19), and SSEA-1 (17) antigens), lack of expression of the major histocompatability H2 and β-microglobulin proteins (11, 12, 14), insusceptibility to infection by various oncogenic viruses (11, 15, 23), and high levels of alkaline phosphatase enzyme activity. Although, as we have seen, pluripotent EC cells differentiate in response to local high density or aggregation, these culture conditions are difficult to control in such a way that the starting population is homogeneous and biochemical changes occur synchronously and reproducibly. Conditions have, therefore, been sought in which EC cells which normally show only a low level of spontaneous differentiation (and are therefore easy to maintain as homogeneous populations of undifferentiated cells) differentiate rapidly in response to a simple chemical signal. At present the most successful of these systems is that first described by Strickland and Mahdavi (20) in which EC cells of the F9 clonal line differentiate into endoderm in response to the addition of retinoic acid, a derivative of Vitamin A. This process occurs over a period of 2-4 days and has been observed in a number of other cell lines (10), although details of the response are not always the same.

In order to discuss the response of F9 cells to retinoic acid it is necessary to refer to the early embryology of the mouse (Figure 1) and, in particular, to the differentiation of the extraembryonic tissues known as parietal and visceral endoderm. There is evidence (5) that these cells are derived from a common percursor population known as the primitive or primary endoderm (Prim End) which, in turn, arises from cells on the blastocoele surface of the inner cell mass during the fourth day of development. Prim End cells which migrate onto the trophectoderm differentiate into parietal endoderm, while cells which remain in contact with the egg cylinder and, later, with the extraembryonic mesoderm, differentiate into visceral endoderm. The two endoderm types have a very different morphology (Figure 1) and also express very different sets of genes. Parietal endoderm cells, for example, synthesise large amounts of the high molecular weight, extracellular matrix proteins laminin and Type IV procollagen (6, 7, 16) which they lay down in the form of a thick basement membrane known as Reichert's membrane. Visceral endoderm cells, on the other hand, make relatively little extracellular matrix (6), but do synthesise and secrete large amounts of alphafoetoprotein (AFP) (4). There is evidence that the phenotype of early endoderm cells can be modulated by interaction with other embryonic tissues. Dziadek (3), for example, has shown that endoderm cells of the 7.5 d embryo which synthesise AFP (as judged by immunoperoxidase staining with specific antiserum) when cultured alone or in contact with embronic ectoderm, are prevented from doing so by interaction with trophoblast-derived cells. Hogan and Tilley (9), on the other hand, have shown that the expression of the parietal endoderm phenotype and the synthesis of laminin and Type IV procollagen is promoted by culture of endoderm cells in contact with trophoblast. While these modulations in phenotype observed in culture make sense in embryological terms (since parietal endoderm is underlaid by trophoblast, and visceral endoderm by ectoderm or mesoderm) the mechanism by which they are achieved are not yet understood. They do, however, illustrate the fact that what a cell makes of itself depends not only on its embryological ancestors but also on its present neighbours.

Returning to F9 EC cells and their response to retinoic acid, it was first reported that the cells appearing in the treated cultures expressed the phenotype of parietal endoderm. This was based on their mophology, synthesis of collagen (20) and basement membrane antigens (18) and absence of AFP (protein or mRNA) production (21). The fact that EC cells differentiate first into endoderm, rather than mesodermal derivatives or nerve, for example, is compatible with other evidence that the cells are biochemically similar to cells of the inner cell mass and embryonic ectoderm of the early post-implantation embryo (12). Experiments in which treatment of F9 cells with retinoic acid was combined with, or followed by, exposure to dibutyryl-cyclic-AMP, showed that virtually all of the cells in a culture could be converted into parietal endoderm over a period of about 4 days (22). The cells have a rounded morphology (see Figure 2C) and synthesise large amounts of laminin and Type IV procollagen (Ref. 22, and see Figure 3A) and do not make AFP (8). Recently, however, evidence has been presented (8) that F9 cultures can give rise to cells which morphologically resemble visceral rather than parietal endoderm and synthesise AFP, provided that the retinoic acid treatment is carried out under certain conditions. One factor favouring visceral endoderm production seems to be close contact between cells which have already differentiated into endoderm and remaining EC cells. This is achieved either when monolayer cultures become locally dense and cells pile up into clumps, or when aggregates of F9 cells are seeded into bacteriological petri dishes. As shown in Figure 2E, such aggregates exposed to retinoic acid differentiate into structures resembling simple embryoid bodies with an outer epithelial layer of visceral endoderm-like cells. These cells synthesise AFP, as judged by immuno-precipitation of radioactively labelled culture medium (Figure 3B) and immunoperoxidase staining using specific anti-AFP immunoglobulin (8). Aggregates exposed to retinoic acid and dibutyryl-cyclic-AMP do not develop this clearly-defined outer layer of visceral endoderm but instead generate rounded, poorly adherent cells resembling parietal endoderm (Figure 2F). On the basis of these observations it has been proposed (8) that F9 cells exposed to retinoic acid differentiated first into a cell-type analogous to the primitive endoderm of the early mouse embryo. In the absence of intercellular contacts (in sparse monolayer cultures, for example) these cells then differentiate further into parietal endoderm, but if they can interact with EC cells (in dense monolayers or in aggregates) then they differentiate into visceral endoderm. In either case, exposure to dibutyryl-cyclic-AMP promotes expression of the parietal endoderm phenotype. Within this framework, further experiments are being carried out to understand the relative roles of retinoic acid, cyclic AMP and cell contact in the differentiation of F9 embryonal carcinoma cells into either parietal or visceral endoderm.

Legend for Figure 1. Stages in Early Mouse Embryogenesis

At the time of implantation the blastocyst consists of an outer vesicle of trophoblast (diagonal lines) surrounding the inner cell mass (ICM). ICM cells nearest to the blastocoele cavity differentiate into primitive endoderm (Prim End) while the remainder become the epiblast or embryonic ectoderm (stippled). Following implantation, trophoblast cells invade the uterus and primitve endoderm cells migrate on to the trophoblast and differentiate into parietal endoderm, laying down a thick acellular basement membrane (Reichert's membrane). As shown in the photograph, parietal endoderm cells have few intercellular contacts and contain a well-developed endoplasmic reticulum filled with material similar to the matrix of Reichert's membrane. At the time of primitive streak formation (6.5-7.0 days p.c.) mesodermal cells (black) differentiate from the ectoderm of the egg cylinder. Extraembryonic mesoderm cells migrate beneath the visceral endoderm and contribute towards the visceral yolk-sac. Visceral endoderm cells (photograph) form a simple cuboidal epithelium, have apical tight junctions, numerous microvilli and pinocytotic vacuoles and only a thin endoplasmic reticulum. Scale bar = 2μ.

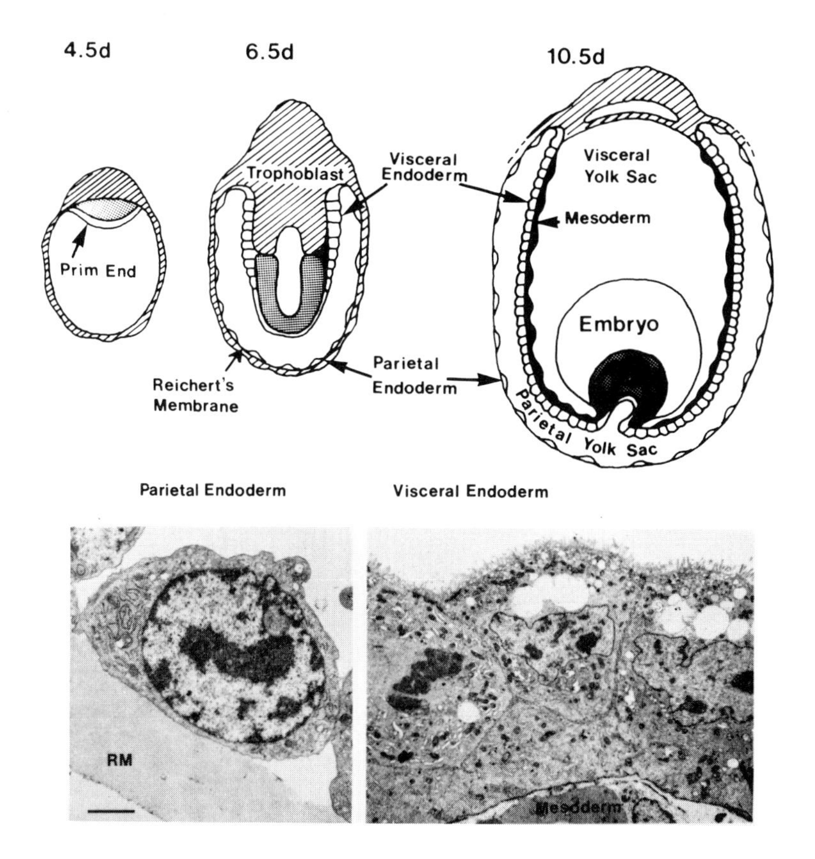
4.5d
6.5d
10.5d
Trophoblast
Visceral Endoderm
Visceral Yolk Sac
Mesoderm
Prim End
Embryo
Reichert's Membrane
Parietal Endoderm
Parietal Yolk Sac
Parietal Endoderm
Visceral Endoderm
RM
Mesoderm

FIGURE 1

Legend for Figure 2.

Morphology of F9 cells in monolayer or suspension culture after exposure to retinoic acid and dibutryl-cyclic-AMP. (A) Control F9 cells in monolayer culture (B) Monolayer culture 4 days after adding 1 x 10^{-7} M retinoic acid (Sigma, all-trans). (C) Monolayer culture 4 days after adding 1 x 10^{-7} M retinoic acid and 10^{-3} M dibutyryl cyclic AMP (Sigma). (D) Control F9 aggregate after 8 days in suspension culture. (E) Aggregate 7 days after treatment with 5 x 10^{-8} M retinoic acid. When viewed in the electron microscope, many of the outer cells resemble visceral endoderm (Figure 1 and Ref. 8). Even after prolonged culture, these embryoid-body-like structures do not differentiate tissues other than endoderm. (F) Aggregate 5 days after exposure to 1 x 10^{-7} M retinoic acid and 10^{-3} M dibutyryl cyclic AMP. In the electron microscope, these outer cells resemble parietal endoderm. Scale bar = 50 μm.

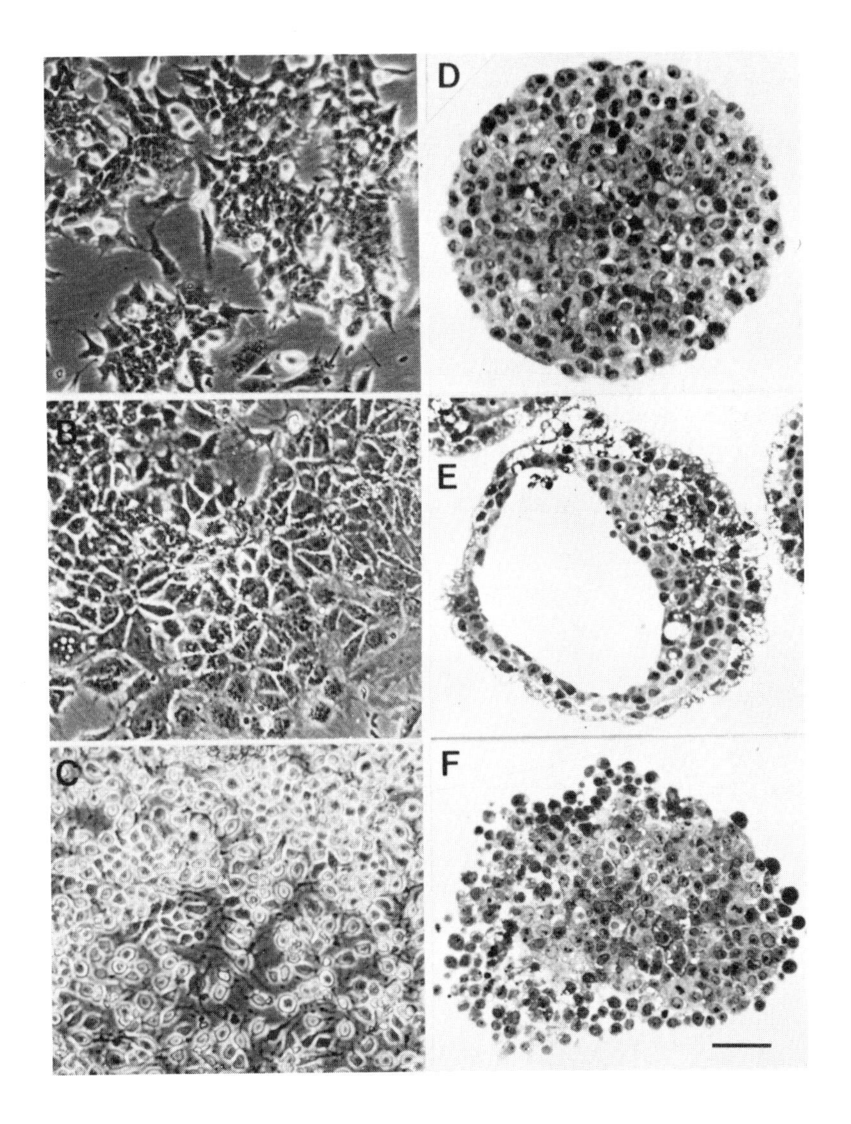

FIGURE 2

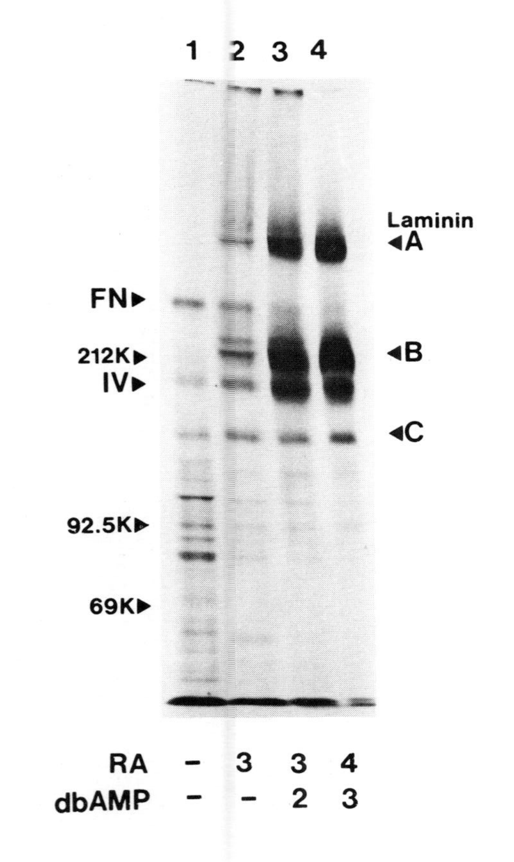
1 2 3 4
Laminin
A
B
C
FN
212K
IV
92.5K
69K
RA - 3 3 4
dbAMP - - 2 3

(A)

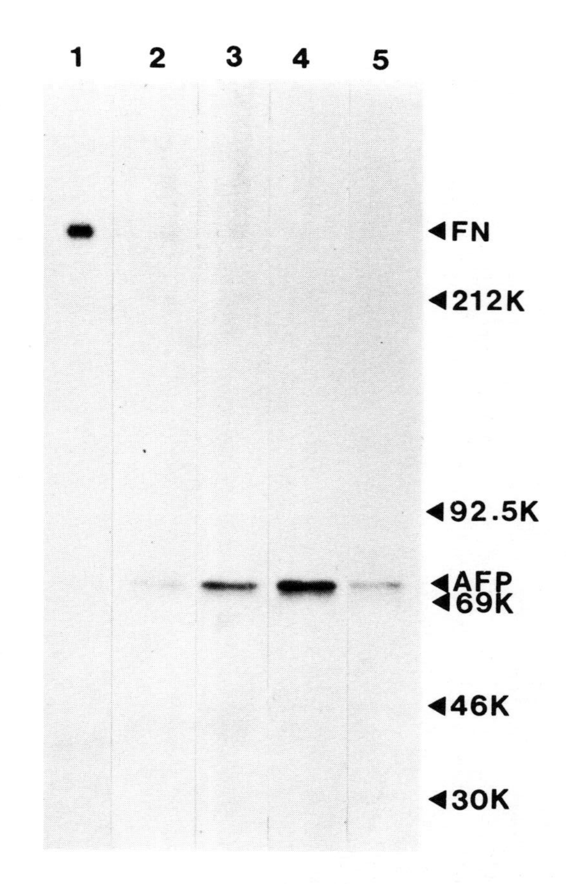
1 2 3 4 5
FN
212K
92.5K
AFP
69K
46K
30K

(B)

FIGURE 3

Legend for Figure 3.

Biochemical changes induced in F9 cells by retinoic acid and cyclic-AMP. (A) Synthesis of extracellular matrix proteins. F9 cells in monolayer culture were incubated for 10 hours with 50 μci/ml ^{35}S Methionine and 0.5 mm $\alpha\alpha$'-dipyridyl, an iron chelating agent which inhibits the hydroxylation of lysine and proline in collagen. The culture medium was harvested and high molecular weight proteins precipitated with a final concentration of 10% (w/v) polyethylene glycol. The proteins were separated on a 5-10% gradient SDS polyacrylamide gel under reducing conditions and radioactive bands visualised by fluorography. Radioactive marker proteins were included in the gel and their position is shown on the left hand side. All methods are described in references 6 and 7. Track 1, proteins synthesised by control F9 cells. The polypeptide with a molecular weight of approximately 265K has been identified as fibronectin (FN) (7) . Track 2, proteins made by F9 cells after three days exposure to 1 x 10^{-7} M retinoic acid. Tracks 3 and 4, proteins made by F9 cells after three and four days exposure to retinoic acid respectively, with 10^{-3} M dibutyryl cyclic AMP being added after the first day. The polypeptides with molecular weight of approximately 450K, 230K and 150K can be immunoprecipitated by antibodies to laminin (9). In these experiments, incubation with $\alpha\alpha$-dipyridyl results in the secretion of underhydroxylated Type IV procollagen (IV) which migrates slightly faster than the B chains of laminin (9). Note that treatment with retinoic acid and cyclic AMP results in the shut-down of fibronectin synthesis.(B). Synthesis of alpha-foetoprotein by F9 aggregates exposed to retinoic acid. Aggregates pre-treated for 5 days with retinoic acid were then incubated for 10 hours with 100 μci/ml ^{35}S Methionine and the culture medium immunoprecipitated with specific rabbit antimouse AFP immunoglobulin and protein A-sepharose. The samples were analysed on 5-10% gradient SDS polyacrylamide gels under reducing conditions and radioactivity visualised by fluorography. All methods are described in Reference 8. Track 1, absence of AFP synthesis by control aggregates not exposed to retinoic acid. Some fibronectin is non-specifically trapped in the protein A-sepharose. Tracks 2-5, AFP synthesised by aggregates exposed to 10^{-6}, 10^{-7} , 5 x 10^{-8} and 10^{-8} M retinoic acid respectively. The radioactive polypeptide of approximately 70K molecular weight co-migrated with AFP immunoprecipitated from the culture medium of 10.5 day visceral yolk-sacs (8).

REFERENCES

1. Artzt, K., Dubois, P., Bennett, D., Condamine, H., Babinet, C., and Jacob, F., 1973. Surface antigens common to mouse cleavage embryos and primitive teratocarcinoma cells in culture. Proceedings of the National Academy of Sciences of the United States of America. 70, 2988-2992.

2. Bernstine, E.G., Hooper, M.L., Grandchamp, S., and Ephrussi, B., 1973. Alkaline phosphatase activity in mouse teratoma. Proceedings of the National Academy of Sciences of the United States of America. 70, 3899-3903.

3. Dziadek, M., 1978. Modulation of alphafoetoprotein synthesis in the early post-implantation mouse embryo. Journal of Embryology and Experimental Morphology. 46, 135-146.

4. Dziadek, M., and Adamson, E.D., 1978. Localisation and synthesis of alphafoetoprotein in post-implantation mouse embryos. Journal of Embryology and Experimental Morphology. 43, 289-313.

5. Gardner, R.L., 1978. The relationship between cell lineages and differentiation in the early mouse embryo. In: Results and Problems in Cell Differentiation (Ed.) Gehring, W.J., Springer Verlag, Berlin. pp. 205-241.

6. Hogan, B.L.M., 1980. High molecular weight extracellular proteins synthesised by endoderm cells derived from mouse teratocarcinoma cells and normal extraembryonic membranes. Developmental Biology. 76, 275-285.

7. Hogan, B.L.M., Cooper, A., and Kurkinen, M., 1980 Incorporation into Reichert's membrane of laminin-like extracellular proteins synthesised by parietal endoderm cells of the mouse embryo. Developmental Biology. 80, 289-300.

8. Hogan, B.L.M., Taylor, A., and Adamson, E., 1981. Cell interactions modulate the differentiation of F9 embryonal carcinoma cells into parietal or visceral endoderm. Nature (in press).

9. Hogan, B.L.M., and Tilley, R., 1981. Cell interactions and endoderm differentiation in cultured mouse embryos. Journal of Embryology and Experimental Morphology (in press).

10. Jetten, A.M., Jetten, M.E.R., and Sherman, M.I., 1979. Stimulation of differentiation of several murine embryonal carcinoma cell lines by retinoic acid. Experimental Cell Research. 124, 381-391.

11. Knowles, B.B., Pan, S., Solter, D., Linnenbach, A., Croce, C., and Huebner, K., 1980. Expression of H-2, laminin and SV40T and TASA on differentiation of transformed murine teratocarcinoma cells. Nature. 288, 615-618.

12. Martin, G.R., 1980. Teratocarcinomas and mammalian embryogenesis. Science. 609, 768-776.

13. Martin, G.R., Wiley, L.M., and Damjanov, I., 1977. The development of cystic embroid bodies in vitro from clonal teratocarcinoma stem cells. Developmental Biology. 61, 230-244.

14. Nicolas, J.F., Avner, P., Gaillard, J., Gennet, J.L., Jacobs, H., and Jacob, F., 1976. Cell lines derived from teratocarcinomas. Cancer Research. 36, 4224-4231.

15. Segal, S., and Khoury, G., 1979. Differentiation as a requirement for simian virus 40 gene expression in F9 embryonal carconoma cells. Proceedings of the National Academy of Sciences of the United States of America. 76, 5611-5615.

16. Smith, K.K., and Strickland, S., 1981. Structural components and characteristics of Reichert's membrane, an extra-embryonic basement membrane. Journal of Biological Chemistry (in press).

17. Solter, D., and Knowles, B.B., 1978. Monoclonal antibody defining a stage-specific mouse embryonic antigen (SSEA-1). Proceedings of the National Academy of Sciences of the United States of America. 75, 5565-5569.

18. Solter, D., Shevinsky, L., Knowles, B.B., and Strickland, S., 1979. The induction of antigenic changes in a teratocarcinoma stem cell lines (F9) by retinoic acid. Developmental Biology. 70, 176-182.

19. Stern, P.L., Willison, K.R., Lennox, E., Galfre, G., Milstein, C., Scher, D., Ziegler Springer, 1978. Monoclonal antibodies as probes for differentiation and tumour-associated antigens: a Forssman specificity on teratocarcinoma stem cells. Cell, 14, 775-783.

20. Strickland, S., and Mahdavi, V., 1978. The induction of differentiation in teratocarcinoma stem cells by retinoic acid. Cell, 15, 393-403.

21. Strickland, S., and Sawey, M.J., 1980. Studies on the effect of retinoids on the differentiation of teratocarcinoma stem cells in vitro and in vivo. Developmental Biology. 78, 76-85.

22. Strickland, S., Smith, K.K., and Marotti, D.R., 1980. Hormonal induction of differentiation in teratocarcinoma stem cells: generation of parietal endoderm by retinoic acid and dibutyryl cAMP. Cell, 21, 247-355.

23. Teich, N.M., Weiss, R.A., Martin, G.R., and Lowy, D.R., 1977. Virus infection of murine teratocarcinoma stem cell lines. Cell, 12, 973-982.

LOSS AND RECOVERY OF THE ABILITY TO DIFFERENTIATE IN CLONED LINES OF MOUSE EMBRYONAL CARCINOMA CELLS

Elizabeth Robertson
Department of Genetics,
University of Cambridge,
Downing Street,
Cambridge CB2 3EH

Introduction

Many embryonal carcinoma (EC) cell lines have been isolated from both spontaneously occurring and induced teratocarcinomas and established as permanent tissue culture lines. EC cells vary considerably in their capacity to differentiate such that distinctions have been made between pluripotent cell lines and nullipotent or non-differentiating cell lines. Pluripotent lines can be further subdivided according to the degree to which they differentiate under in vitro and in vivo conditions. For example, some lines differentiate extensively in tissue culture and as tumours, while others differentiate well only in vivo. Other lines show restricted differentiation capacity forming tumours predominantly of one cell type (5).

Nullipotent lines resemble undifferentiated pluripotential cells morphologically, biochemically and in the expression of certain cell surface antigens. However, they fail to differentiate under normal conditions which optimise differentiation, indeed these cells have been used extensively as controls for the undifferentiated state. It has been assumed that this apparent lack of differentiation capability arises from a stable alteration in gene structure, for example gene deletion or mutation, which prevents cellular differentiation thus blocking nullipotent cells in the stem cell state. Recently, putative nullipotent lines have been examined more carefully for the ability to differentiate.

Sherman and Miller (8) reported that the cell line F9 shows a low level of differentiation both in vitro and in vivo forming small numbers of endodermal cells, and concluded that all cells within the population were capable of forming endoderm with a low probability.

Studies involving the use of biochemical inducers of

differentiation have provided evidence that other nullipotent lines are capable of differentiating. Retinoids and retinoic acid in particular have been shown to cause differentiation of several nullipotent lines to at least one type of differentiated cell (3, 10). The stable phenotypic alteration induced by retinoic acid indicates that nullipotent cells do possess the genetic information that is prerequisite for at least some type of differentiation.

As yet no studies of the differentiation ability of EC cell lines have been directed towards an analysis of the response by individual cells within a culture population. Data presented here from clonal analysis of pluripotent and nullipotent lines reveals heterogeneity in lines previously assumed to be homogeneous with respect to differentiation ability. Detailed cloning studies carried out on the nullipotent cell line Nulli SCC2A shows it to be composed of subpopulations of cells which in turn exhibit characteristic levels of differentiation.

Materials and Methods

When aggregates of certain pluripotent EC cell lines are cultured in suspension they are stimulated to differentiate. The primary event is the formation of a layer of endodermal cells surrounding the core of undifferentiated stem cells (6). Endoderm cells can be discerned after 24 hours culture, indeed alterations in protein synthesis are detected 12 hours after the start of suspension culture (4).

This observation provides the basis of a technique designed specifically to test the differentiation capabilities of cell lines in terms of the competence of constituent cells to differentiate. The method involves screening colonies of single cell origin for ability to form endoderm. Formation of endoderm was used as the criterion for differentiation ability for two reasons; it is the first cell type to appear, facilitating more rapid screening, and also endoderm cells are easily distinguishable from EC cells.

The cell lines used were the pluripotent lines PSMB and PSA4TG (OTT5568) (7, 9), EC10 and EC12, two clonal derivatives of PSA4GT selected for impaired ability to differentiate, and the nullipotent line Nulli SCC2A (6). Pluripotent lines were maintained as undifferentiated cultures on feeder layers of mitomycin treated STO fibroblasts. Nulli SCC2A was maintained on gelatin coated tissue culture plates. Routine culture and cloning experiments were carried our in Dulbecco's Modified Eagle's Medium supplemented with 10% newborn calf serum.

The stages of clonal analysis are outlined in Figure 1. Cell cultures are treated with 0.25% trypsin-EDTA for 5 minutes at room temperature. After this time medium plus serum is added and the cells pipetted vigorously to create a single cell suspension. Cellular aggregates are removed by gravity sedimentation. $3 - 5 \times 10^3$ cells are seeded onto a preformed feeder plate and incubated for 2 days, after which time the cell colonies have reached a diameter of 0.6 - 0.8 mm. Colonies are removed from the feeder layer individually using the sealed end of a finely drawn out pasteur pipette, and transferred into a bacteriological petri dish to which they are unable to adhere. After 4 days in suspension culture colonies are replated onto tissue culture dishes and left undisturbed for 2 days, during which time they reattach and spread. Individual colonies are scored under a x40 binocular microscope for the presence of visceral and/or parietal endoderm cells. Between 5×10^2 and 10^3 colonies are scored per assay.

Clonal lines were established by dissociating individual colonies and plating into 1.6 cm petri dishes together with a suspension of feeder cells. After 7 days the cells are expanded onto 6 cm feeder plates and when confluent retested. Clonal lines were maintained exclusively on feeder layers. Maintenance of EC cells on gelatin coated surfaces alone seems to reduce the ability with which mass aggregates differentiate during long term culture (2) and appears to increase possibilities of alterations in chromosome number.

Results

Differentiation of pluripotent and nullipotent cell lines

Levels of differentiation of five cell lines, assayed by scoring differentiation of primary clones, are summarised in Figure 2a. Pluripotent lines PSMB and PSA4TG display a level of almost 100% of primary clones forming endoderm (over 2×10^3 colonies scored per line). PSA4TG EC10, and EC12, selected for impaired ability to differentiate in vitro, have a reduced frequency of differentiation on clonal

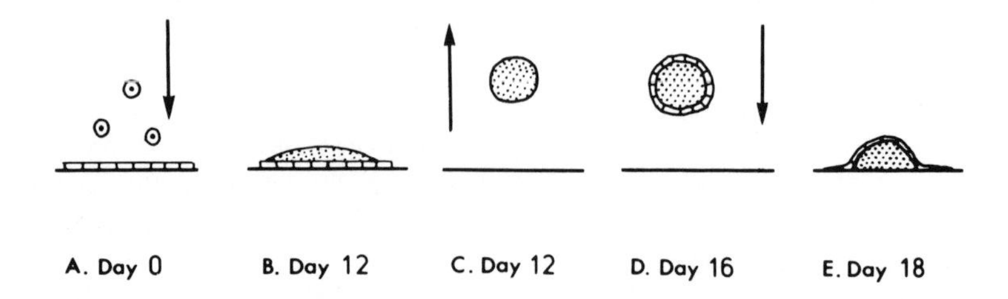

FIGURE 1 - Stages of clonal analysis

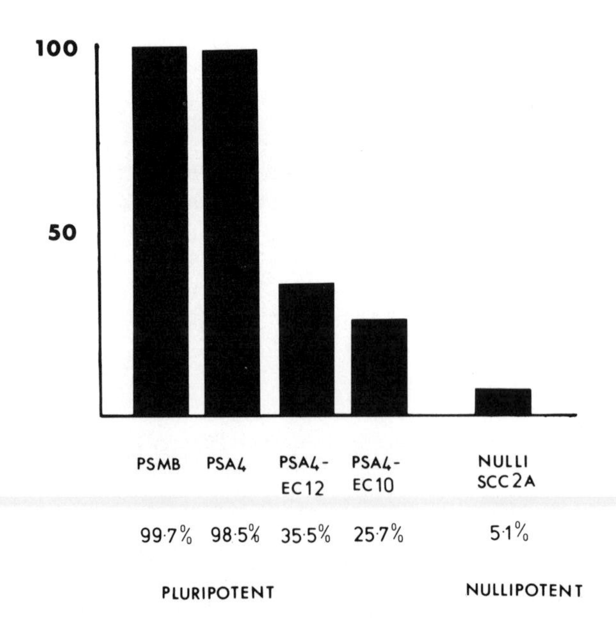

FIGURE 2a - Levels of differentiation in pluripotent and nullipotent cell lines.

analysis; 25% and 35% of clones forming endoderm, respectively. Nulli SCC2A exhibits a level of 5% of primary clones forming endoderm. The level ranges from 2% to 10% on independent testings each of about 10^3 colonies.

Differentiation of Nulli SCC2A

Clonal analysis of the starting population reveals that on average 5% of primary clones are scored as differentiating, the remaining 95% fail to form endoderm and are scored as non-differentiating. There are two possible explanations for this result; firstly, all cells within the population are able to differentiate but only do so with a very low probability, or secondly, the cell population is composed of subpopulations of differentiation competent and incompetent cells. To distinguish between these two possibilities the following cloning strategy was adopted. Primary clonal colonies were scored for endoderm, selected and established as cell lines. These primary clonal lines (distinguished as + or - according to state of differentiation of colony of origin) were retested; subclones of each line scored for the percentage of endoderm forming (dif+) colonies. Thirty-seven primary clonal lines were tested in this manner, twenty-two established from differentiating colonies (termed N+ lines) and fifteen from non-differentiating colonies (termed N- lines). the results of scoring are summarised in Figure 2b. The level of differentiation (5% interval classes) is plotted against the relative frequency of occurence of primary lines whose differentiation level falls within that class. Data from primary N+ lines and N- lines are presented separately. The majority of N- lines exhibit low levels of differentiation; 80% of N- lines have differentiation levels below 10%. The remainder show elevated levels over that seen in the starting population, ranging up to 35 - 40% of subclones forming endoderm. Amongst the primary N+ lines there is a more varied distribution of ability for differentiation. A few lines show levels of less than 10% differentiation while some lines have greatly elevated levels of differentiation (50% - 60%). The majority of N+ lines show intermediate levels ranging between 10 - 30%.

The result of this primary cloning step is to generate clonal lines which display markedly different potentials for differentiation. The stability of the patterns of differentiation of primary clonal lines was retested by a further cloning step. Three primary N+ lines were selected

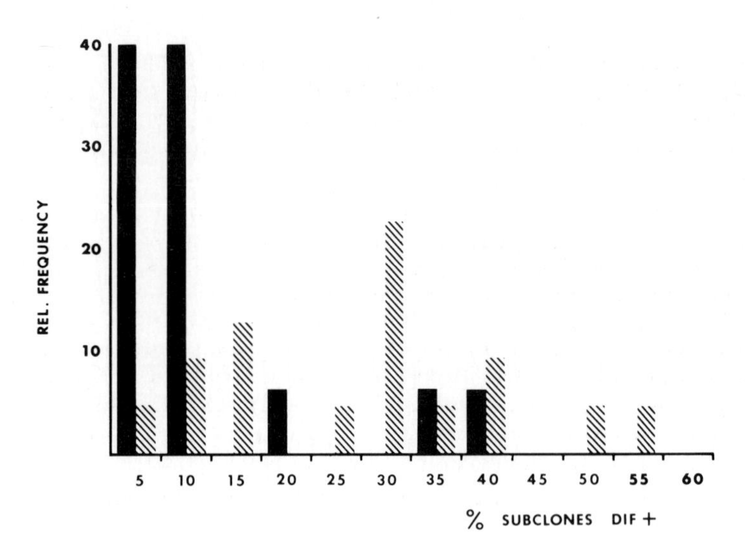

FIGURE 2b - Relative frequency of differentiation levels displayed by Nulli SCC2A primary clonal lines. (hatched bars - primary N+ lines; solid bars - primary N- lines)

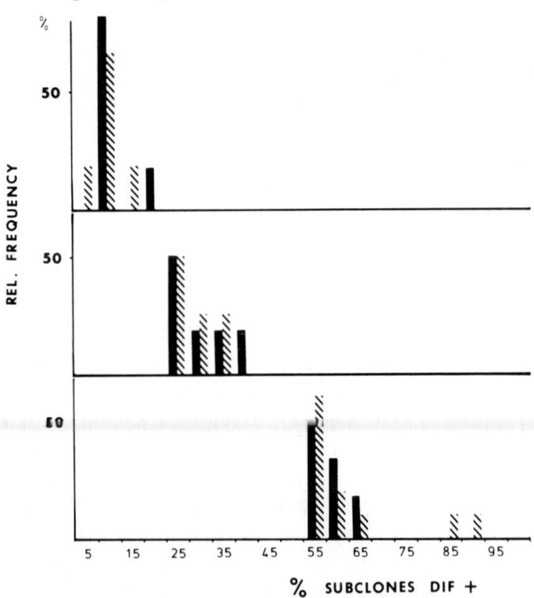

FIGURE 2c - Relative frequency of differentiation levels displayed by Nulli SCC2A secondary clonal lines. (hatched bars - secondary N+- lines; solid bars - secondary N+- lines)

from each of the following differentiation classes; 10 - 15%, 30 - 35%, and 50 - 60%. Equal numbers of colonies of differentiating and non-differentiating phenotype were picked and secondary clonal lines, termed N++ and N+- respectively, established. Approximately twelve lines were established from each primary line and the three sets of lines reassayed (Figure 2c). Data from N++ and N+- lines are presented separately. It would appear that the potential for differentiation exhibited by primary clones is stable; levels of differentiation among each set of secondary clonal lines is a reflection of the level displayed in the primary line, irrespective of whether the lines are N++ or N+-.

The secondary clonal line which showed the best differentiation exhibited levels of endoderm formation comparable with those of pluripotent lines. To test the stability of this reversion a set of twelve tertiary lines were derived from the secondary N++ line, equal numbers being established from differentiating (N+++) and non-differentiating (N++-) colonies. As predicted, on reassaying, all sublines showed patterns of differentiation similar to that of the parental line, the majority falling in the classes 90 - 95% and 95 - 100%. No differences were noted between the distribution of N+++ and N++- lines.

Finally, differentiation of nine primary lines was followed during the course of long term culture. The lines were selected as being representative of the varied levels of differentiation shown by primary N+ lines. The lines were maintained under identical conditions of culture and assayed after 2, 6, 15, 18, and 20 passage generations. The results are shown in Figure 3. It appears that, overall, levels of differentiation are stable although some lines fluctuate to a greater degree. Differentiation levels are sensitive to variations in culture conditions, for example, all lines show slightly elevated levels of endoderm formation after 6 passage generations.

The correlation between levels of endoderm formation in vitro and the potential for differentiation into different cell types was investigated by examination of the extent to which N+ lines differentiate in vivo. Preliminary results show that N+ lines displaying a high level of in vitro endoderm formation (35 - 40%) form tumours in which several differentiated tissues are discernable (neural tissue, cartilage, epithelial cells) in addition to cells of EC morphology. Tumours formed from lines which differentiate poorly (less than 10%) are composed predominantly of EC cells.

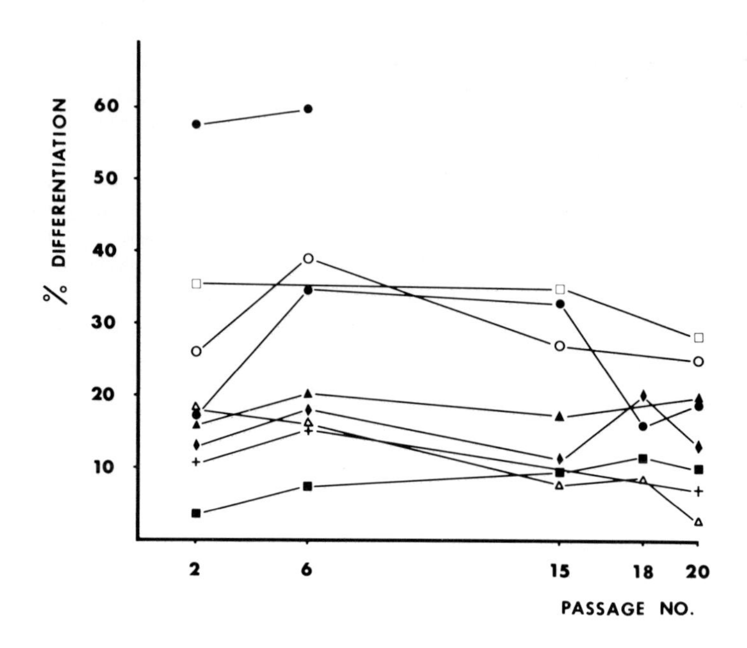

FIGURE 3 - Levels of differentiation of primary Nulli SCC2A lines during long term culture.

Discussion

At present EC cell lines are loosely classified according to the apparent degree to which they differentiate in vitro and in vivo. The technique of clonal analysis outlined here provides a simple method of quantifying differentiation of potential of EC cell lines. There seem to be a good correlation between the level of endoderm formation in culture and the extent to which a given line differentiates under in vivo conditions. Clonal analysis may also facilitate a more sensitive method of monitoring the influence of altered conditions of culture on the differentiation response of a cell population.

The cell line Nulli SCC2A has been reported not to differentiate under normal culture conditions. A more careful scrutiny by a technique of clonal analysis shows the cell population to be composed of subpopulations of variable differentiation potential. These subpopulations are revealed only after a step to establish clonal lines which can then be individually assayed. There is a considerable range in diversity of differentiation ability in primary clonal lines. The majority of constituent cells are revealed as poor differentiators, although all thirty-seven primary clonal lines assayed showed endoderm formation amongst subclones. Deriving primary lines from colonies exhibiting a differentiated phenotype enriches the probability of selecting lines with an elevated capacity for endoderm formation, and it has been possible to recover, at a low frequency, clonal lines whose differentiation ability approaches those seen in pluripotential lines. Procedures involving repeated cloning steps show the pattern of differentiation exhibited by primary clonal lines to be stable, the patterns being repeated in secondary and tertiary clonal lines. Subsequent to the primary cloning step, the state of differentiation of the founding colonies does not influence differentiation in the resulting lines. This conclusion is further verified by long term culture experiments which provide evidence that differentiation potential is inherited stably through successive cell generations.

Given this observed degree of stability, one approach when considering the factors which could govern differentiation ability, is an investigation of chromosome number. The cell line Nulli SCC2A is characterised by a modal number of forty-four, including two major translocations, and is perhaps one of the more karyotypically abnormal EC cell lines. Many EC cell lines are aneuploid, although most lines closely approach the diploid condition (1). Preliminary data from chromosome counts strongly implicate an association between chromosome number and degree of differentiation; the secondary clonal line which shows the highest level of differentiation has a modal number of forty chromosomes. More detailed karyological data are now being collected to test whether loss of chromosomes results in a more balanced karyotype, and, if so, whether loss and recovery of the ability to differentiate can be correlated with loss and recovery of specific chromosomes.

REFERENCES

1. Graham, C.F., 1977 In: Concepts in Mammalian Embryogenesis (Ed.) Sherman, M.I. MIT Press, Cambridge, Massachusetts, pp. 315-394.

2. Hogan, B.L.M., 1976. Changes in the behaviour of teratocarcinoma cells cultivated in vitro. Nature. 263, 136-137.

3. Jettern, A.M., Jettern, M.E., Shaprio, S.S., and Poon, J.P., 1975. Characterisation of the action of retinoids on mouse fibroblast cell lines. Experimental Cell Research. 119, 289-299.

4. Lovell-Badge, R.H., and Evans, M.J., 1980. Changes in protein synthesis during differentiation of embryonal carcinoma cells and a comparison with embryo cells. Journal of Embryology and Experimental Morphology. 59, 187-206.

5. Martin, G.R., 1975. Teratocarcinoma as a model system for the study of embryogenesis and neoplasia. Review Cell. 5, 229-243.

6. Martin, G.R., and Evans, M.J., 1975. Differentiation of cloned lines of teratocarcinoma cells. Formation of Embryoid bodies in vitro. Proceedings of the National Academy of Sciences of the United States of America. 72. 1441-1445.

7. Martin, G.R., and Evans, M.J., 1975. Multiple differentiation of clonal teratocarcinoma stem cells following embryoid body formation in vitro. Cell. 6, 467-474.

8. Sherman, M.I., and Miller, R.A., 1978. F9 embryonal carcinoma cells can differentiate into endoderm-like cells. Developmental Biology. 63, 27-34.

9. Slack, C., Morgan, R.H., and Hooper, M.I., 1978. Isolation of metabolic co-operation-defective variants from mouse embryonal carcinoma cells. Experimental Cell Research. 117, 195-205.

10. Strickland, S., and Mahdavi, V., 1978. The induction of differentiation in teratocarcinoma stem cells by retinoic acid. Cell, 15, 393-403.

PROLIFERATIVE CHARACTERISTICS OF CELL POPULATION IN GERM CELL TUMOURS OF THE TESTIS

Rosella Silvestrini
Cellular Kinetics & Pharmacokinetics,
Istitutio Tumori,
Via Venezian 1,
20133 Milan, Italy

The kinetics of cell proliferation is a biological aspect of testicular tumours which has been little investigated but which is likely to yield important information of considerable clinical relevance.

In the present study, it is intended to analyse the proliferative activity of testicular tumours with a two-fold aim. First, to obtain additional information on the biological character of tumours by determining the kinetics of different histological patterns in pure and mixed forms as well as those of the same histological patterns in differing anatomical sites. Secondly, to assess the relationship of cell kinetics to the biological aggressiveness of the tumour through an analysis of any correlation between proliferative activity and the clinical or pathological stage and to assess the prognostic significance of such analyses.

Proliferative activity was determined by autoradiography using a ^{3}H thymidine labelling index (L.I.). Most patients entered into the study had already been treated by orchidectomy but had not received radiotherapy or chemotherapy at the time of the kinetic determination. The kinetic determinations showed a broad similarity of L.I. values in both embryonal carcinomas and seminomas with partial overlapping of values. Statistical analysis, however, showed a median L.I. value significantly higher ($p < 0.01$) in embryonal carcinomas (44%) than in seminomas (13%). Mature teratomas showed a very low proliferative activity with a median L.I. value of 0.6%. A more detailed analysis of the kinetic behaviour of embryonal carcinomas showed no significant difference in proliferative activity in this type of tumour whether in pure or mixed forms. Similarly, for embryonal carcinomas similar median L.I. value were obtained in primary growths and lymph node metastases. A direct correlation was observed between proliferative activity and clinical or pathological stage in both embryonal carcinomas and seminomas in pure forms. The assessment of the prognostic significance of cell kinetics require a longer follow-up of the patients.

The accepted relevance to prognosis of the clinical and pathological stage and the finding of a relationship between the proliferative activity of a cell population and the bulk of the tumour could make the L.I. a potentially useful marker in monitoring individual tumours and in identifying patients at risk of recurrence.

REGRESSION AND REGROWTH OF TESTICULAR TERATOMA LUNG METASTASES WITH DIFFERENT CHEMOTHERAPY REGIMES

D. Ash
Consultant Radiotherapist,
Cookridge Hospital,
Leeds LS16 6QB

Until a few years ago it was common for patients with metastases from testicular teratoma to be given a number of different chemotherapy regimes, each of which would generally produce some degree of regression of the tumour, but would soon be followed by regrowth and eventual death. A retrospective analysis has been performed on 16 such patients in whom more than one episode of regression and regrowth has occurred under the influence of different chemotherapy regimes. Serial chest x-rays were analysed and the area of clearly defined rounded metastases was measured. These measurements were converted to volume and a graph of changes in volume against time was constructed. From these graphs, it was possible to measure the growth rate of the tumour (doubling time), the regression rate (halving time), the period of regrowth delay between intiation of treatment and regrowth to the pre-treatment size and the percentage volume reduction produced by each treatment. Because of the difficulty of interpreting growth rates from only two points all doubling and halving time measurements have come from growth curves having a minimum of three points.

The mean doubling time was 19.8 days and showed a log-normal distribution. The mean halving time was 13.6 days and did not show a log-normal distribution suggesting that the regression rate of the tumour may not be due entirely to intrinsic properties of the tumour and the host removal factors but might be influenced by the treatment given.

The mean regrowth delay period was 84 days which is equivalent to approximately 5.6 doubling times. The overall mean survival was 442 days from the time of diagnosis of lung metastases to death and this represents a mean of 25.4 doubling times.

No significant correlation was found between doubling time and halving time, regrowth delay and halving time, or survival and doubling time. Comparison of halving times induced by four different chemotherapy regimes and x-ray treatment showed no significant differences. Halving times within patients receiving different treatments showed that the range of halving times for different treatment within the same patient was much smaller than that within patients overall, and suggested that the regression rate induced by successive treatments in the same patient was similar.

Nine patients had more than one measurable deposit and in six the regression and regrowth of metastases within the same patient appeared identical. In three patients, each with three measurable metastases, one showed a pattern of regression and regrowth different from that of the other two confirming the expected heterogeneity within these tumours.

Studies of this type allow much valuable information to be gained about the response of testicular teratomas to chemotherapy and at a time when considerable attention is being paid to animal models and their relevance to human treatment it should not be forgotten that there remains much that can be learnt from observation of tumour responses in patients.

RADIOLABELLED MONOCLONAL ANTIBODIES FOR THE LOCALISATION OF HUMAN TERATOMA XENOGRAFTS IN VIVO

V. Moshakis, R.A.J. McIlhinney and A.M. Neville
Ludwig Institute for Cancer Research
(London Branch,)
Royal Marsden Hospital,
Sutton, Surrey SM2 5PX

A mouse monoclonal antibody, LICR-LON/HT13, raised against cell surface components of a human germ cell tumour xenograft and shown to bind to the cells in vitro, was used to ascertain whether it would localise in this tumour in vivo when growing as a xenograft in immune suppressed mice.

Tumour-bearing mice received intravenously approximatly equal amounts (10-15 μCi) of ^{125}I-labelled specific monoclonal antibody and ^{131}I-labelled non-specific (control) monoclonal IgG simultaneously. At time intervals of up to four days after injection, the animals were sacrificed and radioactivity in blood, tumours and organs was counted in an LKB 1280 ultra-gamma counter. Antibody localisation in the tumour was examined at the cellular level by autoradiography.

The tumour uptake of monoclonal antibody was 17-56 times higher than in normal tissues. The tumour blood ratio reached 10 at four days post-injection. This selective localisation did not occur with the control monoclonal IgG. Localisation was also observed in other human germ cell tumours, but not in non-germ cell tumour xenografts. The degree of localisation was higher in smaller tumours ($<$200 mg).

Autoradiography showed that the injected monoclonal antibody was localised to the viable parts of the tumour and in close association with the tumour cell surface (Figure 1).

Monoclonal antibodies, therefore, are capable of selective tumour localisation, at least in the animal model of human tumour xenografts. On the basis of the above results clinical studies have been initiated.

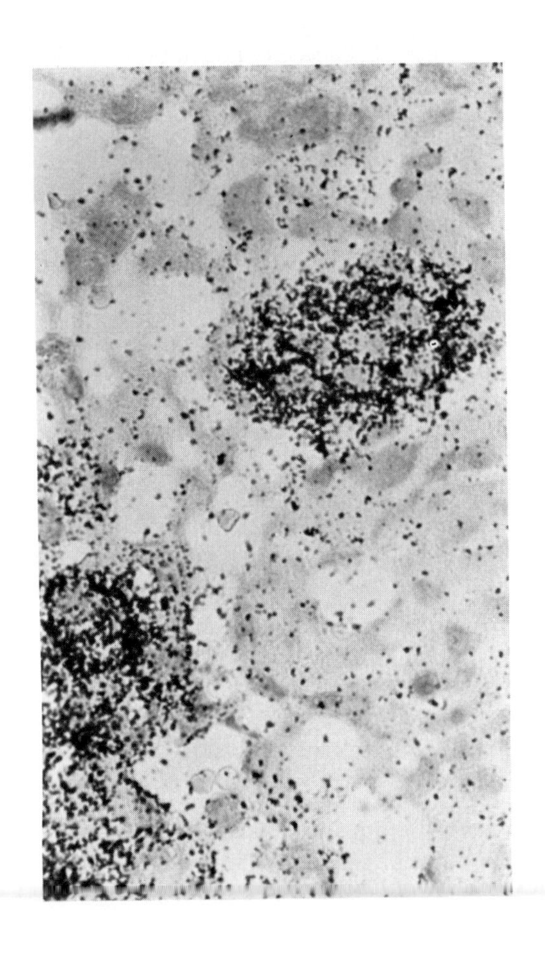

FIGURE 1 - Autoradiograph of a germ cell tumour xenograft labelled in vivo with monoclonal antibody. Note the association of grains (antibody) to the periphery of tumour cells (x 250).

A PEANUT LECTIN BINDING PROTEIN PRESENT ON HUMAN TERATOMA DERIVED CELL LINES

R.A.J. McIlhinney and S. Patel
Ludwig Institue for Cancer Research,
(London Branch),
Royal Marsden Hospital,
Sutton, Surrey SM2 5PX

In the characterisation of human teratoma derived cell lines and the study of their differentiation in vitro it would be useful to have a marker for the undifferentiated embryonal carcinoma cell phenotype. Since murine embryonal carcinoma cell lines have been shown to bind the lectin peanut agglutinin (PNA) and to loose this ability on differentiation (5), the ability of human teratoma derived cell lines to bind PNA has been examined using the cell lines Tera 1, Tera 2, (3), PA1 (6), Susa (4), HX39/7, T_3 B_1 and PL7 (2). All of the cell lines except PL7 and T_3 B_1 bound the PNA as measured by surface fluorescence. These two cell lines were regarded as more differentiated than the others in terms of morphology, ultrastructure and surface characteristics. In order to identify the membrane components responsible for this PNA binding the cell surface glycoproteins from membrane prepared from the cell lines were reported in SDS-polyacrylamide gels, and the position of the PNA binding protein determined by overlaying the gels with iodinated PNA (1). All of the cell lines shared considerable homology in their PNA binding protein composition with Tera 1, Susa and HX39 alone of the teratoma lines having a major PNA binding component of 200,000 molecular weight (Figure 1).

This 200,000 D protein was not present on any of the human tumour cell lines tested derived from differentiated human tissues. Furthermore, neuraminidase treatment of these cell lines did not reveal this protein, indicating that its non-appearance was not simply due to glycosylation differences between the different cell lines. Incubation of the 200,000 D protein positive cell lines with radioactive amino acids, followed by affinity chromatography of cell lysates on immobilised PNA resulted in the isolation of the 200,000 D protein labelled with radioactivity. Therefore the protein is actively synthesised by the cells and not absorbed by them from foetal calf serum in the culture medium. Analysis of the culture supernatents from these radioactive labelling experiments showed that the cells did not secrete the protein.

Given the restricted distribution of this protein on the teratoma derived cell lines it could provide a useful marker for a particular cell type in human teratomas.

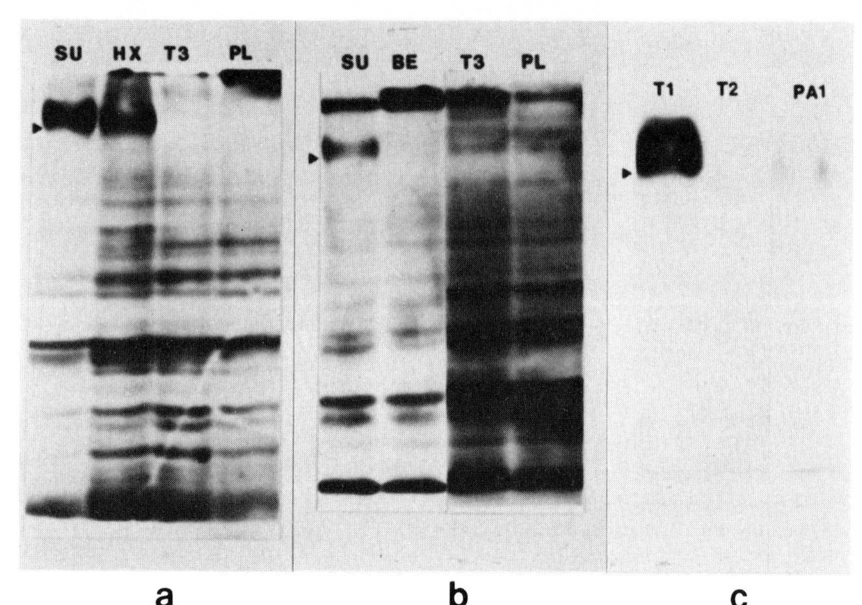

FIGURE 1 - PNA overlay of membrane glycoproteins from the human teratoma derived cell lines, SuSa (Su), HX39/7 (HX), T_3B_1 (T_3), PL7 (PL), Tera 1 (T1), Tera 2 (T2), PA1 and the human choriocarcinoma BeWo (BE). The separation of the proteins was accomplished on 7.5% SDS-polyacrylamide gels and in all cases 100 micrograms of membrane glycoproteins were loaded in each track. The gels a, b and c were run on separate occasions, and (▶) indicates the position of the myosin marker.

REFERENCES

1. Burridge, K., 1976. Changes in cellular glycoproteins after transformation: Identification of specific glycoproteins and antigens in sodium dodecyl sulphate gels. Proceedings of the National Academy of Sciences of the United States of America. 73, 4457-4461.

2. Cotte, C., 1980. PhD Thesis, University of London.

3. Fogh, J., and Trempe, G., 1975. Human Tumour Cells in vitro. In: J. Foch (Ed.) Plenum Press, New York, pp. 115-159.

4. Hogan, B.L.M., Fellows, M., Avner, P., and Jocob, F., 1977. Isolation of a human teratoma cell line which expresses F9 antigen. Nature. 270, 515-518.

5. Reisner, Y., Gachelin, G., Dubois, P., Nicolas, J.F., Sharon, N., and Jacob, F., 1977. Interaction of peanut agglutinin, a lectin specific for non-reducing terminal D-galactosyl residues, with embryonal carcinoma cells. Developmental Biology. 61, 20-27.

6. Zeuthen, J., Norgaard, J.O.R., Avner, P., Fellows, M., Wartiovaara, J., Vaheri, A., Rosen, A., and Giovanella, B.C., 1980. Characterisation of a human ovarian teratocarcinoma-derived cell line. International Journal of Cancer. 25, 19-32.

DISCUSSION

Dr. Evans

Could I ask Miss Robertson if there is any relationship between the growth rate of the lines and the proportion of the cells differentiating and the corollary to that is that if you reduce the growth rate of the lines, do you get an increase in the proportion of cells differentiating?

Miss Robertson

Yes, I think there is a correlation between growth rate and differentiation if you compare nulli-potent lines and pluri-potent lines, but I have not quantified this in any way and do not know what would happen if you did try to do so. I think that there are ways of slowing down the growth rate, for example, by reducing temperature, which I have tried and this does not appear to affect the ability to differentiate, although it is evident that the nulli-potent lines can grow under less stringent conditions, and they show a very rapid rate of growth. I think this probably does indicate that the growth rate does effect the differentiation abilities.

Mr. Corbett

We are probably extremely naive in the way we consider our chemotherapy works. It is very nice to attempt to transfer some of the concept from some of the scientific work into chemotherapeutic terms and I wonder really whether our two speakers consider that when we are giving fractionated chemotherapy are we actually carrying out a sub-cloning experiment? Also whether, in fact, we should be trying to look more deeply at this problem by getting as much surgical tissue as possible and having it examined to try and carry out some assays on some of our patients at various stages of treatment. This to see what, in fact, we are actually doing on these basic levels to our patient's tumours. I wonder whether any of the scientists would tell us whether they would be interested in trying to carry out any such work, if we can supply sequential specimens during treatment?

Dr. Hogan

The trouble is that it is very difficult since working with human tumour tissue, no-one has really succeeded in growing out the stem cells. Until we know more about what kind of cell types there are in human tumours it would be rather difficult to make that analysis. Perhaps one point I did not bring up in my talk was, that Sydney Strickland and others have tried using retinoic acid as a therapeutic agent as it were. He has taken mice, injected them with F9 cells and then treated these mice with massive doses of retinoic acid and he gets temporary regression of the growth of F9 cells compared with control animals and sections of the tumours to do seem to show more differentiated endoderm. Unfortunately, after a period the growth rate increases again and he does not know whether this is because retinoic acid resistant lines develop in the same way as in certain kinds of lymphoma cells where you get steroid resistant lines appearing. I know that Liz Robertson has, in the laboratory, isolated retinoic acid resistant stem cell lines. No one yet knows the mechanism of action of retinoic acid, there are retinoic acid binding proteins in the cytoplasm that maybe move into the nucleus and control gene expression there, but I think the possible use of retinoic acid as a therapeutic agent is unlikely to be relevant to humans.

Dr. Newlands

In relation to retinoic acid, we have actually looked at a limited number of heavily pre-treated patients resistant to chemotherapy whom we have proceeded to treat with a water soluble vitamin A analogue and we have not seen any responses in a small number of patients. The diseases continue to grow throughout treatment and so we have not found any clinical benefit. I was wanting to throw open a question; we are very used to using a serum markers to monitor patients. I was wondering if there was any new information or speculations about AFP, its actual biological function in the normal embryo. Is there any new information on this?

Dr. Hogan

Not that I know of, but people have been extensively studying it's structure and the structure of the genes for AFP and there is very close correlation between the structure of the protein and serum albumin from adults and so presumably it is acting as some kind of foetal albumin. One other point has occurred to me about the effect of chemotherapy is that I suppose from Miss Robertson's presentation, it is obvious that chromosome imbalance may affect the ability of cells to differentiate and these chemotherapeutic agents could well be causing chromosome imbalance. Some of them may, in fact, favour the production of cells which would differentiate poorly, so we need to be a bit careful of this perhaps.

Dr. Evans

It would be possible to see whether the predominant karyotype of the cells changed in tumours as they progressed and in new tumours as they reappeared; they may have different chromosomes. I should think one very strong difference between mouse embryonal carcinoma stem cells and human cells that you grow from tumours, is that the mouse cells have a karyotype closer to the normal 40 chromosome of the mouse (they just have 44 at the most) whereas most human tumour cells lines have about 56 or 60 or more chromosomes, there is a big chromosomal imbalance in the human cell lines.

Dr. Jones

Could I ask the speakers if they have any comments about the phenomenon I mentioned of acceleration of disease following cyto-reductive surgery?

Dr. Pizzocaro

Why not speak of acceleration of disease after chemotherapy?

Dr. Ash

When I compared the growth rate of tumours towards the end of the period of observation with that at the beginning, what I had expected to find was they they might be slower, following a Gompertzian type of growth pattern. In actual fact, for what it's worth, the mean doubling time of the tumours when they were first measured was approximately 20 days and it was about 18 days in their last period of growth, they certainly did not get any slower.

Mr. Corbett

I should like to ask Dr. Silvestrini, your bulk disease had a higher labelling index. Did you look at disease during treatment, or were you always working on pre-treatment specimens?

Dr. Silvestrini

Variation in the course of the disease may occur in a tumour treated with the same drugs according to proliferative activity. As we have observed from the in vitro sensitivity test sometimes the proliferative activity can be stimulated instead of being stopped by some drugs either cyclophosphamide or other alkylating agents but the sensitivity is a peculiar intrinsic characteristic of individual tumours.

Dr. Pritchard

Following on the interesting observation by Dr. Hogan, I should like to ask Dr. Jacobsen whether in her work on human tumours she has noticed any relationship between AFP-producing tissue and adjacent tissues?

Dr. Jacobsen

I did not have this point in mind when I looked at these tumours, but it would be a good idea to see what the neighbouring tissue is like.

Dr. Heyderman

I think that the correlation I found is that there is often a close apposition of the two extra-embryonic types. That one sees yolk-sac elements next to giant syncitial cells of trophoblastic type is a very common association.

Mr. Scott

I would just like to come back to Dr. Jones' point because I think this is a really relevant question to be answered by some of our biologist colleagues and this is the apparent acceleration of growth of quiescent tumour nodules once disturbed, for instance at a second look laparoscopic biopsy. We gynaecologists perhaps are far more experienced than most of our colleagues in this regard, being familiar in the use of a laparoscope and I can confirm the anecdotal experience of many gynaecological colleagues around the country of this very chastening experience of having patients in apparent total remission with perhaps only one or two tiny quiescent nodules disturbed at biopsy and then rapid acceleration. There must be some change in host tumour environment and it seems to me it is for our biological colleagues to answer this, because it is terribly relevant to the development of current practices, especially as regards this whole question of the emphasis on second look surgery. I think there are real dangers here if people do not get these lessons appropriately sorted out.

Mr. Corbett

Would you suggest that perhaps a second look laparotomy where tumours are found should be followed by fairly intensive chemotherapy?

Mr. Scott

I think it should be followed by definitive therapy, it should not be followed by laparoscopic biopsy and an experienced gynaecologist should be capable of telling whether what he sees is tumour or something else.

Dr. Begent

In relation to Dr. Jones' question again. If one follows patients who have malignant teratoma producing HCG or AFP and one does surgery to try and remove drug resistant disease, one almost invariably has to stop the patient's chemotherapy for four weeks at a time when he should be on continuing therapy. You tend then to remove the big mass of disease and you have by this manoeuvre given small metastases that you had not detected before the surgery a free rein to grow over a period of four weeks and sometimes more. In our experience it seems to be this problem, the fact that the metastases were not detected when you undertook the cytoreductive surgery that gives this illusion of accelerated growth but, in fact, it is the unrestricted growth of something that was there all the time but was controlled by previous therapy. It is not actually acceleration as far as we can see.

Mr. Corbett

Well there is some suggestion, I think from Lange's rather anecdotal cases, that in fact it is an acceleration of some of the microscopic deposits of tumour.

Professor Wahlqvist

I would like to ask Dr. Ash, there are two problems, are the big metastases accelerating faster than the small metastases or is it the contrary? When we are discussing the acceleration which takes place after surgery, surely it depends upon the size of the tumours?

Dr. Ash

The size of the majority of metastases that were measured were those with a diameter approximated to 1 to 2 cms. I have not actually made a careful analysis of responses directly in relation to size and, of course, the other thing we cannot forget about these patients in relation to their ultimate prognosis is that they do not just have lung metastases, many of these patients also have intra-abdominal disease and that contributes sometimes even more to their ultimate fate than the lung metastases. I cannot answer your question exactly.

Mr. Corbett

Did you measure all the metastases, or did you just take 2 or 3 selective ones for each patient?

Dr. Ash

You have to concentrate on those which are readily identifiable, measurable and reproducible, that means that you do not measure them all.

EXTRAGONADAL GERM CELL TUMOURS

Chairman of Session
Dr C.K. Anderson

EXTRAGONADAL GERM CELL TUMOURS - BIOLOGICAL AND CLINICAL RELEVANCE

P.J. Corbett
Lecturer,
University Department of Radiotherapy,
Cookridge Hospital,
Leeds LS16 6QB

It is perhaps natural that the majority of the clinical discussions at this conference will revolve around testicular tumours. Although rare, they are by far the commonest site in which malignant germ cell tumours occur. Malignant teratomas and dysgerminomas are also found in the ovary. Gonadal primaries account for some 88% of all germ cell tumours (9). However, histologically identical tumours of many types found in the gonads are also found in a variety of extragonadal sites (Figure 1).

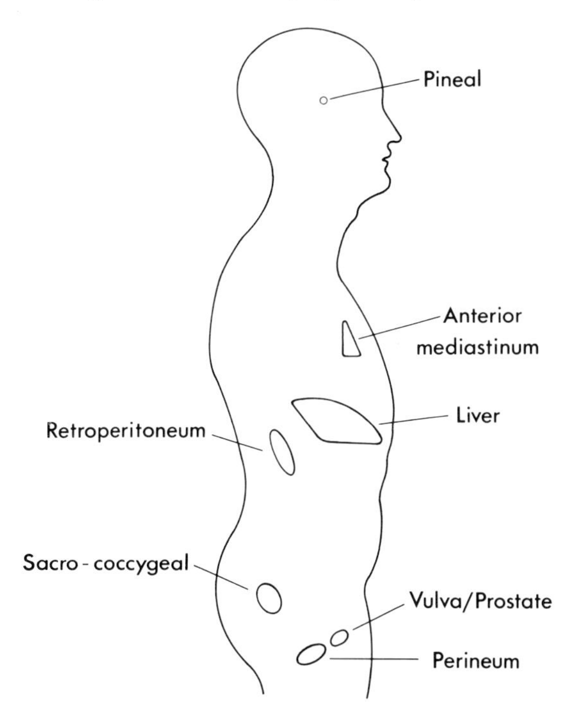

FIGURE 1 - Extragonadal germ cell tumours

These include the retroperitoneum (10), the liver (6), sacrococcygeal region (2), prostate (1) and vulva (11), pineal region (8) and, most commonly described of the extragonadal sites, the anterior mediastinum (7) all basically well defined sites on a median or paramedian axis.

It is easy to conceive of the origin of these tumours in totipotent germ cells residing in their natural habitat, the gonad. Agreement has not yet been reached on the origins of tumours found in the gonads. There are two principal theories - the dualistic theory which suggests that seminomas are of germ cell origin and teratomas are composed of displaced blastomeres which have escaped the influence of the primary organisers during embryogenesis. More popular is the holistic theory which considers all such tumours, including seminomas and teratomas to be of germ cell origin, and embryonal carcinoma to be the stem cell from which all non-seminomatous germ cell tumours are derived. If this is so the irreversibility of malignant transformation and dedifferentiation is called into question.

The origin of these stem cells for tumours occurring in extra-gonadal sites is not known for certain. Suggestions include the failure of migration of primordial germ cells, the dislocation of totipotent cells at an even earlier stage of embryogenesis, and even metastasis from an undetected or regressed primary gonadal germ cell tumour. Under certain circumstances, and in certain sites, all these theories are plausible.

However, the considerable variation in the proportion of tumours exhibiting malignant behaviour, the age variation and predilection for one or another sex found for each of the extragonadal sites suggests that a single aetiology for this group of tumours is unlikely.

Although primary extragonadal germ cell tumours are extremely uncommon they are probably underdiagnosed clinically (3, 5). We have seen cases of mediastinal tumours subsequently diagnosed both histologically and by marker production studies which had originally been labelled as undifferentiated carcinoma of the bronchus, Hodgkin's disease and thymoma.

Clinically these tumours tend to occur in young adult patients, and by reason of their site are often irremovable by primary surgery. In view of their similarity to gonadal tumours, and the advances made recently in the therapy of these, it is natural that similar therapeutic strategems should be utilised when dealing with extragonadal lesions.

Increasing awareness leads to an increased rate of

diagnosis, and multicentre co-operative studies are feasible. The first such study has recently been reported from the South West Oncology Group by Feun et al (4). They studied nineteen patients with assorted histologies, twelve having mediastinal tumours, one lung and six in the retroperitoneum. Interestingly, all their patients were male, as are the majority, but not all, cases reported. These patients were treated with Cis-platinum, Vinblastine and Bleomycin chemotherapy and although nine of sixteen patients responded, three of the responses being complete, the median duration of response was short, being only two months. In all it was concluded that these patients did not fare as well as those with disseminated testicular tumours treated with the same drug regime.

Various reasons why this might be so include those related to diagnosis and inappropriate treatment which may delay appropriate treatment or utilise valuable normal tissue (especially bone marrow) reserves, the site of these tumours (e.g. mediastinal) making debulking surgery either before or after chemotherapy impossible, and also quite possibly tumour related factors such as a different intrinsic drug sensitivity, perhaps related to a different biological origin, development of drug resistance and factors related to tumour vascularisation.

Hence, it may be that if these tumours have a different origin from gonadal tumours the statement made by Feun that "treatment should be the same for disseminated tumours of similar histology" may not be valid.

All histological types of tumour occurring in the testis are found in some or all of these sites. Perhaps numerically most are seen in the anterior mediastinum, and there is a tendency for such tumours to be differentiated along extra-embryonic pathways, especially towards pure yolk-sac tumours, more commonly than those from the testis. Over the last three years we have seen four cases of mediastinal germ cell tumour. Interestingly enough three of these have been pure yolk-sac tumours, two of which have been treated by ourselves, and the fourth was a young man with "undifferentiated carcinoma" though originally to be of bronchial origin involving the mediastinum secondarily, who also had unsuspected intracranial disease at post-mortem and interestingly a small scar in one testis. His tumour produced AFP, but despite the fact that AFP levels declined in response to treatment with platinum, vinblastine and bleomycin (PVB) his disease did not respond and he died from extensive intrathoracic disease on the day his AFP came back as normal. However, he achieved useful palliation and his condition improved sufficiently for him to marry his fiancee, a nurse, a courageous step taken by both of them in the full knowledge of the facts.

Only one of our cases was a female, and the majority of cases arising in this site and all previous pure yolk-sac tumours reported in the literature as arising in the mediastinum, occur in males. This seventeen year old girl presented with respiratory symptoms, a large mediastinal mass and pulmonary deposits. A diagnosis of pure yolk-sac tumour was made after thoracoscopy. As would be expected with a pure yolk-sac tumour there was gross elevation of the AFP level shown plotted on a log-linear scale (Figure 2). Despite a dramatic fall in AFP levels amd marked clinical and radiological improvement following PVB chemotherapy, AFP levels began to rise again. Thoracotomy revealed further active yolk-sac tumour which was not resectable and she succumbed to her disease. Of the three cases of pure yolk-sac tumour of the anterior mediastinum that we have treated all had dramatic initial responses to PVB chemotherapy, resulting in both symptomatic and objective

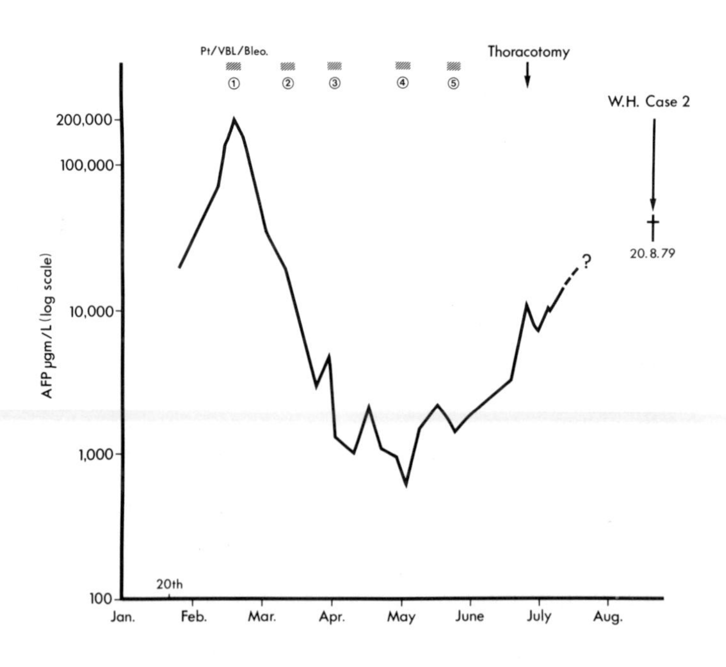

FIGURE 2

improvement, the latter both radiologically and with marked falls in marker levels. However, all the patients relapsed and became unresponsive to further therapy and indeed all died with massive disease. However, in two of our cases very useful palliation was achieved (useful to the patient that is - the only acceptable meaning of the word).

Despite the fact that treatment to date has not produced long-term complete remissions in our hands or in the reports of others, I believe we are on the right track. If we can make others more aware of the existence of extragonadal germ cell tumours, especially when tumours occur in young adults and can encourage them to screen such cases for marker production, rather than accepting a clinical diagnosis of "undifferentiated carcinoma of the bronchus" or even "lymphoma", we will be able to offer appropriate treatment at an earlier stage, before less appropriate treatment has been given, and will be in a better position to adopt an aggressive multimodal approach.

It is also apparent that such cases are best managed by those with special interests in germ cell tumours within specialised centres.

Perhaps the oncological sub-speciality of 'Germ Cell Oncology' is a valid and viable one and it is hoped that discussion of these interesting and unusual extragonadal germ cell tumours will help us both to understand their biology and to improve treatment of all such tumours.

REFERENCES

1. Benson, R.C., Segura, J.U., and Carney, J.H., 1978. Primary yolk-sac tumour of the prostate. Cancer. 41, 1395-1398.

2. Berljung, L., Fagerlund, M., and Hammer, E., 1975. Extragonadal endodermal sinus tumours. Acta Chirurgica Scandinavia. 141, 688-691.

3. Corbett, P.J., Primary mediastinal yolk-sac tumours. Publication pending.

4. Feun, L.G., Samson, B.K., and Stephens, R.L., 1980. Bleomycin and Cis-diamminedichloroplatinum in disseminated extragonadal germ cell tumours. Cancer. 45, 2543-2549.

5. Fox, R.N., Woods, R.L., Tattersall, M.H.N., and McGovern, V.J., 1979. Undifferentiated carcinoma in young men - the atypical teratoma syndrome. Lancet. i, 1316-1318.

6. Hart, U.R., 1975. Primary endodermal sinus tumour of the liver - first reported case. Cancer. 35, 1453-1458.

7. Martini, N., Goldberg, R.D., Haju, S.I., Whitmore, W.F., and Deathe, J., 1974. Primary mediastinal germ cell tumours. Cancer. 33, 763-769.

8. Norgaard-Pedersen, B., Altrechtsen, R., Arendo, J., Windholm, J., and Sandburg, E., 1978. A report of two cases of primary intracranial germ cell tumours. Scandinavian Journal of Immunology. 8, 152-156.

9. O'Hare, M.J., 1978. Teratomas, neoplasia and differentiation: a biological overview. I. The natural history of teratomas. Investigative Cell Pathology. 1, 39-63.

10. Pantoja, E., Llobet, R., and Gonzalez-Flores, B., 1976. Retroperitoneal teratoma - histological review. Journal of Urology. 115, 520-523.

11. Ungerleider, R.S., Donalson, S.S., Warmke, R.A., and Wilbur, J.R., 1978. Endodermal sinus tumour - The Stanford Experience - the first reported case arising in the vulva. Cancer. 41, 1627-1634.

MEDIASTINAL GERM CELL TUMOURS

A. Barrett and M.J. Peckham
The Institute of Cancer Research and
The Royal Marsden Hospital,
Downs Road, Sutton,
Surrey SM2 5PT

Introduction

Primary malignant germ cell tumours of the mediastinum are rare and must be distinguished from metastatic disease from an occult testicular tumour. A mediastinal mass situated anteriorly is characteristic and there should be no evidence of abdominal disease on CT scanning. Only careful post mortem studies can confirm the diagnosis but the clinical picture suggests that mediastinal germ cell tumours are a distinct entity.

Ten patients with mediastinal germ cell tumours have been seen at the Royal Marsden Hospital between 1964 and 1980. Six of these cases have been seen in the last two years which probably reflects a greater awareness of the possibility of germ cell tumours presenting with mediastinal masses.

Nevertheless in five patients, an initial diagnosis of another type of malignancy was made pathologically; in two cases adenocarcinoma, in two thymoma and in one carcinoma of the thyroid.

Estimation of serum AFP and HCG revealed the true diagnosis in one patient with adenocarcinoma who presented with a mediastinal mass and lung metastases. These investigations should be performed routinely in any young patient with an anterior mediastinal mass.

Presenting features in these ten patients are shown in Table 1.

Pleuritic chest pain was very common with a dry unproductive cough but systemic symptoms were also often marked, resolving with successful treatment. Two patients had lung metastases at presentation. The diagnosis of primary mediastinal germ cell tumour was considered only if CT scanning showed a clear anterior mediastinal mass with no evidence of pelvic, paraortic or retrocrural lymphadenopathy

TABLE 1

MEDIASTINAL GERM CELL TUMOURS

Presenting features (10 patients)

SYMPTOM	NUMBER OF PATIENTS
Chest pain	9
Cough	6
Weight loss	5
Fever	4
Lethargy, anorexia	4
Dyspnoea	4
S.V.C.O.	2
Proximal myopathy	1

and both testes were clinically normal.

Patients may be divided into two groups according to the treatment they received - a historical group treated between 1964 and 1976 with various chemotherapy combinations and those treated between 1978 and 1980 with platinum-containing regimes: either vinblastine, bleomycin and platinum (PVB), with BEP where VP 16-213 mg/m^2 for 5 consecutive days is substituted for vinblastine, or BEVIP where both vinblastine and VP 16-213 are given (Table 2).

Patient details for the historical group are shown in Table 3. Only one patient who presented with small volume seminoma is still alive after treatment with cyclophosphamide and mediastinal irradiation but it is of interest that response to chemotherapy was seen in all patients although complete remission was not attained.

From 1978 six patients have been treated. One patient has died. He presented with a very large mediastinal mass which had not changed in size after two courses of PVB in spite of a rapid fall in AFP levels. His chemotherapy was therefore changed and was subsequently ineffective. In

TABLE 2

PVB

Cis Platinum	20 mg/m^2	Daily x 5
Vinblastine	0.2 mg/kg iv	Days 1 and 2
Bleomycin	30 mg iv	Days 2, 9 and 16

(In practice doses of 30 or 40 mg of Cis-Platinum are used daily alternating if necessary)

Repeat every 21 days

BEP

VP-16	120 mg/m^2 (Max 200 mg)	Daily x 5
Bleomycin	30 mg iv	Days 2, 9 and 16
Platinum	20 mg/m^2	Daily x 5

BEVIP

VP-16	120 mg/m^2	Days 1 and 2 Days 3 - 5 if possible
Bleomycin	30 mg iv	Day 2
Platinum	20 mg/m^2	Days 1 - 4
Vinblastine	0.2 mg/kg	Day 1

TABLE 3

HISTORICAL GROUP 1964-1976

PATIENT	HISTOLOGY	CHEMOTHERAPY	R/T	OUTCOME	SURVIVAL
AB	Seminoma	1 G Cyclophosphamide 2 mg Vincristine	3500	A & W	8 years
*BL	Seminoma	Chlorambucil Actinomycin	3700	Died	3 yr 2/12
*IP	MTI	VAM x 1	5000	Died	7/12
PZ	MT	Cyclophosphamide x 1 Platinum x 1 Velbe/Bleomycin x 2 Adriamycin/DTIC x 4	No	Died	7/12

A & W = Alive, no disease

MTI = Malignant teratoma intermediate

VAM = Vinblastine/Actinomycin D/Methotrexate

* = Presented with lung metastases

retrospect it might have been more appropriate to continue treatment for longer, possibly using surgery or radiotherapy for residual disease.

The two patients with seminoma went into complete remission with six courses of PVB and as CT scanning then showed no residual disease no further treatment was given. The patients remain well at 20 months and 14 months respectively. The other three patients had residual masses after chemotherapy. In one (MM) this was of very small volume and has been treated by local irradiation (40 Gy in 4/52). In the other two, surgery successfully removed residual masses, and histological examination showed no evidence of malignancy. Patient CH died in the post-operative period of respiratory failure. Both phrenic nerves were cut at surgery and there was also evidence of severe bleomycin toxicity.

Response of mediastinal germ cell tumours to standard chemotherapy regimes seems to have been good in this group of patients and suggests that the poor prognosis of previously reported patients may now be unjustified, with modern combined approaches.

GERM CELL TUMOURS AND THE YOLK-SAC - A PERSONAL VIEW

B. Norgaard-Pedersen
Sondeborg Hospital,
Post Box 160,
6400 Sondeborg, Denmark

In presenting a personal view of the genesis of germ cell tumours and their relationship with the yolk sac, we must remember the historical development of the concept.

Abelev was the first to describe AFP as an oncodevelopmental protein in patients with teratocarcinoma of the testis. Gitlin demonstrated the origin of AFP in the secondary vitteline channels of the foetal yolk-sac, both in vivo and in tissue culture. Teilum developed the endodermal sinus concept from the study of infantile yolk-sac tumours of testis and, later, from examination of endodermal sinus tumours of ovary. With the advent of immunohistochemistry AFP was localised within the visceral endoderm of the foetal yolk-sac and similar histological components in testicular and ovarian tumours of 'yolk-sac' origin.

More recently the biochemistry of AFP has identified the germ cell tumour AFP as of yolk-sac origin on the basis of the electrophoretic microheterogeneity and Con-A-nonreactivity. Liver produced AFP has a very low Con-A-nonreactive fraction, a pattern also seen in hepatocellular carcinoma. In yolk-sac AFP the Con-A-nonreactive fraction is high, and this pattern is also seen in germ cell tumour produced AFP.

DISCUSSION

Dr. Oliver

Most of the patients described by Mr. Corbett, and ten to sixteen patients with mediastinal germ cell tumours seen by the South West Oncology Group, and classified as poor responders, had been heavily treated by a variety of modalities before the true diagnosis was known. I would suggest that they are in no way different in behaviour or response to germ cell tumours of the testis but that the apparent different behaviour is a response to previous, inappropriate therapy.

Mr. Corbett

This may be so. Now that patients are being referred earlier, before inappropriate therapy is instituted, we may begin to see an effect on prognosis.

Dr. Anderson

The main point, which cannot be emphasised too strongly, is that germ cell tumurs occur in the mediastinum.

Dr. Oliver

The use of marker assays in patients with anterior mediastinal masses is an appropriate mode of investigation but the value of this depends largely on the marker service available.

Dr. Milford Ward

There is no reason why marker results should not be available within three to four days.

Mr. Corbett

Provided that Hodgkin's Disease can be excluded, there is little to be lost by initiating appropriate germ cell tumour chemotherapy on the basis of a high degree of clinical suspicion whilst awaiting assay results from the marker laboratory.

FEMALE GERM CELL TUMOURS

Chairman of Session
Dr C. K. Anderson

GERM CELL TUMOURS IN FEMALES

E. Wiltshaw
Medical Oncologist,
Royal Marsden Hospital,
Fulham Road,
London SW3

In the ovary 94% of tumours are of epithelial origin and the remaining 6% include a variety of neoplasia such as germ cell tumours, sex and stromal tumours and sarcoma not specific to the ovary. This paper reviews the present position regarding the management of germ cell tumours only.

Table 1. Ovarian Germ Cell Tumours

Dysgerminoma
Immature (Malignant) Teratoma
Embryonal Carcinoma
Choriocarcinoma
Endodermal sinus tumour
Mixed tumours

Dysgerminoma

The male equivalent of dysgerminoma is the testicular seminoma and the successful management of this tumour follows similar lines. Dysgerminoma spreads primarily to the lymph nodes and the paraortic nodes will be those first involved. However, Stage I disease does occur and may be successfully treated by surgical removal of the single ovary (1). If the lymph node metastases have occurred then radiotherapy to the ipsilateral pelvic area plus paraortic radiotherapy should be given (9). Some physicians treat more widely with radiotherapy in these cases applying whole abdominal fields and mediastinal and supraclavicular radiotherapy as well. However, there is good evidence that many patients can still be cured if conservative treatment is given and further radiotherapy or chemotherapy (vincristine, actinomycin D and cyclophosphamide) are administered only with overt recurrence.

With this plan of management minimal toxicity and damage is caused by treatment while a survival rate of 85-90% can be expected.

Immature (Malignant) Teratoma

In a large study of malignant teratoma of the ovary Norris (12) emphasised three important features associated with prognosis, size of the tumour, stage and degree of differentiation.

Most primary tumours were large at diagnosis but those measuring more than 20 cm were usually associated with Stage II or III (FIGO) disease. Their cases were treated almost entirely by surgical means and only occasional patients had radiotherapy or any form of chemotherapy. Nevertheless, of forty patients presenting with Stage 1 disease, twenty-seven were alive and well for a median time of seven years. For later stage disease the prognosis was much worse only 38% of sixteen patients surviving more than five years.

Grade of tumour was also important in a prognostic sense especially if related to the stage at presentation. Thus the better differentiated tumours (Grade 1) in forty patients with Stage 1 disease showed a 100% survival rate while Grade II had 70% and Grade III only 33% survivors. Grading of teratoma histology is even more important today when apparently inadequate chemotherapy may be associated only with residual tumour of Grade 0 type. This type of histology can be accompanied by long term survival and possibly cure (4, 13).

Despite the good prognosis of some teratomas there is plenty of evidence that others are highly malignant and in Norris' series, twenty of fifty-six patients died of disease two to forty-eight months after diagnosis while a further four patients were alive but with evidence of active tumour at one month to more than six years. They noted however, that no deaths had occurred between five and ten years disease free follow up suggesting that at that stage, cure might be anticipated.

Recently a better prognosis has been seen in cases of advanced teratoma with the additional use of chemotherapy while the usefulness of radiotherapy remains in doubt. Curry (3) reported a series of patients treated with chemotherapy (modified VAC) post-operatively and compared them with patients given other drugs or no chemotherapy.

Table 2. Malignant Teratoma

Variable Prognosis

Prognosis depends on size, degree of differentiation (Grade) and Stage.

Stage 1	Grade I and smaller tumours (<10cm) Surgery only	95-100% Cure.
Stage I	Grades II and III and larger tumours.	Oophorectomy + VAC.
Stages II & III all grades		Surgery ('debulking') + VAC Cures Stage I near 95% Cures Stage II and III 85%

Eight patients, receiving no drugs at all, died within forty months of diagnosis. Five patients were given methotrexate, actinomycin D and cyclophosphamide (MAC) in combination and three died within twenty-three months. Lastly, twelve patients were treated with their modified VAC regime. Two have died at three and twenty-six months while ten are alive and free of disease at sixteen to sixty-eight months with a median of forty-three months follow up.

Modified VAC is probably the most successful chemotherapy so far tested in ovarian teratoma and consists of:

Vincristine 1.5mg/m^2 (up to 2mg) IV weekly for twelve weeks

Actinomycin D 0.5mg IV daily for five days) repeated
Cyclophosphamide 5-7mg/Kg IV daily for five days) every
) 4-6 weeks

Curry (3) gave the treatment for as long as two years but other series suggest that a shorter period of one year or even less may be curative.

Embryonal Carcinoma

Embryonal carcinoma is basically a teratomatous lesion in which the tumour tissue is very undifferentiated and morphologically analogous to the structures present in the earliest stages of embryonic development. It can be regarded pathologically as similar to the embryonal carcinoma of the testis of young males.

In a study of twenty-seven cases Neubecker and Breen (10) showed that in 52% of cases the tumour was limited to one ovary at laparotomy but that apparent localised disease had no significance in relation to prognosis. Twenty-one cases had follow up data and eighteen were dead while three were living with disease. Of those who died survival in 78% was less than ten months. Death was due to extensive intra-abdominal disease and radiotherapy following initial surgery, or for recurrence, failed to reverse progressive tumour growth.

Table 3. Embryonal Carcinoma

Poor Prognosis

Stages I-IV	Surgery alone 80% dead ten months
	Surgery + chemotherapy ? survival

VAC probably only partially successful, Adriamycin or other drugs may be necessary.

Unfortunately, I could find no data on the use of chemotherapy in this particular tumour type and it is probably that these cases have been included with other malignant teratomas in most series. However, the Norwegian data (Davey, M. - personal communication) suggests that VAC is not sufficient to produce cures and that the addition of adriamycin to the regimen may be of value.

Choriocarcinoma

Again this sub-group of teratoma is extremely rare and only occasional patients have been reported. Wider et al, (15) were perhaps the earliest to show that such a tumour, when arising in the ovary, may respond to chemotherapy in a similar way to chorionic tumours of the uterus. They treated four patients with a combination of methotrexate, actinomycin D and chlorambucil. Three were living and free of disease between ten and fifty-five months later. The fourth patient died after twenty months with resistant tumour while on chemotherapy. In these patients βHCG levels can be used to follow progress.

Table 4. Choriocarcinoma

Poor prognosis βHCG
Surgery alone almost 100% dead in twelve months
Surgery + MAC (Mtx, Actino D, chlorambucil) ¾ A/W
>10-55 months
Best to treat as for chorionic tumours of the uterus

Endodermal sinus tumour of Teilum

Endodermal sinus tumour or yolk-sac tumour is associated with the production of alpha-fetoprotein and is seen in both males and females arising in the gonads or in extra-gonadal areas.

Table 5. Endodermal Sinus Tumour

Survival and chemotherapy (Literature 1976-79)

	No Cases	Surviving >2 years
Surgery ± Radiotherapy	111	9 (8%)
Surgery + Chemotherapy	47	22 (47%)

Before the widespread use of chemotherapy, survival for two years was unusual and additional radiotherapy was of no value, nor was radical surgery. Combining six (6, 8) large series from the literature only eight per cent of one hundred and eleven cases were surviving more than two years, while with additional chemotherapy, usually VAC (vincristine, actinomycin D and cyclophosphamide), forty-seven per cent of forty-seven patients were alive at two years (2, 14). The two year figure is important since the vast majority of all relapses from endodermal sinus tumour (EST) have already occurred within that period. The chemotherapy figures are impressive since the majority of patients treated by chemotherapy had advanced disease at operation (usual FIGO Stage III).

EST in the ovary spreads to the paraortic nodes then to the liver and then to the lungs. AFP levels are very useful in assessing therapeutic effects but must be followed closely, probably weekly, if the correct treatment is to be

TABLE 6

ENDODERMAL SINUS TUMOUR - R M H 1965-1981

Patient	Stage	Surgery	Chemo	AFP	Months Survival
JW	III	Biopsy	-	-	5
CH	III	BSO	5FU Cyc Vcr	-	5
BG	Ia	SO	-	-	7
JWK	III	SO.Om.	-	-	4
AW	III	BSO.TH	chlor	-	38
ZM	Recurr.	-	VAC/VAC/VA	1,300	39
ACS	III	SO.Om	CHAMOMA	6,000	36+
SC	Ic	SO	Einhorn	211	19+
SE	III	SO	Einhorn	340	15+
LB	III	SO	Einhorn	2,900	13+
AG	IV	Biopsy	VAC→Einhorn	93,000	4
MM	III	Biopsy	Einhorn	170	11+
KT	Ic	SO	Einhorn	240	4+
Sh.C	Ic	SO	Einhorn	2,500	2+

applied. If the level does not continue to fall at a constant rate then the type of chemotherapy, which should be given to all patients postoperatively, must be changed at once. At the present time there are three relatively successful forms of chemotherapy. First, the modified VAC regimen mentioned earlier for teratoma of the ovary, second the VBC or Einhorn regimen commonly given to patients with testicular teratoma (vinblastine, bleomycin and cis-platin) and lastly the rather more complicated regimen of the Charing Cross Hospital group in which three to four drugs are used in combination for several courses followed by other combinations in a set sequence provided that response is continuing in a satisfactory manner (11). It is not yet known which one of these schemes is the best since EST is very rare in women. In fact several large centres in Europe, including the UK, are collaborating to try out VBC and, if the response is not satisfactory, will go on to VAC and then to VP16 plus Adriamycin as third line treatment. The experience gained in the study will help to define the best approach to treatment and it is hoped that they can then go on to tackle other rare tumours in the same co-operative manner.

Our own experience in treating EST shows that before chemotherapy four out of four patients died of disease within seven months. Since chemotherapy has become routine treatment, three out of ten patients have died at four, thirty-eight and thirty-nine months. Of the dead cases chlorambucil alone was given to one and VAC to another. The third case had a very large tumour load with 93,000 μg/l of AFP and the tumour rapidly became resistant to both VAC and VBC, she survived only four months (Table 6).

In the follow up of patients with this tumour it is important to realise that a normal pregnancy has occurred in at least one patient following successful treatment and thus rising AFP level does not always signify recurrence (5).

The importance of initial treatment with effective chemotherapy cannot be emphasised too strongly since, despite modern treatment, there has been no case reported so far which has survived a recurrence.

Mixed Tumours

While some germ cell tumours show only one histopathological pattern, many others show a mixed picture (e.g. dysgerminoma + teratoma). Thus it is advisable to look for both AFP and

βHCG in all cases, it is also imperative to treat the patient for the most aggressive lesion which is seen histologically. Again appropriate therapy must be applied to all patients with a raised AFP or βHCG as these are undoubted indicators of residual aggressive tumours.

Some workers do not distinguish between the various sub-groups of germ cell tumour but in the female this may be important since the aim now is to give sufficient treatment for cure with the least possible toxicity and psychological trauma.

Summary

A review of the present position regarding treatment of germ cell tumours in the female shows many features similar to those in the male. Dysgerminoma may be treated more conservatively than other germ cell tumours and oophorectomy is recommended for early smaller tumours. With widespread or recurrent disease, cure is still possible with radiotherapy with or without chemotherapy.

Some teratomas are also curable by conservative surgery alone but this decision is less easy since recurrent disease is rarely curable and thus chemotherapy postoperatively is usually recommended.

In the most malignant tumours, that is those containing elements of embryonal carcinoma, choriocarcinoma or endodermal sinus tumour, vigorous chemotherapy is essential probably in all cases.

Radiotherapy is only recommended in the case of dysgerminoma. Radical surgery appears to have no place in the management of these tumours, although there may be a case for 'debulking' of very large, widespread endodermal sinus tumours.

Unfortunately these recommendations are based, for the most part, on small series of cases treated in a variety of ways and it is imperative that prospective studies be done on larger numbers of cases. For this reason referral to centres involved in such trials is to be encouraged.

REFERENCES

1. Assadourian, L.A., Taylor, H.B., 1969. Dysgerminoma. An analysis of 105 cases. Obstetrics & Gynecology. 33, 370-379.

2. Cangir, A., Smith, J., van Eys, J., 1978. Improved prognosis in children with ovarian cancers following modified VAC (vincristine sulfate, dactinomycin and cyclophosphamide) chemotherapy. Cancer. 42, 1234-1238.

3. Curry, S.L., Smith, J.P., Gallagher, H.S., 1978. Malignant teratoma of the ovary: Prognostic factors and treatment. American Journal of Obstetrics & Gynecology. 131, 845-859.

4. Di Saia, P.J., Saltz, A., Kapan, A.R., Morrow, C.P., 1977. Chemotherapeutic retroconversion of immature teratoma of the ovary. Obstetrics & Gynecology. 49, 346-350.

5. Duncan, I.D., Young, J.L., 1980. Endodermal sinus tumour of the ovary: serum alpha-fetoprotein levels before and after treatment and during pregnancy. British Journal of Obstetrics & Gynecology. 87, 535-538.

6. Flamant, F., Gaillou, B., Pejovic, M.H., Gerard-Marchant, R., Gout, M., Lemerle, J., Sarrazin, D., Zucker, J.M., Schweisgath, O., 1978. Prognostic factors in malignant germ cell tumours of the ovary in children excluding pure dysgerminoma. European Journal of Cancer. 14, 901-906.

7. Greasman, W.T., Felter, B.F., Hammond, C.B., Parker, R.T., 1979. Germ cell malignancies of the ovary. Obstetrics & Gynecology. 53, 226-230.

8. Kurman, R.J., Norris, H.J., 1976. Endodermal sinus tumour of the ovary. A clinical and pathologic analysis of seventy-one cases. Cancer. 38, 2404-2419.

9. Lucraft, H., Mann, J.R., Pearson, D., 1980. Malignant ovarian tumours in children. In: Ovarian Cancer, (Ed.) Newman, Ford and Jordan, Pergamon Press.

10. Neubecker, R.D., Breen, J.L., 1962. Embryonal carcinoma of the ovary. Cancer. 15, 546-556.

11. Newlands, E.S., Begent, R.H.J., Kaye, S.B., Rustin, G.J.S., Bagshawe, K.D., 1980. Chemotherapy of advanced malignant teratomas. British Journal of Cancer. 42, 378-384.

12. Norris, H.J., Ziskin, H.J., Benson, W.L., 1976. Immature (malignant) teratoma of the ovary. Cancer. 37, 2359-2372.

13. Piver, M.S., Sinks, L., Barlow, J.J., Tsakada, Y., 1976. Five year remissions of metastatic solid tumour of the ovary. Cancer. 38, 987-993.

14. Slayton, R., Hreshchyshyn, M.M., Silverberg, S.G., Shington, H.M., Park, R.C., Di Saia, P.J., Blessing, J.A., 1978. Treatment of malignant ovarian germ cell tumours. Cancer. 42, 390-398.

15. Wider, J.A., Marshall, J.R., Bardin, C.W., Lipsett, M.B., Ross, G.T., 1969. Sustained remissions after chemotherapy for primary ovarian cancers containing choriocarcinoma. New England Journal of Medicine., 280, 1439-1442.

ALPHA-FETOPROTEIN AND HCG APPARENT HALF LIVES IN THE CLINICAL MANAGEMENT OF MALIGNANT OVARIAN TERATOMA

I.V. Scott, A. Milford Ward, A.R. Bradwell and A. Wilson
Derbyshire Hospital for Women,
Derby
Royal Hallamshire Hospital,
Sheffield
and Queen Elizabeth Hospital,
Birmingham

Comparison of the apparent serum half life (AHL) of alpha-fetoprotein with known physical half life of 3.4 days reveals three categories of patients. One group showed rates of fall of serum AFP consistent with an AHL of less than five days following surgical removal of the tumour and adjuvant chemotherapy. This group had clinical Stage 1 disease or in vitro cytotoxicity tests on cultured tumour cells indicating sensitivity to safely attainable therapeutic levels of anti-cancer drugs. Such patients

TABLE 1

AFP "AHL" AND PROGNOSIS

	Case No	AHL	Recurrence (R) Death (D) Tumour Free (TF)
No Fall or AHL>10 days	1	>>10	D 8 months
	2	"	D 8 "
	3	"	D 2 "
	4	"	D 4 "
	5	16.8	R 5 "
AHL <10 days >5 days	6	6.3	TF >24 "
	7	6.2	R 23 "
	8	5.5	R 9 "
	9	5.1	R 11 "
AHL <5 days	10	4.9	TF >24 "
	11	4.5	TF 24 "
	12	4.2	TF 24 "
	13	4.0	TF 24 "

appear to have a good prognosis and remain in remission after completion of six months chemotherapy (Table 1).

A second group showed an AHL of greater than five days but less than ten days. This was associated with disease of Stage II or more, or in vitro tests showing poor sensitivity to safe therapeutic levels of cytotoxic drugs. These patients showed failure to remit with adjuvant chemotherapy or early relapse after apparent remission. Figures 1 and 2 are typical of such a case where recurrence occurred after a six months marker free interval. Attempts were made in one

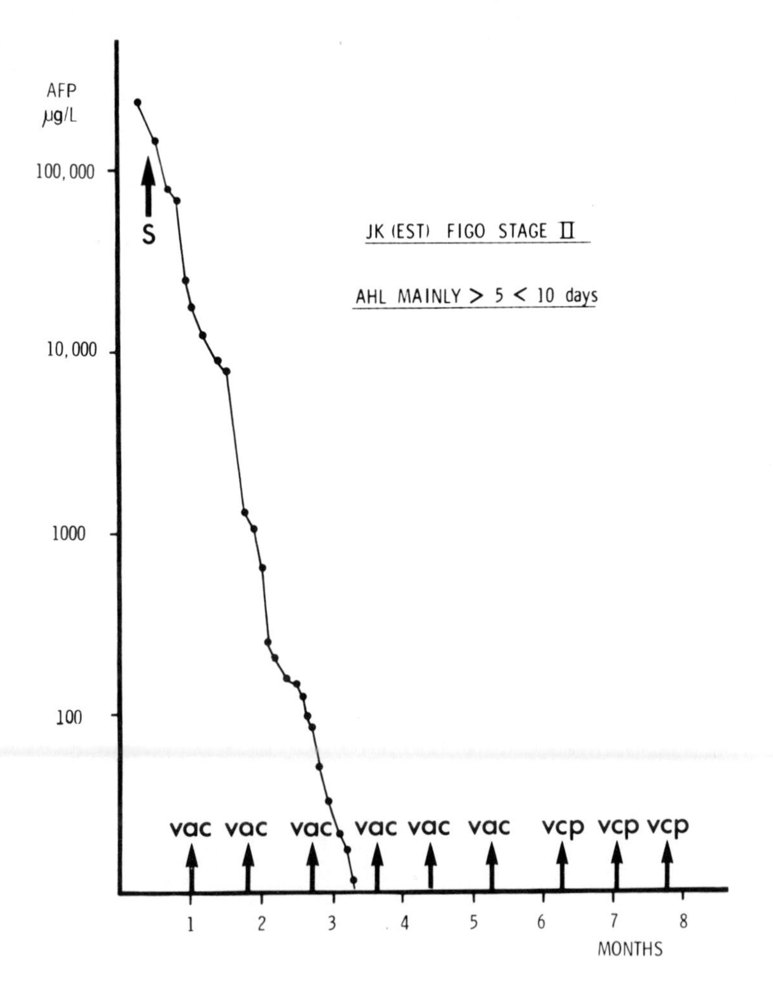

FIGURE 1

case to locate tumour residuum using ^{131}I labelled anti-human AFP serum and scanning with a gamma camera. This was of limited value but might prove useful in the future if higher specificity anti-sera could be produced. Grey scale ultrasound examination followed by laparoscopy seems more effective for the present and as good as CAT scanning. It seems possible that some patients in this second group might be cured if the tumour residuum could be located and removed given adequate sensitivity to chemotherapy.

Patients showing an AHL of more than ten days all have extensive disease and appear to be incurable by any means.

Although a similar approach may be taken with βHCG due to inadequate specificity of antisera and cross reactivity with L.H.

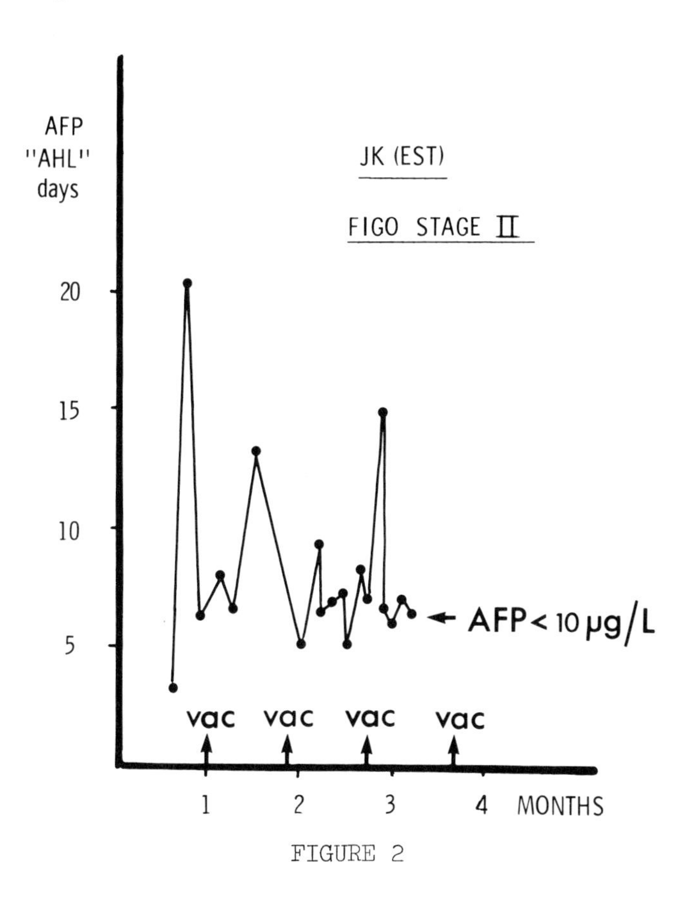

FIGURE 2

AFP PRODUCTION BY OVARIAN CYSTADENOCARCINOMA

P.K. Buamah, G. Bates, A. Milford Ward
Freeman Hospital,
Newcastle upon Tyne and
Royal Hallamshire Hospital,
Sheffield

AFP is well described as a marker of germ cell and endodermal sinus tumours of ovary, but elevated levels may also be seen in patients with cystadenocarcinoma, a tumour of 'epithelial' origin.

We have measured serum AFP in forty patients referred for treatment with advanced Stage III/IV non germ cell tumour of ovary, (ovarian cystadenocarcinoma). The histological appearances of all these tumours were carefully reviewed and considered consistent with cystadenocarcinoma. Four distinct cytological patterns were identified:

i. Serous
ii. Endometroid
iii. Mucinous
iv. Undifferentiated

Each patient was given cis-platinum 100mg/m^2 alone as a bolus intravenously at three weekly intervals for five courses under conditions of forced diuresis. Serial AFP estimations were determined before and during the course of treatment. Three serum AFP profile patterns were distinguished:

i. Elevated levels at the beginning of therapy falling to baseline (<10 μg/l) during therapy.
ii. AFP negative - (<10 μg/l)
iii. Normal levels at the beginning of therapy for variable periods of time, then rising above baseline with tumour recurrence.

Moderately raised levels of AFP above 65 μg/l were noted in 14/40 (35%) of the patients. Although slight to moderate elevations may be observed in patients with regenerative liver disease, serum AFP is considered to be within normal limits in all malignancies other than those of hepatic or

TABLE 1

AFP IN EPITHELIAL TUMOURS OF THE OVARY

		n	% raised AFP	
Talerman 1980	Epithelial	44	0	
Donaldson 1979	Adenocarcinoma			
	Mucinous	4	25	
	Serous	17	47	41%
	Endometroid	1	0	
This series	Adenocarcinoma			
	Mucinous	4	100	
	Serous	25	28	35%
	Endometroid	5	20	
	Undifferentiated	6	30	

germ cell origin. (Table 1).

Talerman (2) found no raised AFP in forty-four cases of ovarian cystadenocarcinoma, but our results support the observations of Donaldson (1) who found elevated AFP in patients with various ovarian tumours of non-germ cell origin. This finding could be interpreted as questioning the epithelial origin of ovarian cystadenocarcinoma.

REFERENCES

1. Donaldson, E.S., Von Nagell, J.R., Gay, E.C., Purcell, S., Meeker, W.R., Kashmir, R., Hunter, L., and Van de Voorde, J., 1979. Alpha-fetoprotein as a biochemical marker in patients with gynaecologic malignancy. Gyncaecologic Oncology. 7, 18-24.

2. Talerman, A., Haije, W.G., and Baggerman, L., 1980. Serum AFP in patients with germ cell tumours of the gonads and extragonadal sites. Cancer. 46, 380-385.

GERM CELL TUMOURS IN CHILDHOOD

Chairman of Session
Dr C.K. Anderson

YOLK-SAC TUMOURS IN CHILDHOOD

Jillian R. Mann
Paediatric Oncologist,
Birmingham Children's Hospital,
Ladywood Middleway,
Birmingham B16 8ET

Each year about twenty children in the United Kingdom develop yolk-sac tumours. They may arise in the testis, ovary, sacroccocygeal/pelvic/presacral/vaginal sites, retroperitoneum, thorax or pineal. Those in the testis typically present in boys age less than three years and 60-75% are cured by radical orchidectomy. Ovarian tumours usually present at the end of the first or in the second decade, are seldom cured by surgery and radiotherapy and may be associated with other germ cell elements. Tumours in the sacroccygeal/pelvic sites are more common in girls, usually presenting before the age of three years. They may be associated with other germ cell elements and, in general, have a very bad prognosis.

A retrospective survey of sixty-one children treated for yolk-sac tumours by members of the United Kingdom Children's Cancer Study Group (UKCCSG) confirmed these features, and revealed that five children with unresectable or metastatic disease had been apparently cured by the addition of chemotherapy with VAC, with or without added adriamycin to surgical and radiotherapeutic measures. Published reports of successful chemotherapy with VAC and other agents are scanty and offer little immediate encouragement.

In 1979 the UKCCSG embarked upon a study of the treatment of yolk-sac tumours in children in which the value of AFP monitoring, and chemotherapy with VAC, adriamycin, vinblastine, bleomycin and cisplatinum is being assessed. Twenty-three patients have been treated to date. Children with testicular tumours are treated by orchidectomy and their AFP levels reviewed. In Stage I chemotherapy is only initiated when the AFP fall is unsatisfactory or if there is recurrent disease. In Stages II-IV VAC therapy is continued for two years with addition of second-look surgery, radiotherapy or other agents as necessary. For tumours in other sites resection or resection biopsy is followed by VAC with such other therapeutic manipulations as necessary.

The initial VAC regimen was of low dose and was soon shown to be ineffectual. This was replaced by an intensive high dose regimen which unfortunately causes a degree of myelosuppression in most children.

The preliminary results show a good clinical response in six of eight children with testicular primaries, two with metastatic disease and in five of eight children with ovarian primaries.

The tentative conclusions that can be drawn at this early stage of the study are:

1. AFP is an extremely valuable marker in assessing Stage and response to therapy.
   βHCG has been of rather less value.
2. The low dose VAC regimen was ineffective.
3. High dose is effective in controlling metastatic disease but causes myelosuppression.
4. In children failing treatment with VAC, all other therapeutic procedures, including radiotherapy, appear ineffective.
5. The Einhorn regimen may be appropriate as a primary treatment in children with sacrococcygeal and pelvic primaries.

TESTICULAR TUMOUR IN CHILDREN

P.J.B. Smith
Consultant Urologist,
Bristol Royal Infirmary,
Bristol BS2 8HW

Testicular tumour in children is a rare condition. In the United Kingdom between 1967 and 1972 of the 1,446 patients who died from testicular cancer only thirty-six were under the age of fourteen at the time of death. Testicular tumours can be divided into germ cell tumours, which include yolk-sac tumour, teratoma and seminoma, and the non-germ cell tumours of Sertoli and Leydig cells, together with the connective tissue tumours of the tunica and epididymis - rhabdosarcoma.

Over a fifteen year period some eight children with testicular tumour have been traced in the Avon area. These include three with yolk-sac tumours, two of whom survived following orchidectomy alone. One child died with multiple metastases despite adjunctive radiotherapy. One child had a teratoma and is alive and well after orchidectomy alone, as also is one child with a seminoma. Three children presented with rhabdosarcomas, all were treated by orchidectomy, radiotherapy, and in one, chemotherapy. Two are alive but the third child died with multiple metastases.

The small numbers involved, and the time span of this study, are such that the recent developments of chemotherapy are not immediately relevant but it does seem that yolk-sac tumour, the commonest of the germ cell tumours in children, can be cured by orchidectomy alone. Metastatic disease in these tumours is manifest by a rise in alpha-fetoprotein and when this occurs chemotherapy should be instituted. Teratoma also responds to orchidectomy alone, though argument exists as to whether lymphadenectomy, radiotherapy or chemotherapy are required for metastatic disease. Seminoma occurs at puberty and should carry the favourable prognosis of the adult disease. Rhabdosarcomas are aggressive tumours and require both radiotherapy and chemotherapy in addition to orchidectomy.

ALPHA-FETOPROTEIN IN INFANTILE GERM CELL TUMOURS

B. Norgaard-Pedersen, J. Kamper, E. Sandberg Nielsen,
A. Sell, H. Sogaard, H. Hertz, I. Tygstrup and M. Yssing
Sonderborg Hospital,
Post Box 160, 6400 Sonderborg,
Denmark

Serum-alpha-fetoprotein (AFP) has been examined in nineteen infants (all less than four years of age) with germ cell tumours. The series consists of 1) Eleven infants with testis tumours including seven cases of endodermal sinus tumours (EST) producing AFP, and four cases of teratoma (TT) not producing AFP. 2) Six infants with sacrococcygeal tumours all producing AFP including five cases of EST combined with TT and one case of pure EST. 3) One infant with ovarian EST and TT producing AFP and 4) One infant with vaginal embryonal carcinoma with EST elements producing AFP. All patients with testicular tumours except one, are alive without tumour recurrence. This patient died thirteen months after orchiectomy of complications of chemotherapy without tumour disease. Two of six patients with sacrococcygeal tumour are alive, both more than three years after initial disease. One infant with ovarian and one infant with vaginal tumour both died after one and two years respectively. Serum AFP levels were a sensitive indication of tumour recurrence and disease activity. AFP half-life curves for seven patients with infantile testicular EST are shown in Figure 1.

It is important that half-life curves are assessed by regression analysis from multiple estimates rather than on two point AFP determinations. The value of multiple marker estimations cannot be stressed too strongly.

In studying decay curves, one occasionally sees apparently stable levels (as in patient 5). This was associated with an episode of fever and was probably due to tumour necrosis with release of intracellular AFP.

Both infants who died, ovarian and vaginal primaries, demonstrated marked AFP elevation some months before there was clinical evidence of tumour recurrence.

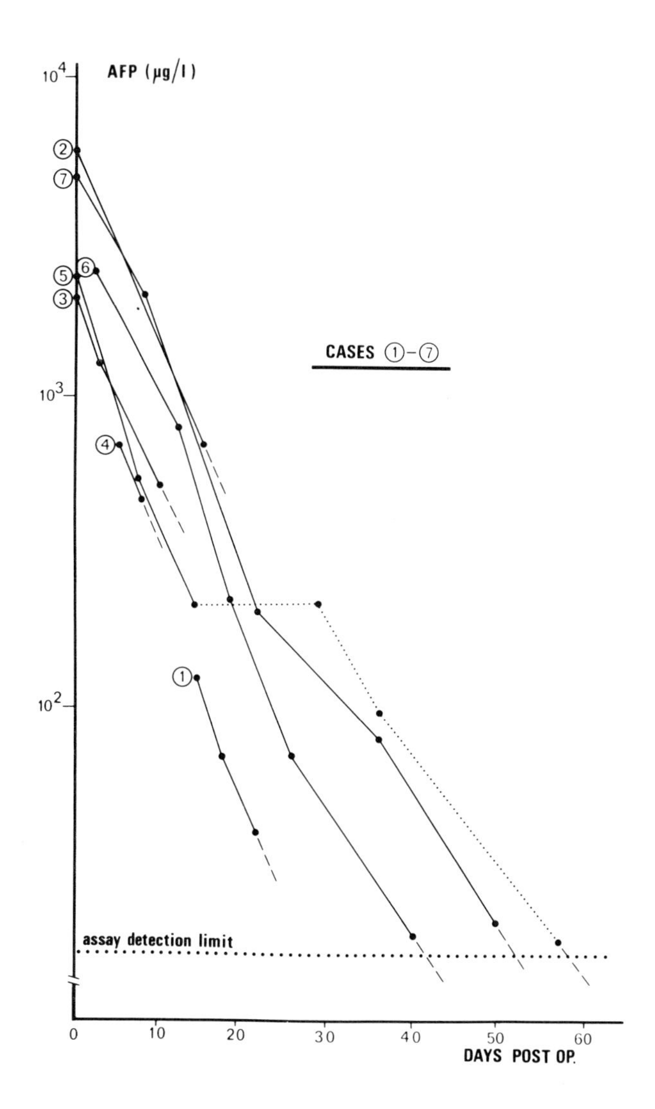

FIGURE 1 - Pre- and postoperative serum AFP concentrations in infantile testis tumour patients. Apparent half-life curves have been calculated by regression analyses to 4.3 - 5.8 days (Mean: 4.9 days). All patients had normal serum AFP values (less than 20 micrograms/l) 2 months after orchidectomy.

DISCUSSION

Dr. Begent

Could I ask Dr. Norgaard-Pedersen if he has any experience of AFP in the neonatal period. We have occasionally had problems with children soon after delivery who would be expected to have a small amount of circulating AFP clearing from the serum over the next few weeks or month. If that patient has also had yolk-sac tumour excised we have found it very difficult to determine whether there is residual yolk-sac tumour or whether it is a normal foetal product that we are seeing. Has he any suggestions for dealing with the problem.

Dr. Norgaard-Pedersen

We certainly have met these cases because some of the tumours are present at birth. One answer is to compare samples from day to day or week to week; the other is to determine what type of AFP it is. The usual type of AFP in the new-born is of liver origin, the Con-A non reactive fraction of the AFP is less than 1%. If you have a germ cell tumour at the same time you will have a Con-A non-reactive AFP which is at least about ten and maybe up to 50%. In that way you will be able to say whether the AFP is of tumour or foetal liver origin.

Dr. Pritchard

I think I should just comment, as a paediatric oncologist, on the natural history of the sacrococcygeal tumours. Some members of the audience may be puzzled if they remember from their medical student days that most of these lumps appearing in the sacrococcygeal region can be successfully removed by a surgeon and be cured. So that the patients coming to the paediatric oncologist are a selected group. We recently surveyed just under thirty children presenting to Great Ormond Street with sacrococcygeal masses, age ranges from birth to two to three years, and about 60% of these patients were cured by surgery alone, 20% have overt malignant elements in their tumour at diagnosis. A greater proportion of the older patients have a malignancy at diagnosis - very few of the neonates - however, some of the neonates do and three babies with histologically benign tumours subsequently developed metastatic disease and even scrutinising the tumours in retrospect no malignant elements could be identified. My question to Dr. Norgaard-Pedersen would be, are you going to have the very small amount of AFP coming from the tumour swamped out by the very much larger amount coming from the normal foetal infant liver?

Dr. Norgaard-Pedersen

AFP clearance after birth is taking place within the first two or three weeks. From a mean level of 50,000 μg/l at birth, the AFP at one month is less than 1,000 μg/l. It is right, however, that during the first year of life you might see AFP elevations due to other causes, intercurrent disease especially liver disease and you might see levels as high as 1,000 μg/l. In infantile testicular tumours or sacrococcygeal tumours initial values are in the order of 6,000 or 7,000.

I think it would be nice if we can conclude that orchidectomy should be the only treatment for testicular tumours in infancy if they are monitored correctly.

Dr. Mann

Yes - I think it is very important to do further investigations. We do take X-rays of the chest and do a CT scan of the abdomen if we can. We have seen retroperitoneal lymph nodes involvement. I think that, providing the AFP is carefully monitored and that the initial staging procedures are done, I would agree that orchidectomy is the only treatment necessary.

TUMOUR MARKERS
INCLUDING
RADIO-IMMUNO-LOCALISATION

Chairman of Session
Dr W.G. Jones

MARKERS IN GERM CELL TUMOURS: THE CURRENT STATE OF THE ART, AFP, βHCG AND AHL KINETICS

A. Milford Ward
Director,
Supraregional Protein Reference Unit,
Royal Hallamshire Hospital,
Sheffield

Tumour markers, as distinct from tumour antigens, can be defined as serum or body fluid constituents found in inappropriate concentrations in tumour bearing patients. They may be synthetic, produced by the tumour, or reactive, produced by host tissues in response to the presence of the tumour. Either category may be used with varying degrees of specificity in the diagnosis of malignancy or the monitoring of tumour burden.

Table 1. Tumour Markers

Synthetic	tumour specific	Oncofetal proteins/antigens Proteins
	non specific	Proteins Hormones Enzymes Nucleosides
Reactive	non specific	Acute phase proteins Enzymes

With notable exceptions, specificity of the markers for particular tumours and indeed for the presence of malignancy is poor; diagnosis of malignancy based on screening for the presence of markers in the serum or other body fluid is rendered impractical by an unacceptable level of false positive and false negative reactions. The more rational use of tumour markers is in the monitoring of tumour burden in patients diagnosed by clinical and histopathological examination. In this context the markers may be used reliably to indicate response to therapy and the recurrence of tumour after a period of apparent remission. It is in this context that the tumour markers have contributed to the changed management and prognosis of the germ cell tumours.

Although various markers have been described in association with germ cell tumours, and the use of lactic dehydrogenase and ferritin, probably synthetic and reactive markers respectively, are described elsewhere in this symposium, the principle markers are still alphafetoprotein (AFP) and the β subunit of human chorionic gonadotrophin (βHCG).

Alpha fetoprotein (AFP)

AFP is a normal serum protein of the human foetus, being synthetised from the 10th gestational week in the foetal liver and gastointestinal tract and in the yolk-sac (1). Maximal serum levels of 3mg/ml are found in the 13th gestational week and fall thereafter during intrauterine and early neonatal life to less than 10ng/ml from the age of three months. Neosynthesis of this protein after the age of three months indicates the presence of increased hepatocyte turnover, as a regenerative process, or the induction of neoplasia involving tissues which possessed AFP producing potential in embryonic life.

AFP is, therefore, a serum marker for tumours of the hepatic parenchyma, a small proportion of tumours of the foregut, and of germ cell tumours containing yolk-sac or endodermal sinus elements (8). Such tumours may be found in a variety of extragonadal sites as well as in both male and female gonads.

Increased AFP levels are also seen in the maternal serum during pregnancy, foetal AFP diffusing across the foetal membranes and placenta into the maternal circulation.

AFP is a glycoprotein, molecular weight 65000 daltons, with a biological half life of 3.5 - 4.0 days. The glycosylation of hepatic and yolk-sac produced protein differs in as much as they can be distinguished by their differential Conconavallin-A binding properties despite immunological identity.

β subunit of Human Chorionic Gonadotrophin (β HCG)

Human chorionic gonadotrophin is a placental glycoprotein, molecular weight 38000 daltons, with a biological half life of less than twenty-four hours. HCG is composed of two dissimilar polypeptide chains. The α chain is antigenically identical to the α chains of human luteinising hormone (LH) and follicle stimulating hormone (FSH). The β chain, whilst distinct from that of FSH, shares eighty-six identical residues with that of LH, differing only in the extra C-terminal domain of twenty-six residues. Assay specificity for HCG is achieved by attention to the β subunit and the use of an antiserum specific for the C-terminal domain (9).

HCG is produced by the syncytial trophoblast and by non-gestational trophoblastic elements in gonadal and extragonadal germ cell tumours. Inappropriate hormone secretion by a variety of tumours may result in the presence of βHCG as a serum marker in a wide variety of neoplasms (7).

Marker Kinetics - Apparent Half Life (AHL)

Whilst the serum concentrations of a marker may be of value in assessing the tumour burden, the rate of elimination of the marker following treatment gives an indication of the adequacy of that treatment. By observation of the marker kinetics it is not necessary to await basal values before expressing a verdict of complete tumour removal (2, 4). With complete excision of marker producing tissue, the marker level will fall within the biological half life of the marker. The apparent half life (AHL) is the observed marker decay in the patient following tumour excision or treatment. With complete tumour elimination the AHL equates with the biological half life.

AHL values greater than five days for AFP and one day for βHCG are considered abnormal, and indicative of residual disease.

Extending AHL values during post surgical monitoring of germ cell tumours give an even earlier indication of active tumour recurrence than conventional marker recurrence as signalled by increasing serum levels.

Clinical use of AFP and βHCG in germ cell tumours

The value of serial estimates of AFP and βHCG in the diagnosis, detection of recurrence and monitoring of therapy in patients with germ cell tumours has been extensively discussed (3, 5).

The incidence of marker positivity can be illustrated by reference to samples submitted to the Sheffield marker laboratory from the Yorkshire and Trent regions. Table 2

TABLE 2

MARKER POSITIVITY IN ALL PATIENTS, INCLUDING POST TREATMENT FOLLOW UP, WITH TESTICULAR TUMOURS

	n	marker positive	%	
MTU	23	15	65	
MTI	27	16	60	
CT	13	9	79	65%
YST	6	3	50	
MTT	2	2	100	
S	74	30	40	
ACT	2	0	0	
Others	9	0	0	
	156			

TABLE 3

MARKER POSITIVITY IN PREOPERATIVE SAMPLES FROM PATIENTS TESTICULAR TUMOURS

	n	marker positive	%	
MTU	12	11	90	
MTI	14	14	100	
CT	5	4	80	94%
YST	1	1	100	
MTT	1	1	100	
S	35	18	50	
ICT	1	0	0	
Others	9	0	0	

details one hundred and fifty-six patients in whom the histology of the primary tumour has been reviewed by the Regional Testicular Tumour Panel. The incidence of marker positivity is modified by the fact that many of the patients underwent primary treatment some years before marker estimation was initiated.

This group illustrates the universatility of AFP and βHCG for teratomatous tumours of the testis and the value of preoperative samples for marker estimation. The marker positivity amongst the seminomas is higher than usually recorded and is examined in more detail in Table 4.

It has been held that AFP elevations are inconsistent with the diagnosis of seminoma, and that in such cases the teratomous element of a combined tumour should be sought assiduously. In all seven cases in which the AFP was elevated extensive histological examination of the primary tumour failed to reveal a combined tumour although, in one case, yolk-sac elements were detected in an iliac lymph node at necropsy. One case, where a raised AFP was detected in a preoperative sample, would appear to be similar to that

TABLE 4

TUMOUR MARKERS IN PATIENTS WITH SEMINOMA

		marker elevation			
	n	AFP	βHCG	both	marker negative
all patients	74	3	24	3	44
preoperative only	35	1	17	-	17

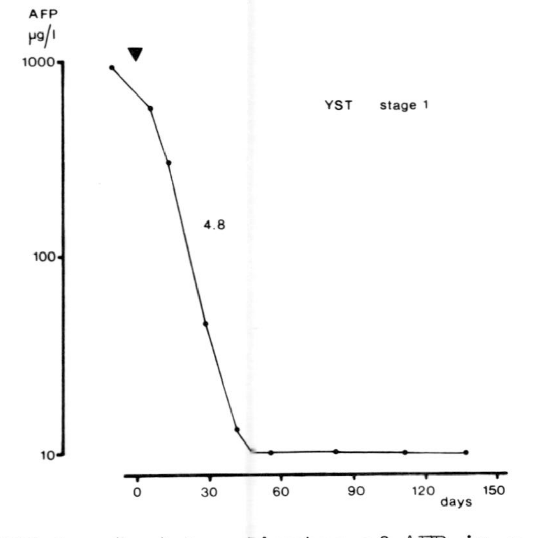

FIGURE 1 - Serial estimates of AFP in a patient with a Stage 1 YST of testis. Complete excision of tumour at operation. No adjuvant therapy.

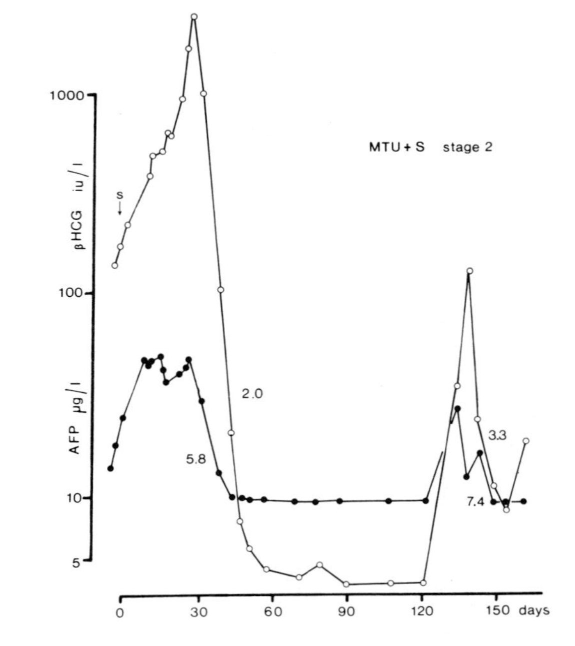

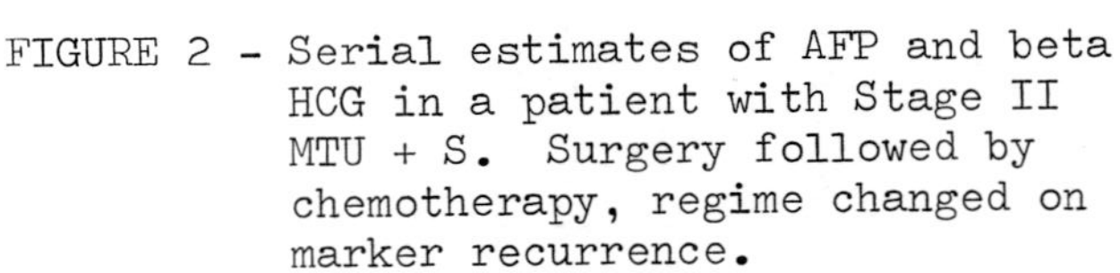

FIGURE 2 - Serial estimates of AFP and beta HCG in a patient with Stage II MTU + S. Surgery followed by chemotherapy, regime changed on marker recurrence.

described by Raghaven (6).

Marker kinetics in relation to tumour eradication and prognosis are illustrated in Figure 1 and Figure 2. Complete excision of a Stage I (Figure 1) is followed by marker decay on an AHL of 4.8 days. The residual disease in the Stage II MTU + S (Figure 2) can be predicted by the prolonged AHL values for both AFP and βHCG seventy days before marker recurrence becomes evident.

Conclusions

The regular use of serum markers in patients with germ cell tumours has been shown to be of value both in diagnosis and management. Preoperative estimation of AFP and βHCG shows marker positivity in over 90% of malignant teratomas and in over 40% of seminomas. Initial marker level is

TABLE 5

SHEFFIELD SCHEDULE FOR MARKER STUDIES IN PATIENTS WITH GONADAL AND RELATED TUMOURS

1. on presentation
2. prior to operation or primary treatment
3. twice weekly after primary treatment : to continue for eight weeks or until advised by the marker laboratory
4. weekly samples for six months : if no recurrence
5. monthly samples

related to tumour burden and indirectly to overall prognosis. Marker kinetics, following the initial treatment of excision, are related to the effectiveness of therapy; prolonged AHL values indicate residual disease and a high probability of recurrence.

The optional use of serum markers in the diagnosis and management of patients with germ cell tumours is dependent on a close liaison between clinician and marker laboratory and on a frequent sampling rate. The protocol recommended by the Sheffield marker laboratory (Table 5) allows for frequent sampling in the first eight weeks after initial treatment, reducing to weekly and monthly samples when basal levels are achieved. Recurrence or change in treatment regime reverts the sample protocol to that following initial treatment.

REFERENCES

1. Gitlin, D.A., Perricelli, A., and Gitlin, G.M., 1972. Synthesis of AFP by liver, yolk-sac and gastrointestinal tract of the human conceptus. Cancer Reasearch. 32, 979-982.

2. Kohn, J., 1978. The dynamics of serum alphafetoprotein in the course of testicular teratoma. Scandinavian Journal of Immunology. 8, Suppl. 8, 103-107.

3. Lehmann, F.G., (Ed.) 1979. Carcinoembryonic Proteins. Elsevier/North Holland, Amsterdam.

4. Milford Ward, A., and Bates, G.E., 1979. Serum AFP and apparent half life estimates in the management of endodermal sinus tumours. Protides of the Biological Fluids. 27, 365-368.

5. Norgaard-Pedersen, B., and Axelson, N.H., (Ed.) 1978. Carcinomembryonic Proteins. Scandinavian Journal of Immunology. 8, Suppl. 8.

6. Raghavan, D., Heyderman, E., Monaghan, P., Gibbs, J., Ruoslahti, E., Peekham, M.J., and Neville, M., 1981. Hypothesis: when is a seminoma not a seminoma? Journal of Clinical Pathology. 34, 123-128.

7. Rosen, S.W., 1975. Placental proteins and their subunits as tumour markers. Annals of Internal Medicine. 82, 71-83.

8. Talerman, A., and Haije, W.G., 1974. AFP and germ cell tumours; a possible role of yolk sac tumour in the production of AFP. Cancer. 34, 1722-1726.

9. Vaitukaitis, J.L., Braunstein, C.D., and Ross, G.T., 1972. A radioimmunoassay which specifically measures human chorionic gonadotrophin in the presence of human luteinising hormone. American Journal of Obstetrics and Gynecology. 113, 751-758.

CLONING OF HUMAN TESTICULAR CANCER IN SOFT AGAR: POTENTIAL DIAGNOSTIC AND THERAPEUTIC APPLICATIONS

R.F. Ozols, Brenda J. Foster and Nasser Javadpour
National Cancer Institute,
Bethesda,
Maryland, 20205
U.S.A.

Introduction

An in vitro assay for the growth of human tumour stem cells (HTSCA) in soft agar has recently been described (4, 5). The initial results using the assay to aid in the individual selection of chemotherapy have demonstrated a strong correlation between in vitro sensitivities and clinical results (14). Malignant cells from a variety of tumour types (including melanoma, ovarian cancer, neuroblastoma, oat cell lung cancer) and a variety of tissues (solid metastases, malignant effusions, and cytologically positive bone marrow) have been reported to form colonies in agar (3, 12). We have used this technique to grow ovarian cancer colonies from 70% of malignant effusions as well as from 92% of cytologically malignant peritoneal washings (9).

We have recently been successful in growing colonies from testicular carcinoma specimens. The HTSCA as applied to testicular carcinoma has potentional applicability in the selection of chemotherapy for relapsed patients as well as in the search for new agents with activity in testicular cancer. Furthermore, the ability to form tumour colonies in agar may correlate with the malignant potential of a tumour and thus aid in the diagnosis and management of some patients in whom the histology of tumour mass is equivocal.

Testicular Cancer Colony Growth in Soft Agar

The technique used for the growth of human testicular cancer cells in agar is similar to that initially described by Hamburger and Salmon (5) with certain modifications as previously described (9), Figure 1. A single cell suspension of cells was made by mechanical dissocation of the solid specimen by mincing, passing through needles and filtration

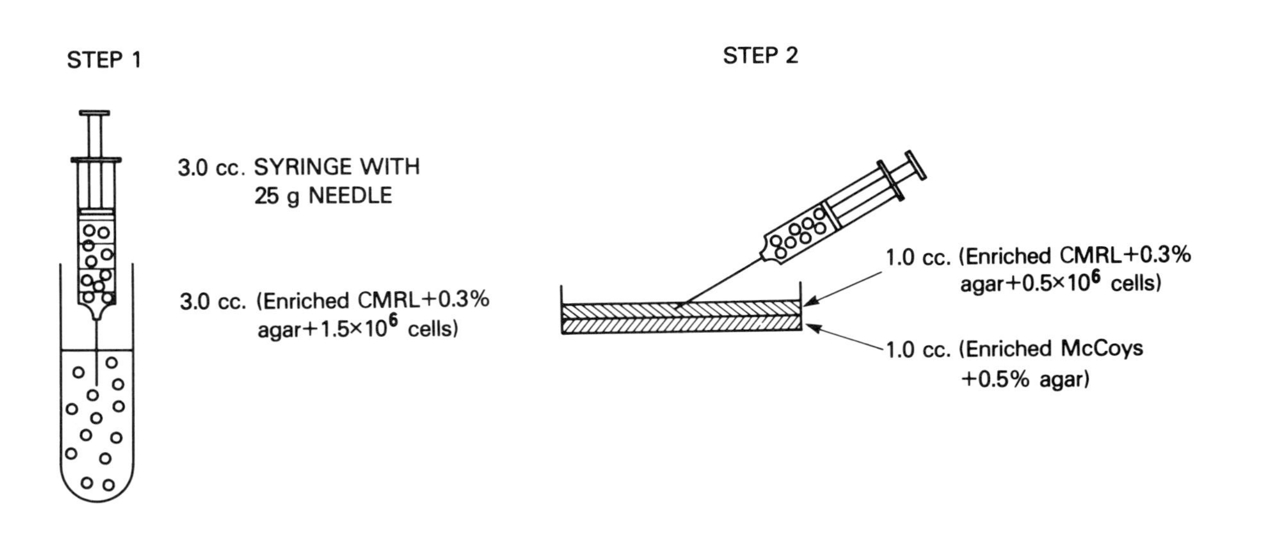

FIGURE 1 - Plating of tumour cells in soft agar

through a 20 micron gauge nylon mesh. Malignant effusions were centrifuged, and the cells washed as previously described. The total number of nucleated cells was determined with a haemocytometer without any attempt to determine a differential count of malignant or non malignant cells such as macrophages and polymorphonuclear cells.

The cells were cultured in 35 mm petri dishes using two layers of selective media. The underlayer consisted of 1.0 ml of McCoy's medium with 10% heat inactivated foetal calf serum, various nutrients and 0.5% agar (5). Conditioned media was not used. The cells were resuspended in enriched CMRL media and 0.3% agar at a final concentration of 500,000 cells per plate. The cells were incubated at 37°C in 6% carbon dioxide. The plates were examined using an inverted microscope after seven, fourteen and twenty-one days of incubation.

For determination of drug sensitivities, the cells were incubated with chemotherapeutic agents for one hour at 37°C. The cells were washed with McCoy's medium and plated in triplicate as described above. The percent survival of

TABLE 1

TESTICULAR CARCINOMA CLONING IN SOFT AGAR

Histological Type	Colony Growth	>25 colonies per plate	Drug Testing
Choriocarcinoma	1/1	1/1	0/1
Embryonal or Mixed	5/7	5/5	3/5
Seminoma	2/3	2/2	1/2
Total	8/11	8/11	4/11

tumour colony forming units was then compared to untreated controls.

Results

Colony Growth

Eleven specimens of metastatic testicular cancer were placed in the soft agar culture system and colony formation was observed in eight (73%), Table 1. Cells from the following subtypes of germ cell cancers produced colony formation: seminoma, embryonal cell, and choriocarcinoma. The median colony forming efficiency (number of colonies x 100%/500,000 nucleated cells) was 0.02% (range 0.005 - 0.03).

Characterisation of Colonies

Cell aggregates containing thirty or more cells were scored as colonies. The colonies grew as compact aggregates of cells, Figure 2, and reached their maximum size by day fourteen. The upper layers of agar which contained the tumour colonies were fixed as described by Salmon and Buick (13) and stained with hematoxylin and eosin. Immunoperoxidase staining for alphafetoprotein (AFP) and human chorionic gonadotrophin (HCG) was performed using previously described immunocytochemical techniques (7, 8). In tumour colonies derived from embryonal cell carcinoma immunoperoxidase staining demonstrated the presence of colonies which stained for either AFP or HCG. Not all colonies on a plate were positive for a marker but within a positive colony most of the cells stained for either HCG or AFP. In colonies derived from a patient with seminoma neither AFP nor HCG could be detected.

One specimen was obtained from a patient with embryonal cell carcinoma who had a residual mass following chemotherapy. Pathologic examination revealed a mature teratoma with no viable embryonal cells identified. However, immunoperoxidase staining of the fresh specimen as well as of the colonies revealed AFP and the patient also later became seropositive for AFP.

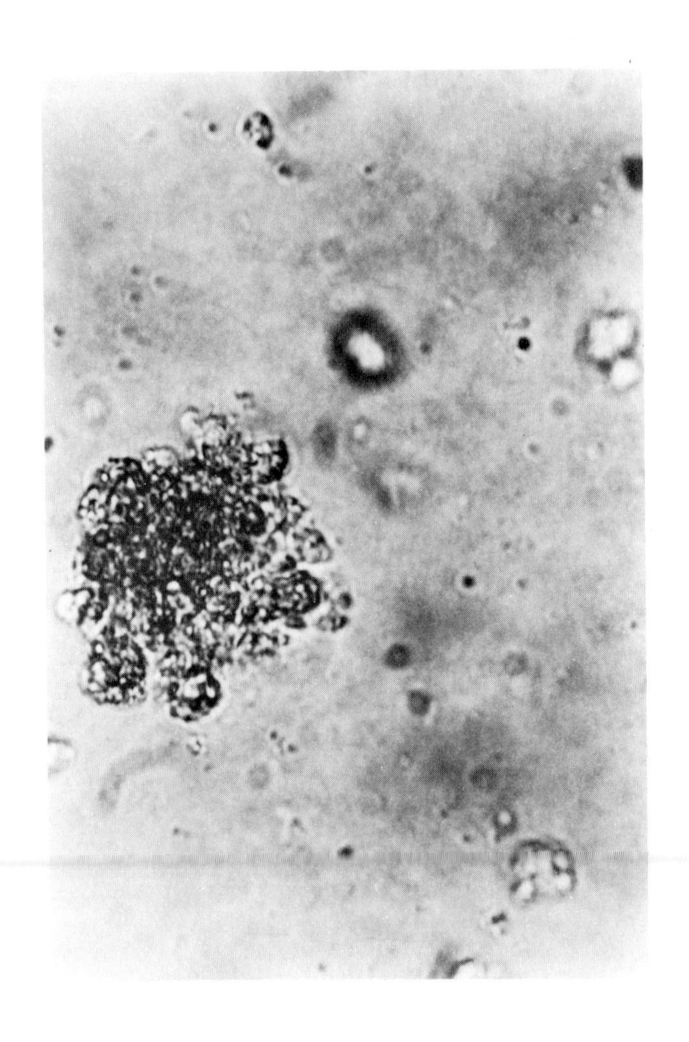

FIGURE 2 - Human testicular cancer colony grown in soft agar. Day 14 of culture (x 200).

Drug Sensitivity Studies

Enough single cells were obtained from six specimens to perform drug sensitivity studies. However, only four of these specimens produced colonies. The limiting factor for drug sensitivity studies was the number of viable single cells obtained from the tumour specimens. The specimens weighed between 2-4 grams and as only $1\text{-}2 \times 10^6$ cells were usually obtained per gram of tumour, multiple drug screening was not routinely possible.

Figure 3 depicts the effect on colony formation of exposure to adriamycin, VP-16 and platinum in a patient with seminoma who had relapsed after treatment with platinum, vinblastine and bleomycin. A moderate drug response in vitro was observed only at a concentration of VP-16 which was ten times greater than the peak attainable plasma level.

Discussion

The results of this and other studies (16) indicate that human testicular carcinoma specimens can be cloned in soft agar. The HTSCA has potential utility in the selection of a patient's chemotherapy as well as an aid in the diagnosis of residual masses following chemotherapy.

However, the results also point out some major problems with the assay as currently performed which need to be resolved before the assay can be routinely applied to clinical use. Foremost of these is the preparation of a single cell suspension from a solid tumour mass (11). We have used mechanical disaggregation followed by filtration through a 20 μ mesh. Without the use of enzymes (colleganase and/or DNAase), this has usually resulted in only $1\text{-}2 \times 10^6$ cells/gram of tissue. In most instances this has only allowed for plating of control specimens and has not allowed us to test for drug sensitivity against a panel of chemotherapeutic agents which would be required before the assay is clinically useful. Additionally, the cloning efficiency of the specimens is low. While eight of eleven specimens produced greater than twenty-five colonies per plate only six of eleven produced more than fifty colonies per plate. These two problems, preparation of single cell suspension from solid tumours and low cloning efficiency, are major areas of continued investigation. Among the many questions to be answered is whether the use of digestive enzymes in the preparation of a single cell

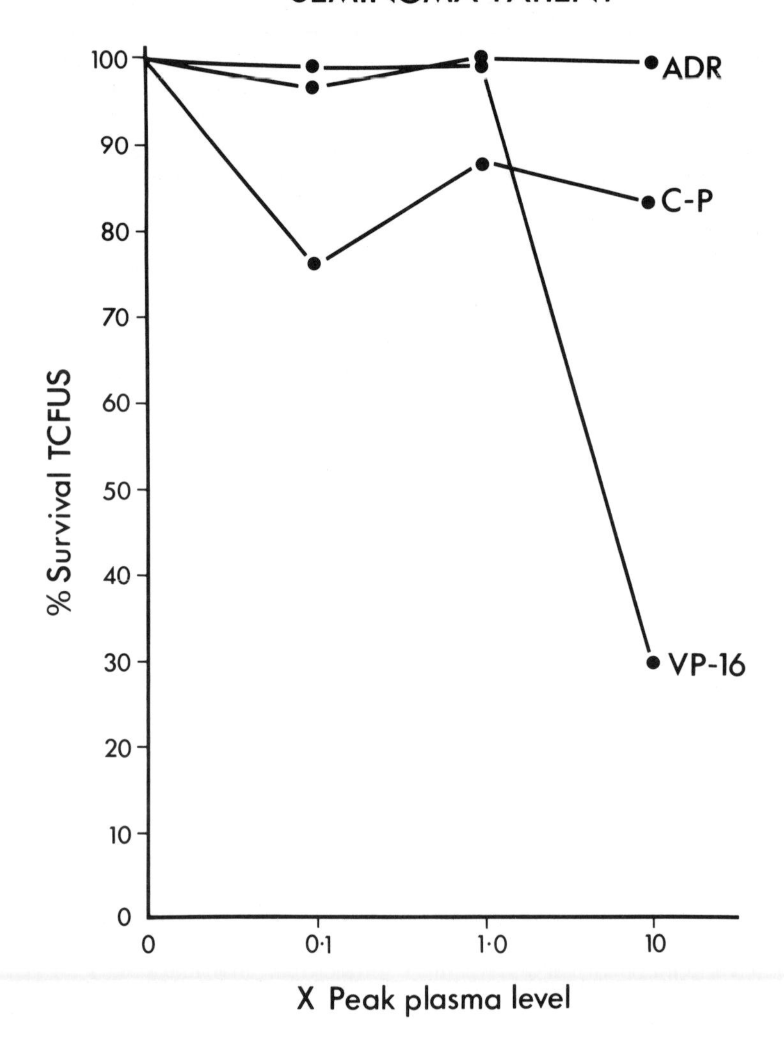

FIGURE 3 - Percent survival of tumour colony forming units after a one hour exposure to adriamycin (ADR), cis-platinum (C-P) and VP-16. The concentrations of the drugs were multiples of the peak attainable plasma levels.

suspensions will alter the chemotherapy sensitivity pattern.

These problems may be overcome if tumour amplification can be produced by short term suspension culture in selective serum-free media. If such a medium could be defined for testicular carcinoma, then it may allow for a rapid increase in the number of cells available for drug testing. Furthermore, the addition of conditioned media and/or growth factors to the soft agar culture system needs careful examination since the optimum culture conditions in soft agar for testicular cancer have not yet been defined.

Once these problems are overcome, then the HTSCA as applied to testicular cancer will have significant clinical utility. The most obvious application would be in the selection of chemotherapy for an individual patient. Salmon has reported this assay to have an accuracy of greater than 95% in predicting clinical resistance and 67% accuracy rate in predicting for response in relapsed ovarian cancer patients (1, 14). Since there are active agents which are not part of current primary chemotherapeutic induction regimens in testicular cancer, this assay could potentially be used to design individualised salvage regimens in relapsed testicular cancer patients. Any beneficial role of the HTSCA in the selection of initial chemotherapy regimens will be more difficult to establish since platinum containing combinations currently have an approximately 90% response rate. Furthermore, the assay could be used to aid in the search for new active agents in testicular cancer.

Other potential chemotherapeutic applications of the HTSCA to testicular cancer include the determination of dose-response relationships to specific drugs. It would be beneficial in testicular carcinoma to determine dose-response relationships to drugs such as VP-16, platinum, and cyclophosphamide. High dose chemotherapy trials with autologous bone marrow infusion are underway in the treatment of refractory testicular cancer patients (15). It would be useful to determine in vitro whether there is an associated increase in cytotoxicity with increasing dose before subjecting patients to such intensive therapeutic measures. Analogous in vitro studies have been performed by us in ovarian carcinoma (10). In specimens obtained from patients who had relapsed on adriamycin containing combination regimens there was not significant inhibition of ovarian tumour colony formation at exposure to adriamycin concentrations ten times greater than achievable by systemic therapy. This suggests that in some patients with ovarian carcinoma the resistance to adriamycin is of such a degree that even if adriamycin could be safely administered at 2-5 times higher does it would not be likely to produce any increase in cytotoxicity.

This assay also has potential application for the investigation of the mechanisms of drug resistance (6) and the nature of drug-drug interactions. The HTSCA also may be useful in the diagnosis of residual masses following chemotherapy. If cells obtained from a residual mass form colonies in soft agar this may be indirect proof of the malignant potential of the tumour regardless what the histologic nature of the mass is since cloning in agar is a characteristic of malignant cells, although non-malignant cells may under appropriate conditions also form colonies (2). Further evidence for malignant potential could be inferred from peroxidase immunochemical staining of colonies for AFP and HCG. The growth in agar may allow for detection of cells which produce these tumour markers at a time when immuno-staining in the primary specimen is not observed.

In summary, the HTSCA as applied to testicular cancer is a promising research area which when the technical problems are surmounted may result in an assay which will be of both diagnostic and therapeutic importance.

REFERENCES

1. Alberts, D.S., Salmon, S.E., Chen, H.S.G., Surist, E.A., Soehnlen, B., Young, M., and Moon, T.E., 1980. In vitro clonogenic assay for predicting response of ovarian cancer to chemotherapy. Lancet. ii, 340-342.

2. Bradley, E.C., Reichert, C.M., Brennan, M.F., and von Hoff D.D., 1980. Direct cloning of human parathyroid hyperplasia cells in soft agar cultures. Cancer Research. 40, 3694-3697.

3. Carney, D.N., Gazdar, A.F., and Minna, J.D., 1980. Positive correlation between histologic tumour involvement and generation of tumour cell colonies in agarase in specimens taken directly from patients with small cell carcinoma of the lung. Cancer Research. 40, 1820.

4. Courtenay, V.D., Selby, P.J., Smith, I.E., Mills, J., and Peckham, M.J., 1978. Growth of human tumour cell colonies from biopsies using two soft agar techniques. British Journal of Cancer. 38, 77-81.

5. Hamburger, A.W., and Salmon, S.E., 1977. Primary bioassay of human tumour stem cells. Science. 197, 461-463.

6. Hogan, W.M., Ozols, R.F., Grotzinger, K.R., et at 1981. The effect of amphotericin B on the cytotoxicity of adriamycin and L-phenylalamine mustard in L1210 leukaemia in murine ovarian cancer and on in vitro Phase II trial using the human tumour stem cell assay. Proceedings of the American Association for Cancer Research. 22, (In Press).

7. Javadpour, N., 1979. The value of biologic markers in diagnosis and treatment of testicular cancer. Seminars in Oncology. 6, 37-47.

8. Javadpour, N., McIntire, K.R., and Waldmann, T.A., 1978. Immunochemical determination of human chorionic gonadotrophin (HCG) and alpha-fetoprotein (AFP) in sera and tumours of patients with testicular cancer. National Cancer Institute Monographs. 49, 209-213.

9. Ozols, R.F., Willson, J.K.V., Grotzinger, K.R., and Young, R.C., 1980. Cloning of human ovarian cancer cells in soft agar from malignant effusions and peritoneal washings. Cancer Research. 40, 2743-2747.

10. Ozols, R.F., Willson, J.K.V., Weltz, M.D., Grotzinger, K.R., Myers, C.E., and Young, R.C., 1980. Inhibition of human ovarian cancer colony formation by adriamycin and its major metabolites. Cancer Research. 40, 4109-4112.

11. Rupniak, H.T., and Hill, B.T., 1980. The poor cloning ability in agar of biopsies of primary tumours. Cell Biology International Reports. 4, 479-486.

12. Salmon, S.E., 1980. Cloning of human tumour stem cells. Alan R. Liss, Inc., New York.

13. Salmon, S.E., and Buick, R.N., 1979. Preparation of permanent studies of intact soft agar colony cultures of haematopoietic and tumour stem cells. Cancer Research. 39, 1133-1136.

14. Salmon, S.E., Hamburger, A.W., Soehnlen, B., Durie, B.G.M., Alberts, D.S., and Moon, T.E., 1978. Quantitation of differential sensitivity of human tumour stem cells to anticancer drugs. New England Journal of Medicine. 298, 1321-1327.

15. Spitzer, G., Dicke, K.A., Litam, J., Verma, D.S., Lanzotti, V., Valdivieso, M., McCredie, K.B., and Samuels, M.L., 1980. High-dose combination chemotherapy with autologous bone marrow transplantation in adult solid tumours. Cancer. 45, 3075-3085.

16. von Hoff, D.D., 1980. New leads from the laboratory for treating testicular cancer. In: Therapeutic Progress in Ovarian Cancer, Testicular Cancer and the Sarcomas. (Ed.) A.T. van Oosterom, et al. Martinus Nijhoff Publications, The Hague, pp. 225-234.

TUMOUR MARKER LEVELS AND PROGNOSIS IN MALIGNANT TERATOMA OF THE TESTIS

R.H.J. Begent, E.S. Newlands, J.R. Germa-Lluch
and K.D. Bagshawe
Charing Cross Hospital,
Fulham Palace Road,
London W6 8RF

The relationship of tumour bulk to prognosis in malignant teratoma is widely recognised. However, tumour dimensions as estimated by a variety of methods are not readily quantified and do not necessarily represent the volume of viable tumour. Human chorionic gonadotrophin (HCG) and alphafetopotein (AFP) are found separately or together in the serum of more than 75% of patients with disseminated malignant teratoma. Their concentrations in serum are related to viable tumour mass. The relationship of pre-treatment concentrations of these tumour markers to prognosis has been investigated and compared with other potential prognostic indicators.

Differences between survival curves constructed according to the various parameters were tested for statistical significance. The most consistent indicator of poor prognosis was serum AFP above 1×10^3 μg/l or serum HCG above 1×10^5 IU/l. A tumour mass in excess of 5 cms diameter or the presence of more than eight pulmonary metastases were also prognostic indicators but gave a less satisfactory discrimination between good and poor prognostic groups than the tumour marker concentration (1).

Since 1977 the cytotoxic chemotherapy regimen used at Charing Cross Hospital has given improved survival in malignant teratoma (2), including the group of patients with HCG exceeding 1×10^5 IU/l or AFP exceeding $1 \times 10^3 \mu$g/l. However, the latter group still had a significantly worse prognosis than those with low tumour marker levels (p=0.008) (Figure 1). When tumour bulk was judged by the system of Peckham et al (3), differences between the 'good' and 'poor' prognostic groups do not reach statistical significance. Thus it appears that as the prognosis of disseminated malignant teratoma is improved by chemotherapy, measurement of pre-treatment tumour marker

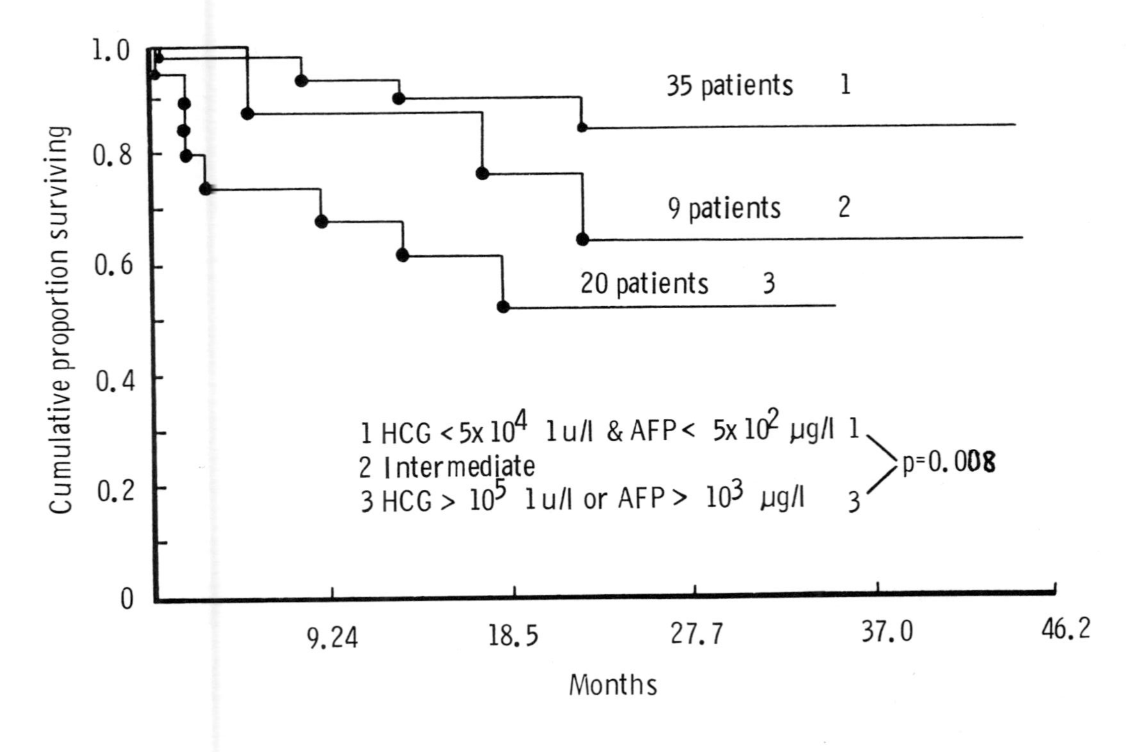
Cumulative proportion surviving
1.0
0.8
0.6
0.4
0.2
0
35 patients 1
9 patients 2
20 patients 3
1 HCG < 5x 10^4 Iu/l & AFP < 5x 10^2 µg/l 1
2 Intermediate
3 HCG > 10^5 Iu/l or AFP > 10^3 µg/l 3
p=0.008
9.24
18.5
27.7
37.0
46.2
Months

FIGURE 1

concentrations continues to provide the most satisfactory prognostic indicator. The assays for these tumour markers are readily standardised between centres and results should be stated in reports of chemotherapy of malignant teratoma in order to provide a basis for comparison between series.

REFERENCES

1. Germa-Lluch, J.R., Begent, R.H.J., and Bagshawe, K.D., 1980. Tumour marker levels and prognosis in malignant teratoma of the testis. British Journal of Cancer. 42, 850-855.

2. Newlands, E.S., Begent, R.H.J., Kaye, S.B., Rustin, G.J.S., and Bagshawe, K.D., 1980. Chemotherapy of advanced malignant teratomas. British Journal of Cancer. 42, 378-384.

3. Peckham, M.J., McElwain, T.J., Barrett, A., and Hendry, W.F., 1979. Combined management of malignant teratoma of the testis. Lancet. ii, 267-270.

ENDOCRINE EFFECTS OF TESTICULAR GERM CELL NEOPLASMS

H. Fox
Department of Pathology,
Stopford Building,
University of Manchester,
Oxford Road,
Manchester M13 9PT

We often tend to think of testicular seminomas and teratomas as being endocrinologically inert, and in any large series of such tumours there is always a small number of patients who have clinical signs of hormone disturbance either in the form of gynaecomastia or of reduced libido. In this communication I will show that a very considerably proportion of these patients have a sub-clinical endocrine disorder which amounts to a standardised abnormality and I would suggest that this is due to the presence within these neoplasms of tissue of functional trophoblast even though it may not be histologically recognisable as such. This report is not original endocrinological work but the result of considerable researching of the literature and the assembly of widely scattered information. Although rare, pure testicular choriocarcinoma does provide the blue-print for the endocrine profile of all patients with neoplasms of the germ cell group. They obviously have a high plasma HCG. Rather less commonly they also have a high circulating level of HCS, or, to use its older and incorrect term, HPL. This indicates that the tumour trophoblast is functioning as it if were first trimester trophoblast rather than third trimester trophoblast. High levels of circulating HCG should be associated with increased levels of testosterone due to HCG stimulation of the Leydig cells in the non-neoplastic portions of the testis. This is not the case, and the testosterone values within these patients are either in the low normal range or they are sub-normal. It has been shown that the Leydig cells of the patient with an HCG secreting neoplasm are resistant to the effects of HCG and they are also resistant to LH. In vitro culture techniques have shown that HCG, even in relatively low concentration, causes a marked reduction in LH receptors on the Leydig cells. Accompanying this low testosterone there is a markedly elevated osetrogen level in both the urine and the plasma. At first sight it might appear that the oestrogen is coming from the Leydig cells, but there are two objections to the hypothesis. The first is that although it is suspected, it is certainly not proven, that Leydig cells can synthesise oestrodiol. Secondly, if Leydig cells do synthesise oestrodiol, they must do so under the control of LH because Leydig cells have LH receptors but do not have FSH receptors. (Sertoli cells are the only cells in the testis with FSH receptors.) In view of the available evidence it is unlikely that the high levels of oestrogens are produced by the Leydig cells in the non-neoplastic part of the testis. It would seem reasonable to suggest that the high oestrogen levels are due to the ability which is shared by both normal and neoplastic trophoblast to convert circulating adrenal C19 steroids, particularly dehydroandrosterone, dehydroepiandrosterone and dehydro-epiandrosterone sulphate, to oestrogen. The ability of testicular choriocarcinoma to convert C19 steroids to oestrogen has been demonstrated in tissue culture. FSH values are low because of the raised oestrogens and LH levels are high because of the fall in androgen levels. Prolactin levels are also elevated in many of these patients and this is almost certainly due to the effects of oestrogen on prolactin release from the pituitary. Prolactin secretion has never been shown in these neoplasms.

The information on seminoma is not as complete or as marked as it is in choriocarcinoma. Twenty to 40% of patients with seminoma do have circulating HCG and the endocrinological profile of these patients shows the same type of abnormality as is seen in choriocarcinoma. They have low testosterone values, elevated or markedly elevated oestrogen levels together with increased levels of prolactin and LH and reduced levels of FSH.

The available data on teratoma is even more sparse but again 30 to 40% have significant increases in HCG, testosterone levels are low, whilst oestrogen and prolactin levels are increased.

A significant proportion of patients with seminoma or teratoma exhibit the same endocrine abnormality as is seen in choriocarcinoma. This endocrine abnormality is characteristic of biologically functioning trophoblast. It is well known that HCG containing cells can be demonstrated in these tumours but it is a debatable point whether HCG production constitutes evidence of the trophoblastic nature of a cell. HCG producing cells are found in a wide variety of neoplasms which are not of trophoblastic origin or association. I would, however, suggest that the HCG producing tissue in germ cell neoplasms is not only histological trophoblast but behaving as trophoblast in the full biological sense of the word, particularly in the ability to convert C19 adrenal steroids into oestrogen. It is the presence of this functional trophoblast which is responsible for a sub-clinical spectrum of endocrine abnormality which in a small proportion of patients may increase to a level to produce physical symptoms. It may also be responsible for the decrease in size that often occurs in the non-neoplastic testis.

NEW MARKER POSSIBILITIES

Frances Searle
Department of Medical Oncology,
Charing Cross Hospital,
London W6 8RF

The value of human chorionic gonadotrophin (HCG) and alpha-fetoprotein (AFP) as tumour markers is unquestioned. The requirement I am to discuss, is whether we will be able to identify, by equally useful markers, those silent germ cell tumours which do not express AFP and HCG. Germ cells are well classified from the pathologists' point of view, but the correlate the chemist needs to find is a cellular biochemical function related to their proliferation and tendency to metastasise. For the purpose of this paper, I would prefer to consider the cells in terms of a series of chemical factories, and see what we know about their production lines and if we can make educated guesses about their likely biochemical specialities.

The germ cell is mobile in early life, though its normal development and proliferation, as described by Gondos (16), seems to be closely integrated with that of its more specialised progeny, which suggests a predominant role for receptors and possible feedback mechanisms between the cells, i.e. specific mechanisms which may be partially retained in the malignant cell. The host involvement of embryo-derived tumours in some mouse strains and the cellular interactions in promoting or switching off of alpha-fetoprotein synthesis bear witness to this. The potential for mobility leads one to consider cell surface enzymes which may help the cell to "bed-in"; for example, plasminogen activator. The receptors lead one to a consideration of glycolipids, which play an integral role in defining cell-surface anchorages for hormones, etc., and may therefore contain specific molecular sequences. I will return to these in a moment.

It is tempting to work back from the yolk sac tumour cell, where the alpha-fetoprotein is presumably expressed because a foetal type of m-RNA is functioning (by analogy with Tamaoki & co-workers (24) on hepatoma m-RNA), and

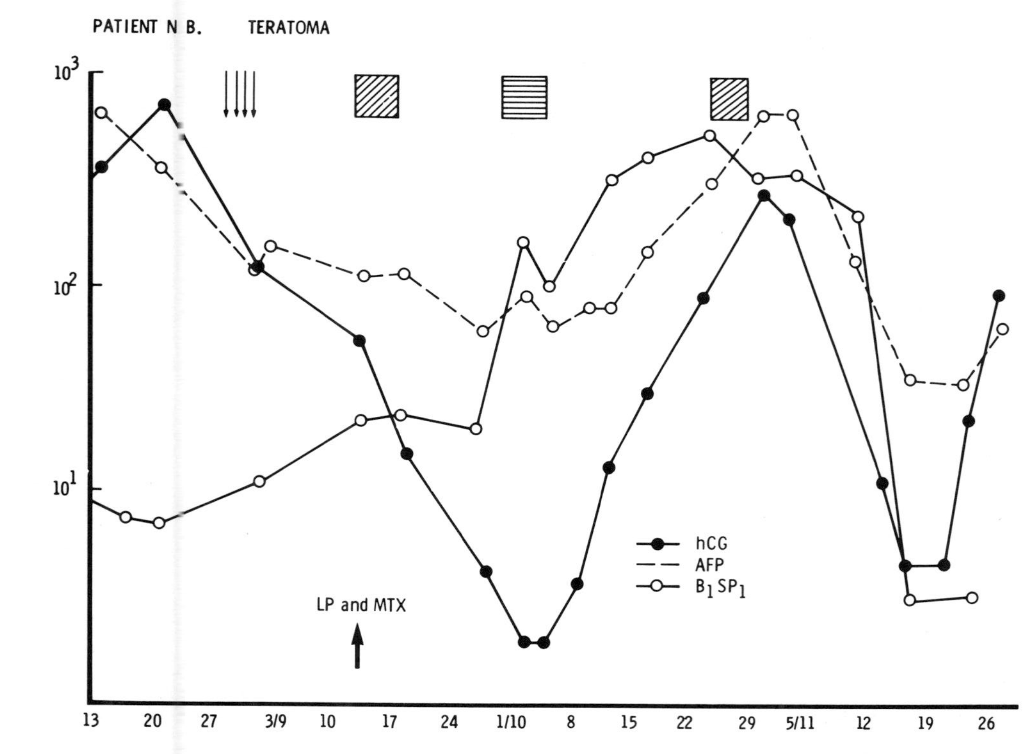

FIGURE 1 - Comparative serum levels of human chorionic gonadotrophin (IU/l), AFP (microgram/l) and pregnancy-specific-beta-glycoprotein (microgram/l) for a patient with malignant teratoma containing trophoblastic elements.

speculate that there would be other identifiably foetal m-RNA's. Equally it is tempting to extrapolate back from the fact that HCG production in chorio cell lines can be stimulated by dibutyryl-cyclic AMP (19) or some cytotoxic drugs (7), to wonder if there is a distortion, such as those reviewed by Hunt & Martin (18), in cyclic nucleotide balances common to the germ cell tumours, which will be reflected in other increased secreted products if these cells are so stimulated. However the evidence from the pathologists seems to contain arguments for and against how directly the trophoblast, yolk-sac, and teratoma types of cells progress from the germ cell. It is therefore safest for the biochemist to assume that the identifiable smoke signals such as AFP and HCG are not necessarily indicative of partial dominant pathways in the "silent" germ cell tumours though they may be.

Let us briefly consider the status of pregnancy-specific-β-glycoprotein (SP_1) as a marker in trophoblastic tumours. When we assayed 581 samples collected serially from 38 patients during and after treatment for choriocarcinoma, or after evacuation of a hydatidiform mole, there were few instances in which the conclusions of response to treatment drawn from the two assays would have been discordant (Figure 1), but analysis of the ratio of numerical values in the sera increasingly began to suggest that the two markers were charting cells which could potentially respond differently to the treatment schedules (Figure 2). The effect was more marked in 7 patients with teratoma containing trophoblastic elements, where the SP_1 remained elevated in 114/505 samples when the HCG was apparently normal. However the interpretation of these results is balanced by three patients with choriocarcinoma who showed elevated SP_1 for at least 5 months while remaining clinically disease-free and without HCG elevations. One could speculate that the SP_1 expression is concerned with a less aggressive state than rising HCG but whether this argues for different cells, or the same cell in a more quiescent state, or rather the preferential vulnerability of the SP_1 synthetic metabolic pathway to drug treatment, it would be impossible to determine from this type of data alone. It is interesting that Rosen et al (36) have found immunoperoxidase staining of fibroblasts with antisera to SP_1, and Javadpour et al (21) a discordance in staining of cells in regard of HCG, AFP and SP_1.

A marker which is being considered currently for germ cell tumours is LDH, (particularly in advanced tumours) e.g.

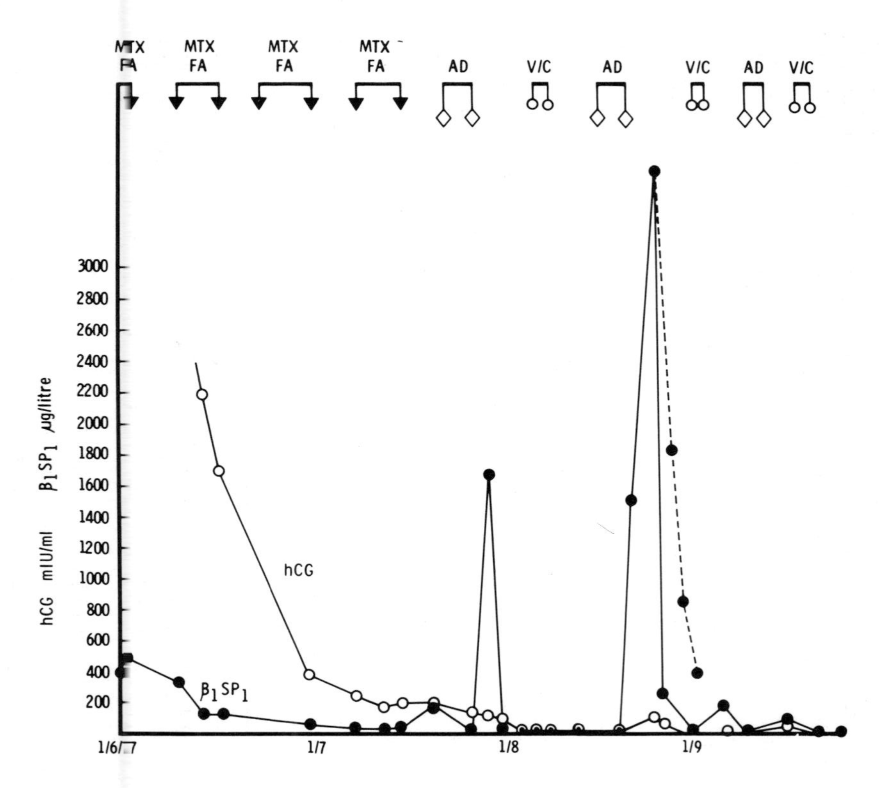

FIGURE 2 - Occasional discordance in the serum levels of human chorionic gonadotrophin and pregnancy-specific-beta-glycoprotein in a patient with choriocarcinoma, under treatment.

by Edler von Eyben (13) and ourselves (Figure 3). Some interesting work that may lead to associated markers is coming out of the consideration of LDH isozymes. Thus dibutyryl-cyclic AMP increases levels of LDH-3 and its m-RNA in certain glioma cells (12), while LDH's in muscle and other tissues and fluids of tumour-bearing mice can have what is referred to as a "changed pattern" (20, 4).

What seems to have been established so far is that the LDH isozyme distribution shifts gradually towards the isozyme pattern of the tumour (in mouse muscles located distally from a transplanted mammary carcinoma) while in the human, 2 peptides have been identified whose kinetics indicate a specific interation in vitro with either LDH-H_4 or LDH-M_4 at the level of monomer, dimers or a transition state of the tetramer accessible in the proces of oligomerisation. It cannot be excluded that in vivo conditions exist where the two peptides could exert similar inhibitary mechanisms, and strong allosteric effects might be induced by peptides.

In all, while the overall host LDH pattern would probably have to be quite distorted to be outside the variations of normal fluctuations, if the isoenzyme pattern in surrounding normal tissues can be altered, as demonstrated, possibly there will be a discernable variation in their determinant peptide levels at an earlier stage of disease than marked LDH variations.

Earlier, I referred to considering glycolipids in the search for germ cell markers, since such molecules are closely connected with receptors for hormone regulators. A sulphogalactoglycerolipid (SGG) has been purified and chemically characterised (26). There is strong evidence that SGG is present only in germ line cells. The precursor galactoglycerolipid has also been isolated from testis and its conversion to SGG by a sulphotransferase is believed to occur during the early spermatocyte stages. Now it has been claimed that antibodies which cross-react with sperm antigens can be found in the serum of patients with teratoma (41). Certainly a number of patients with teratoma do produce immune complexes in their sera. These results will be published in detail elsewhere (3). It should be possible to check with overlay techniques whether within these complexes are contained the antibodies to SGG or its precursor glycolipid. Since it is possible to label the percursor glycolipid with ^{35}S utilising the sulphotransferase, the basis for an assay exists for the precursor, at least.

Following on with carbohydrate chemistry, it is

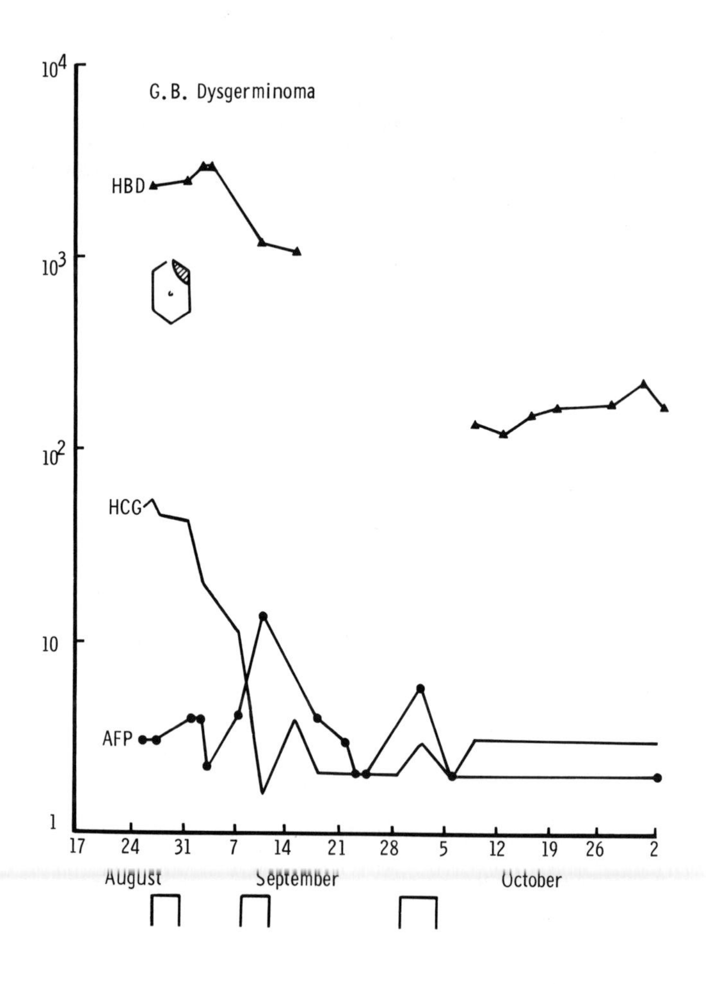

FIGURE 3 - Serum levels of lactate dehydrogenase (LDH 1 and 2) of a patient presenting with a dysgerminoma.

appropriate to discuss the F9 antigen, which, according to Fraley et al (14), is already being considered as a candidate for localising studies with radiolabelled antibodies. Now the F9 antigen is defined by the 129/SV mouse anti-F9 antiserum (1), and in a recent paper Morello et al (29) have managed to distinguish three independent antigenic determinants within this grouping, by isolating the IgM, IgG, and IgG2ab by protein-A-sepharose affinity chromatography of the antiserum. The interest of the F9 antigen lies in the fact that, in the mouse, according to Gachelin (15), it disappears as the major histocompatibility antigens appear. In man, there is evidence from Holden (17) that, in cell lines derived from metastatic teratocarcinoma, the expression of the F9 antigen or a cross-reacting substance, seems to correlate with a lack of β_2-microglobulin. It would seem a hopeful candidate for reasonable specificity to some undifferentiated teratocarcinomas. Reisner (34) found that the two cell lines which were positive for the F9 antigen also bound strongly to peanut agglutinin (28). If these experiments can be interpreted to mean that the binding is to the same, rather than concomitant, structures, it suggests an interesting relationship with the antigen discussed in Dr. McIlhinney's paper which also reacts with peanut agglutinin. To complicate the situation, Ostrand-Rosenberg et al (31) found an H2-precursor antigen, called by them Ag-1, which was density dependent on the human teratocarcinoma line Tera-2, being present on subconfluent but not confluent cells, so the state of growth of the cells from which the membrances are fractionated for antigen purification may be critical.

At the beginning of this paper, I referred to the enzyme plasminogen activator. Plasminogen activators are serine proteases which exist extracellularly and also as membrane-associated cellular enzymes. Plasminogen activator activity is found in various tissues, including heart, macrophages, erythrocytes, leukocytes, blood vessels, lung and kidney. Therefore at first sight it might appear to be an unlikely candidate for a marker for a particular cell. However, it is an enzyme which is raised markedly in malignancy.

Sherman (38) has recently assayed plasminogen activator in trophoblast cells by its fibrinolytic activity and concluded that his results are consistent with the view that plasminogen activator is involved in trophoblast invasiveness. Granulosa cells, during ovulation, produce and secrete plasminogen activator (2). Plasminogen activator (in the mouse) can be found in embryoid bodies

(43). Mouse F9 embryonal cells when induced to differentiate became capable of producing the enzyme (39). The authors suggest plasminogen activator production may be characteristic of an invasive or migratory capacity.

It is when one considers the enhancers and inhibitors of plasminogen activators that the possibilities become rather interesting. 12-O-teradecanoyl-phorbol-13-acetate is able to enhance only one of two immunologically distinguishable forms of plasminogen activator in a hamster lung clone (44); it is believed that the TPA mimics an endogenous growth control substance binding to the cell receptor. Various forms of plasminogen activator have been found in ovarian cancers (46). Specific inhibitory responses (10) to a range of chloromethyl ketones have been found. If the side range in individual sera, as indicated by Rifkin (35), reflects the concentration of serum inhibitors rather than fibrinolytic activities, so that the basal enzyme level is fairly consistent, it is possible to envisage that a subtraction assay, in which an inhibitor particularly suitable for the germ-cell-derived plasminogen activator could be used to deplete that activity preferentially in the assay, would provide a useful indication of invasive cell load.

So far in this paper I have considered potential markers which are coming through in the literature recently. Now suppose we want to branch out again. I would like to briefly outline two types of approach which would appeal to me, one based on xenografts, the other a speculative extrapolation from known chemical pathways in malignant transformations.

The difficulty with evaluating a new candidate marker isolated by virtue of its antigenic behaviour is that its limitations are usually found at the stage of sensitive assays in human sera, after a good deal of investment of time and effort has gone into at least semipurifying the antigen. This can be due to two factors: absorption experiments on the preliminary antisera may fail to demonstrate a low but fluctuating level in normal sera, so that the range displayed in the malignant disease is not great enough to be significantly elevated above normal, or alternatively the control tissue chosen to assess the original differential between the normal and malignant cells is not really sufficiently similar. An antigen which appears to be present in 10-100 fold greater quantities in the malignant tissue may be associated with the preponderance of, for example, an epithelial cell type rather than genuinely correlatable with malignancy, so that, against the background of other cell turnover in the whole host, the differential is not sustained.

The xenograft model should eliminate some of these difficulties, since, as the tumour grows, there should be a time-dependent increase of a circulating product, if there is one to be found.

The snag with the xenograft model is the lack of human host modulating cells. To establish a useful clinical assay with current radioimmunoassay sensitivity, pragmatically we need a differential of some 500 ng/ml between the level in human sera from normal or non-malignant disease patients, and patients with advanced disease. The xenograft model will be favourable in the sense that the tumour burden carried by the nude mouse can be high. The challenge for the chemist is to identify a relevant circulating protein or polypeptide which increases in concentration in the mouse serum over serial assays and decreases rapidly after removal of the tumour. It is possible to identify sub-microgram amounts of proteins quantitatively on SDS gels (40). I would like to suggest that one compares two tumour xenografts, for example an appropriate germ-cell tumour and another human tumour (say a colonic cancer) in the same nude mouse strain. One could take serial samples of the sera of each and compare under the most highly resolving system available (SDS; perhaps ampholyte for the particular area of interest) and look for a protein band of increasing importance which occurs in the germ-cell derived xenograft sera but not in the other sera. If there is a hint of such a band, it is reasonable to suppose a degree of glycosylation, and feeding the animals with labelled galactosamine may enhance definition. It might well be profitable to explore whether production can be enhanced by serum preparations of the smaller molecular weight ranges where growth factors are to be found.

One snag is that an "organ-associated" rather than malignancy associated product may be picked up. This will need a second screening device. Since only a few micrograms of a protein are necessary to immunise a mouse, use could be made of those litter mates which are heterozygous and therefore do not show the full characteristics of the nude mouse. This should eliminate the probability of raising antibodies to accompanying mouse proteins isolated from a slag-gel procedure. Although the amount of antibody achieved will be small, it should serve as a reference and be sufficient for preliminary immunoperoxidase or cell-binding tests to ensure that is neither particular to the cells originally injected, nor produced as a response between xenograft and mouse, nor a generally widely-distributed protein dependent on cell cycle for example.

If the antigen survives these criteria, it might be time to move into the production of a monoclonal antibody towards it, as this gives immense advantages in purification by affinity chromatography.

It could be argued that the dual tumour screen advocated would actually prevent homing-in on a general malignancy-associated product, if such exists. This leads me to the second line of approach that I think it would be interesting to explore, though it depends very much on whether the germ-cell tumour is predominantly a "misplaced normal" cell, or whether it has itself genuine cell surface modifications.

Phosphate metabolism is an area of considerable interest in regard to malignant transformation. There is the role of 53,000 protein, associated with the SV 40 T antigen, to be established; there is the question of tyrosine phosphosylation, particularly in the phosphorylation of membrane proteins in response to growth factors, and the interdependence of cytoskeletal architecture and kinases generally. Are there any hints that these mechanisms play a dominant part in germ-cell tumours? Is calcium implicated in a mediatory role, particularly since mithramycin, though highly toxic, produces its best responses in some germ-cell tumours, whilst also producing a decrease in serum calcium and in phosphorus (22). Since the drug reduces the rate of bone resorption without affecting the rate of bone accretion (23) and is believed to affect RNA synthesis, is the effect in the tumour cell some form of modulation of calcium transport within the cell to which the tumour is exceptionally sensitive?

I believe that pursuing phosphate metabolism might lead to a marker for the following reasons: recent work, stemming from original observations by Collett and Erikson (11), has indicated that certain proteins associated with the transformation of virally infected cells to the malignant state may be protein kinases which have the novel property of catalysing the phosphorylation of tyrosine residues in their substrate proteins (25). Similar kinases may participate in normal regulation of cell proliferation. Opperman (30) attributes the neoplastic transformation by src to overloading the host cell with a normal cellular gene product. Linzer et al (27) demonstrated by partial peptide map analysis the 54K phosphoproteins from SV40-transformed cells and uninfected embryonal carcinoma cells were very similar if not identical. Phosphotyrosine has been identified as a product of an epidermal growth factor-activated protein kinase in A-431 cell membranes. Todaro (42) discusses the production of growth factors which stimulate achorage-independent growth of tumour and normal cells. These factors may be analogous to peptide growth factors expressed early in normal embryonic development.

Now it is a big jump to suggest that there is a picture building up of inter-linked phosphorylated protein kinases which can be stimulated by growth factors to produce tyrosine-phosphorylated membrance proteins, and that amongst all these pathways one molecule will contribute towards identifying the malignant germ-cell-derived tumour. However, we can bear in mind the work of Ushiro & Cohen (45), who have identified phosphotyrosine as a product of epidermal-growth factor-activated protein kinase in cell membranes of the human epidermoid carcinoma line A-431, and also of Carpenter et al (8), where the binding of EGF to the membranes rapidly activates a cyclic-AMP-independent phosphorylating system. We can take into consideration the work of Scott & Dousa (37) on defective cyclic AMP-dependent phosphorylation of low molecular weight membrane proteins in transformed BALB/T3 cells, alongside Cho-Chung's work (9) on rat mammary tumours, where a 56,000 dalton cyclic-AMP-receptor protein increases in regressing tumours. It seems it would probably be a worthwhile gamble to expose the cells of an appropriate germ-cell tumour line to γ-^{32}P-ATP and check for phosphorylated tyrosine molecules on membrane proteins. Pretreatment of the cells with retinoids (32) has been used to inhibit the sarcoma-growth factor induced cell growth of rat kidney cells. This manoeuvre, which could help to identify proteins in the sequence which are particularly concerned with a growth-factor pathway may be unavailable, owing to the effect of retinoic acid on the cells' differentiation. Adenylate cyclase activity is of course closely concerned, witness the synergistic effect between epidermal growth factor and cholera toxin observed in mice mammary cells by Yang (47). Cyclic AMP receptors could be usefully blocked by the photoaffinity label 8, azido-cyclic AMP (33), in order to help sort out the destination of the phosphate group from the ATP.

It is my belief that the study of a growth-factor associated membrane phosphorylation pathway, with one eye open to possible modulation by calcium metabolism, (vide the complicated interaction of cyclic AMP, GTP binding proteins and calcium ion mobilising systems referred to by Bennett

(5)), could lead to identifying a germ-cell membrane component which could be shed into the circulation as part of the proliferative process of the germ cell tumour (6) and thus act as a marker of its presence. Since its very stimulus may depend on host interactions, the model will be difficult.

REFERENCES

1. Artzt, K., Dubois, P., Bennett, D., Condamine, H., Babinet, C., Jocob, F., 1973. Surface antigens common to mouse cleavage embryos and primitive teratocarcinoma cells in cluture. Proceedings of the National Academy of Sciences, New York. 70, 2988-2992.

2. Beers, W.H., Strickland, S., Reich, E., 1975. Ovarian plasminogen activator relationship to ovulation and hormonal regulation. Cell. 6, 387-393.

3. Begent, R., Chester, K., (in preparation 1981).

4. Bengtsson, G., Anderson, G., 1981. The effect of Ehrilich Ascites tumour growth on lactate dehydrogenase activities in tissues and physiological fluids of tumour bearing mice. International Journal of Biochemistry. 13, 53-62.

5. Bennett, J., 1980. Drug receptors and their effectors. Nature. 285, 192-193.

6. Black, P.H., 1980. Shedding from cell surface of normal and cancer cells. Advances in Cancer Research. 32, 75-199.

7. Browne, P., Bagshawe, K.D., Enhancement of human chorionic gonadotrophin production by anti-metabolites. (Submitted for publication, 1981).

8. Carpenter, G., King, L., Cohen, S., 1979. Rapid enhancement of protein phosphorylation in A-431-cell membrane preparations by EGF. Journal of Biological Chemistry. 254, 4884-4891.

9. Cho-Chung, Y.S., 1980. Cyclic AMP and its receptor protein in tumour growth regulation in vivo. Journal of Cyclic Nucleotides Research. 6, 163-177.

10. Coleman, P., Kettner, C., Shaw, E., 1979. Inactivation of plasminogen activator from HeLa cells by peptides of arginine chloromethyl ketone. Biochimica Biophysica Acta. 569, 41-51.

11. Collett, M.S., Erikson, R.L., 1978. Protein kinase activity associated with avian sarcoma virus src gene product. Proceedings of the National Academy of Sciences, New York. 75, 2021-2024.

12. Derda, D.F., Miles, M.F., Schweppe, J.S., Jungmann, R.A., 1980. Cyclic AMP regulation of lactate dehydrogenase. Isoproterenol and N_6 O_2 -dibutyryl cyclic AMP increase levels of LDH-5 isozyme and its MRNA in rat glioma cells. Journal of Biological Chemistry. 255, 1, 112-121.

13. von Eyben, E., 1978. Biochemical markers in advanced testicular tumours. Cancer. 41, 648-652.

14. Fraley, E.E., Lange, P.H., Kennedy, B.J., 1979. Germ cell testicular carcinoma in adults. New England Journal of Medicine. 25, 1370-1377.

15. Gachelin, G., 1978. Cell surface antigens of mouse embryonal carcinoma cells. Biochimica Biophysica Acta. 516, 27-60.

16. Gondos, B., 1974. Differentiation and growth of cells in the gonads. In: Differentiation of growth of cells in vertebrate tissues. (Ed.) G. Goldspinte, Chapman & Hall.

17. Holden, S., Bernard, O., Artzt, K., Whitmore, W.F., Bennett, D., 1977. Human and mouse embryonal carcinoma cells in culture share an embryonic antigen F9. Nature. 270, 518-520.

18. Hunt, N.H., Martin, T.J., 1979. Cyclic nucleotide metabolism in tumours. Australia and New Zealand Journal of Medicine. 9, 584-599.

19. Hussa, R.O., Story, M.T., Patillo, R.A., Kemp, R.G., 1977. Effect of cyclic 3', 5'-AMP derivatives, prostaglandins and related agents on human chorionic gonadotrophin secretion in human malignant trophoblast in culture. In Vitro. 13, 443-449.

20. Ibrahim, G.A., Abbasnezhad, M., Yasmineh, W.G., Theologides, A., 1980. Changes in lactic dehydrogenase isoenzyme pattern in muscle of tumour-bearing mice. Experimentia. 36, 1415-1417.

21. Javadpour, N., 1980. Immunocytochemical discordance of B_1 SP_1, AFP and HCG in testicular cancer. Journal of Urology. 124, 615-616.

22. Kennedy, B.J., 1970. Mithramycin in testicular neoplasms. Cancer. 26, 755-766.

23. Kiang, D.T., 1979. Mechanisms of hypocalcaemic effect of mithramycin. Journal of Clinical Endocrinology and Metabolism. 48, 341-344.

24. Koga, K., O'Keefe, D.W., Iio, I., Tamaoki, T., 1974. Transcriptional control of alphafetoprotein synthesis in developing mouse liver. Nature. 252, 495-497.

25. Langan, T., 1980. Malignant transformation and protein phosphorylation. Nature. 286, 329-330.

26. Lingwood, C.A., Murray, R.K., Schachter, H., 1980. Preparation of rabbit antiserum specific for mammalian testicular sulphoglactoglycerolipid. Journal of Immunology. 124, 769-774.

27. Linzer, D.T.H., Maltzman, W., Levine, A.J., 1979. Characterisation of a murine cellular 80/40 T antigen in SV 40 transformed cells and uninfected embryonal carcinoma cells. Cold Spring Harbour Symposia on Quantitative Biology. 44. Part 1, 215-225.

28. Lotan, R., Skuttelsky, E., Danon, D., Sharon, N., 1975. Purification composition and specificity of anti-T lectin from peanut. Journal of Biological Chemistry. 250, 8518-8523.

29. Morello, D., Condamine, H., Delarbre, C., Babinet, C., Cachclin, G., 1980. Serological identification and cellular distribution of three F9 antigen components. Journal of Experimental Medicine. 152, (5), 1497-1505.

30. Opperman, H., Levinson, A.D., Varmus, H.E., Levintow, L., Bishop, J.M., 1979. Uninfected vertebrate cells contain a protein that is closely related to the product of the avian sarcoma virus transferring gene (src). Proceedings of the National Academy of Sciences, New York. 76, 1804-1808.

31. Ostrand-Rosenberg, S., Edidin, M., Jewett, M.A.S., 1977. Human teratoma cells share antigen with mouse teratoma cells. Development of Biology. 61, 11-19.

32. Paranje, M.S., de Larco, J.E., Todaro, G.J., 1980. Retinoids block ornithine decarboxylase induction in cells treated with tumour promoter TPA or peptide growth hormones EGF and SGF. Biochemica Biophysica Research Communications. 94, 586-591.

33. Pomerantz, A.H., Rudolph, S.A., Haley, B.E., et al 1975. Photoaffinity labelling of a protein kinase from bovine brain with 8-azidoadenosine-3' 5'-monophosphate. Biochemistry. 14, 3858-3862.

34. Reisner, Y., Gachelin, G., Dubois, P., Nicolas, J.E., Sharon, N., Jacob, F., 1977. Interaction of peanut agglutinin, a lectin specific for non-reducing terminal galactosyl residues with embryonal carcinoma cells. Developmental Biology. 61, 20-27.

35. Rifkin, D.B., Pollack, R., 1977. Production of Plasminogen activator by established cell lines of mouse origin. Journal of Cell Biology. 73, 47-55.

36. Rosen, S.H., Kaminska, J., Calvert, I.S., Aaronson, S.A., 1979. Human fibroblasts produce "pregnancy-specific" β-1-glycoprotein in vitro. American Journal of Obstetrics and Gynaecology. 134, 734-738.

37. Scott, R.E., Dousa, T.P., 1980. Defective cyclic AMP-dependent prosphorylation of membrane proteins. Cancer Research. 40, 2860-2868.

38. Sherman, M.I., 1980. Studies on temporal correlation between secretion of plasminogen activator and stages of early mouse embryogenesis. Oncodevelopment Biology and Medicine. 1, 7-10.

39. Sherman, M.I., Strickland, S., Reich, E., 1976. Differentiation of early mouse embryonic and teratocarcinoma cells and in vitro plasminogen activator production. Cancer Research. 36, 4208-4216.

40. Smith, B.J., 1980. Quantitative staining of submicrogram amounts of histone and high mobility group proteins on SDS polyacrylamide gels. Journal of Chromatography. 240, 200-209.

41. Teodorczkinejayam, J.A., Jewett, M.A., Burke, C.A., and Ostrand-Rosenberg, S., 1980. Detection of circulating antibodies to teratoma-defined antigens in patients with testicular tumours. Clinical and Experimental Immunology. 40, 438-444.

42. Todaro, G.J., Fryling, C., de Larco, J.E., 1980. Transforming growth factor produced by certain human tumour cells : polypeptides that interact with epidermal growth factor receptors. Proceedings of the National Academy of Sciences, New York. 77, 5258-5262.

43. Topp, W., Hall, J.D., Marsden, M., Teresby, A.K., Rifkin, K., Levine, A.J., Pollack, R., 1976. In vitro differentiation of teratomas and distribution of creatine phosphokinase and plasminogen activator in teratocarcinoma-derived cells. Cancer Research. 36, 4217-4223.

44. Tucker, W.S., Kirsch, W.M., Martinez-Hernandez, A., Fink, L.M., 1978. In vitro plasminogen activator activity in human brain tumours. Cancer Research. 38, 297-302.

45. Ushiro, H., Cohen, S., 1980. Identification of phosphotyrosine as a product of epidermal growth factor-activated protein kinase in A-431 cell membranes. Journal of Biological Chemistry. 255, 8363-8365.

46. Vetterlein, D., Young, P.L., Bell, T.E., Roblin, R., 1979. Immunological characterisation of multiple molecular weight forms of plasminogen activator. Journal of Biological Chemistry. 254, 575-578.

47. Yang, J., Guzman, R., Richards, J., Imagawa, W., McCormick, K., Nandi, S., 1980. Growth factor and cyclic nucleotide-induced proliferation of normal and malignant mammary epithelial cells in primary culture. Endocrinology. 107, 35-41.

SERUM LACTATE DEHYDROGENASE (LDH) AND LDH ISOENZYMES IN MEN WITH MALDESCENDED TESTES

F.E. von Eyben, G. Skude and S. Krabbe
Department of Oncology, Malmo General Hospital,
S-314 01 Malmo,
University Department of Paediatrics,
Copenhagen,
and Department of Clinical Chemistry,
Malmar Lasarett, Kalmar, Sweden.

One hundred and thirty men previously treated for maldescended testes were screened for testicular neoplasia. Four had neoplasia (1). One had raised levels of serum alphafetoprotein (AFP) and serum human chorionic gonadotrophin (HCG) (3). As serum lactate dehydrogenase (LDH) is often raised in advanced testicular germ cell tumours (2), we measured serum LDH and serum LDH isoenzymes in these men. The upper limit of normal serum LDH isoenzymes was determined in another study (4).

One patient with endodermal sinus tumour and carcinoma-in-situ had normal levels of serum LDH and serum LDH isoenzymes in serum from an arm vein, whereas LDH and LDH-1, LDH-2, and LDH-3 activities were raised in serum from the testicular vein at orchiectomy. One man with microinvasive seminoma and two with carcinoma-in-situ had normal levels of serum LDH and serum LDH isoenzymes. Of one hundred and twenty-six men without testicular neoplasia, two had raised serum LDH and serum LDH-1, LDH-2, and LDH-3 levels, two raised serum LDH and serum LDH-3 and LDH-4 levels, and one hundred and twenty-two normal levels of serum LDH and serum LDH isoenzymes. Serum LDH and serum LDH isoenzyme determinations, as well as serum HCG and serum AFP determinations, were less sensitive compared with testicular biopsy to detect testicular germ cell neoplasia in this group of men. Serum LDH, serum LDH isoenzymes, serum AFP, and serum HCG determinations should not be used as the only screening procedures for testicular neoplasia in patients at risk.

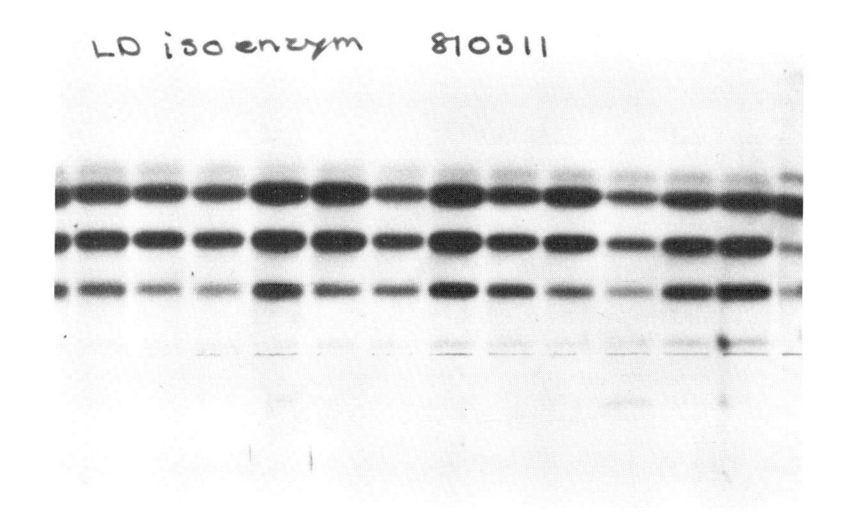

FIGURE 1 - LDH isoenzyme activity in patients with maldescended testes from left to right: 1 and 2 carcinoma in situ, 3 seminoma, 4 endodermal sinus tumour - arm vein determination, 5 endodermal sinus tumour - testicular vein determination, 6-12 patients without testicular neoplasia: 6 & 8 raised levels of LDH-1, LDH-2, LDH-3, 9 & 12 raised levels of LDH-3 and LDH-4.

TABLE 1

MEN WITH RAISED LEVELS OF SERUM LDH AND SERUM LDH ISOENZYMES

	LDH	LDH-1	LDH-2	LDH-3	LDH-4	LDH-5
Man with endodermal						
sinus tumour arm vein	250	83	100	58	5	5
testicular vein	560	218	196	134	6	6
Patients without						
testicular neoplasia	650	260	228	143	13	7
	890	303	329	321	18	9
	540	173	184	162	16	5
	530	170	180	164	11	5
Upper limit of normal	480	180	184	131	10	8

REFERENCES

1. Krabbe, K., Skakkebaek, N.E., Berthelsen, J., von Eyben, F., Volstedt, P., Mauritzen, K., Eldrup, J., and Nielsen, A.H., 1979. High incidence of undetected neoplasia in maldescended testes. Lancet. ii, 999-1000.

2. von Eyben, F.E., 1978. Biochemical markers in advanced testicular tumours : serum lactate dehydrogenase, urinary chorionic gonadotrophin and total urinary oestrogens. Cancer. 41, 648-652.

3. von Eyben, F., Krabbe, S., and Skakkebaek, N.E., 1980. Alphafetoprotein and human chorionic gonadotrophin in men with maldscended testes. British Journal Cancer. 42, 156-157.

4. von Eyben, F.E., Skude, G., Klepp, O., and Fossa, S.D., 1981. Total serum lactate dehydrogenase and its isoenzyme pattern in testicular germ cell tumour patients. (Unpublished).

SERUM FERRITIN AS A THIRD MARKER IN MALIGNANT GERM CELL TUMOURS

B.W. Hancock, Ann Grail, Gillian Bates, W.G. Jones* and A. Milford Ward
Royal Hallamshire Hospital,
Sheffield and
*Cookridge Hospital,
Leeds.

It is well established that alphafetoprotein (AFP) and human chorionic gonodotrophin (βHCG) are excellent disease status markers in malignant germ cell tumours. AFP is produced by yolk-sac and HCG by trophoblastic elements of these tumours but since these elements are absent in some untreated cases and may be eliminated by therapy with the persistence of non-marker producing tumour cells there is clearly a need for other markers. We have assessed serially, together with serum AFP and HCG, serum ferritin as a third marker in twelve patients with malignant germ-cell tumours.

Elevated levels of serum ferritin were detected in ten of the twelve patients; AFP was increased in ten and βHCG in six. One patient with seminoma showed no elevation of any marker. In two patients ferritin levels rose with disease progression. In four patients levels fell, though somewhat slowly, with remission; in one of these ferritin was the only assessable marker. In a further two patients ferritin levels were never greatly elevated, though there were considerable fluctuations in association with treatment. In the three remaining patients ferritin levels remain significantly elevated; persistent tumour is likely in two of these (figure 1).

The cause of elevation of serum ferritin in germ cell tumours is multifactorial. Undoubtedly there may be disturbances in reticuloendothelial iron storage with malignant disease and its treatment; tissue damage, inflammation and infection also play an important role. There is also evidence of increased or altered ferritin production by tumour cells. Indirect immunofluorescence with ferritin antisera shows the presence of ferritin in cell smears from malignant teratomas (3). Tumour related ferritin abnormalities have also been observed in malignant lymphoma (2) and other neoplasms; isoelectric profiles show abnormal bands present in a variety of neoplasms (1).

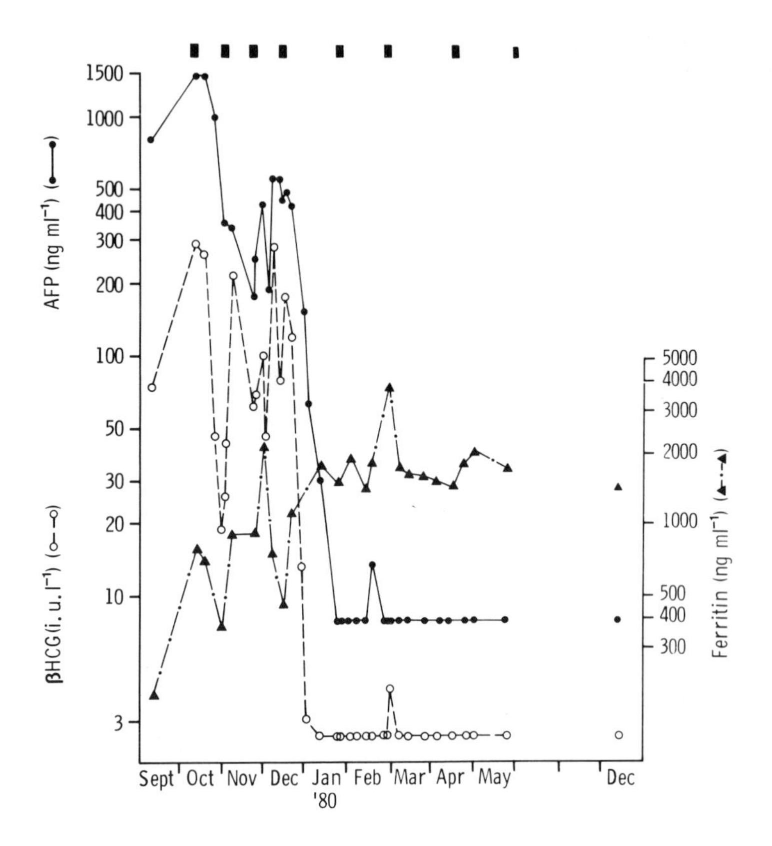

FIGURE 1 - Male patient aged 25 with mixed teratoma/seminoma. Following cytoxic chemotherapy beta HCG and AFP levels after considerable fluctuation, returned to normal in association with apparent clinical remission; serum ferritin, however, increased and has remained elevated. The nature of a posterior mediastinal mass remains, as yet, uncertain.

Serum ferritin when used in conjunction with AFP and HCG measurements may give additional information on disease status in germ cell tumours; continued elevation suggests persistent disease and falls to, and maintenance within the normal range, are associated with a good response to treatment.

REFERENCES

1. Arosio, P., Yokota, M., and Drysdale, J.W., 1976. Structural and immunological relationships of isoferritins in normal and malignant cells. Cancer Research. 36, 1735-1739.

2. Hancock, B.W., Prince, L., May, K., and Richmond, J., 1979. Ferritin, a sensitising substance in the leucocyte inhibition test in patients with malignant lymphoma. British Journal of Haematology. 43, 223-233.

3. Wahren, B., Alpert, E., and Espisti, P., 1977. Multiple antigens as marker substances in germinal tumours of the testis. Journal of the National Cancer Institute. 58, 489-498.

DISCUSSION

Dr. Norgaard-Pedersen

In contrast to the marker incidence figures presented by Dr. Milford Ward, of the two hundred and ninety-five patients in the Danish Testicular Cancer project with non-seminomatous germ cell tumours who fulfil the claim for being monitored with AFP and HCG in pre-operative samples, the positivity for either AFP or HCG was certainly lower than the figure you presented and is only about 50%.

Since in the Danish study all the testicular cancer patients are included it might be that your marker incidence figures reflect a degree of selection. Inadventent selection by Stage could explain this difference. One hundred and fifty-one of our two hundred and ninety-five patients were Stage I and in these the positive AFP and HCG rate was 45% and 55% respectively. I am not able to present the complete review but this will be completed soon for the first one thousand Danish Testicular Cancer Project patients and at that time we will have considerably more patients monitored ideally by pre-operative and post-operative markers.

Dr. Milford Ward

As far as AFP and HCG positivity in the teratoma group is concerned, I believe your figures ought to be very much higher than that. I do think our 90+% positivity is in any way unusual or different from what one would expect. Two areas of selection certainly exist: the surgeons must be motivated to send a sample before operation and the surgical specimen, the histology, must be reviewed by a pathologist I know or by the Regional Testicular Tumour Panel. I have not, at this juncture, attempted to review the data by clinical stage.

As far as the seminomas are concerned from the Greater Yorkshire series - yes, the marker positivity is higher than I have seen reported elsewhere. Our other colleagues have been talking about 30%, 35% maybe as an upper limit. I am sure there is a degree of selection because the surgeon does not always bother to send serum for markers when he thinks seminoma is the likely diagnosis. The message that I want to get across to our surgical colleagues is that all testicular masses should have serum markers estimated pre-operatively, whatever the likely diagnosis.

Once the diagnosis of hydrocoele or spermatocoele is made one can stop doing markers, but all tumours need their marker levels doing regularly, and I believe twice weekly is not inappropriate for the first two months after surgery, and weekly thereafter for a considerable period of time. I do not think that monthly samples are really adequate monitoring.

Dr. Oliver

I want to return to the problem of markers in prognosis in relation to the data presented by Dr. Begent. You had forty-two large volume abdominal disease or large volume pulmonary disease in your group, whereas you only had twenty in the poor prognostic group on the basis of markers. So that in fact you have half of your large volume abdominal disease in the low marker category and half in the high. When you take that sub-group which is defined for a standard volume of disease, is there a clear distinction on the basis of markers?

Dr. Begent

If they have a big lump and low markers they have a good prognosis if treated at Charing Cross. I would be very glad if somebody else would repeat this study and see if that applies elsewhere. That is why I say the markers give you more accurate split if you want to define a group who require different therapy.

Professor Wahlqvist

Dr. Ozols discussed the reliability of the chemosensitivity assays, and he mentioned the questionable point of whether enzyme treatment changes the chemosensitivity. Dr. Edelstein in our Institute has found for an experimental tumour that if it is treated with trypsin to make a cell suspension, you achieve a much higher adriamycin level in the cells than if you make just a chemical suspension.

Dr. Heyderman

I would like to ask Dr. Hancock two questions. Is it the acidic or basic ferritin that you are measuring? Some years ago Warren showed that ferritin could be localised by immunofluorescence in germ cell tumours of the testis. I wondered if any of your group had tried to do this or have you actually extracted any of the tumours for ferritin?

Dr. Hancock

We have been working mainly with malignant lymphoma rather than teratoma and undoubtedly we can show ascidic isoferritin in this condition. We have no experience with the testicular tumours acidic isoferritin, but are hoping to do so.

Dr. Ashley

Two or three people in this session have talked about metastases consisting of benign tumour. When I see a testicular tumour from an adult which apparently contains benign tissue I am still prepared to accept that it is a malignant tumour, but well differentiated. We should never talk of benign metastases. I was brought up to believe that seminoma was a relatively curable tumour; that the patient needed orchidectomy and a limited amount of radiotherapy and that the results from this treatment were so good that one should not contemplate chemotherapy unless there was a very good reason.

We now hear that the patients with seminoma may have AFP in the serum and they may have raised levels of HCG. Does this mean that the tumour contains yolk-sac elements in which case it is a mixed tumour; or does it mean that this is a seminoma which will behave clinically as a seminoma but happens to have cells producing AFP? The more common situation is the seminoma which produced HCG. Are we to treat these as seminoma or are we to treat them as mixed tumours?

Dr. Ozols

To answer the comment about 'benign' teratoma where 'mature' teratomatous elements are resected from a patient following chemotherapy, there is a growing body of evidence to suggest that the patient has a good long term prognosis. Benign refers to the overall prognosis and 'mature' would be a more precise description of the tumour.

I would consider any testicular tumour patient with an elevated AFP as having a mixed tumour and treat him accordingly. The questions of HCG production is not so clear cut and will be discussed in detail in a later session.

ISOTOPE LOCALISATION OF GERM CELL TUMOURS

A.R. Bradwell, D.S. Fairweather and P.W. Dykes
Department of Immunology,
University of Birmingham Medical School,
Vincent Drive,
Birmingham B15 2TJ

The idea of localising tumour deposits with isotopes is not new but has only recently become a practical proposition with the advent of subtraction scanning (1). This manipulation is required because although the tumour contains up to five times the radiolabelled antibody, compared with the surrounding normal tissue, this represents only about 0.1% of the total injected antibody (2). The vast quantities of radiolabelled antibodies that are in the vascular and extra vascular compartments are "subtracted" by comparing the emission from an isotope which localises in these fluid spaces. This is achieved by injecting ^{99m}Tc-Pertechnetate, which locates in the extravascular space and ^{99m}Tc-labelled human albumin which remains in the intravascular pool. The subtraction of the technetium emissions from the ^{131}I antibody emission constitutes the specific antibody on the tumour.

Initial trials were conducted with carcinoembryonic antigen-producing tumours and satisfactory localisation of primary and metastatic deposits was demonstrated. However, the localisation of such tumours is of little practical use because current therapy is inadequate. Therefore we have attempted to localise germ cell tumours using the alphfetoprotein marker because useful treatment is available.

Antibody preparation

High titre, high affinity antibody to human AFP was raised in sheep. the antibody was solid phase absorbed to remove antibodies cross reacting with normal tissues and subjected to ion exchange chromatography to yield the IgG fraction. Specific antibody represented between 5-10% of the total. The specific antibody fraction could be increased to 70% by affinity chromatography but in our experience this has not been a benefit as one tends to lose the highest affinity antibodies which are needed for the ultimate tumour localisation. For injection into the patient the antibody was sterilised by filtration and toxicity tested for pyrogens. The antibody was iodinated with ^{131}I by the chloramine-T method and free iodide removed. Soluble complexes were removed by centrifugation and after a final filtration, the labelled antibody was mixed with 1% human albumin to give a concentration of 100 μg of antibody to 20ml total volume.

Scanning procedure

Thyroid uptake of ^{131}I was blocked with potassium iodide. 600 μCi of ^{131}I anti-AFP was given intravenously and scans were recorded 24 hours after injection and again at 48 hours if the first scan was negative. ^{99m}Tc-pertechnetate (500 μCi) and ^{99m}Tc-labelled human albumin (500 μCi) were given, 30 minutes and 5 minutes respectively before each scan.

To date twelve patients with germ cell tumours have been investigated by this technique, five on presentation as part of their initial staging of which three were re-scanned during remission. Six further patients with rising AFP concentrations from tumour- recurrence were also scanned. One patient with a seminoma was included as a marker negative control.

Results

High serum AFP concentrations did not seem to limit the effective localisation of the tumour deposits. The results are shown in the table. All patients with raised serum AFP concentrations had positive scans as did two patients who had normal AFP concentrations but had tumour deposits demonstrable clinically. Three patients (3, 5 and 10) were scanned before and after treatment. The scans became negative with successful treatment (Table 1).

Overall antibody scans were positive in ten patients with tumour localisation in fifteen sites. The antibody scans identified all eight para-aortic node metastases compared with 5/8 by CT scanning. Two pelvic deposits were positive by antibody scanning but negative by CT scan. Two pulmonary deposits were, on the other hand, detected by CT scan but missed by antibody scan.

TABLE 1

ANTIBODY SCANS IN PATIENTS WITH GERM CELL TUMOURS

	Diagnosis	S1AFP IU/ml.	Scan
1	MTT	5341	+
2	MTT	1498	+
3	MTT	1231	+
	MTT	<3	-
4	EST	263	+
5	MTT	187	+
	MTT	<3	-
6	MTT	117	+
7	MTT	98	+
8	MTT	59	+
9	MTO	<3	+
10	MTT	<3	+
	MTT	<3	-
11	MTT	<3	-
12	S	<3	-

MTT = Malignant Teratoma Testis

EST = Endodermal sinus tumour

MTO = Malignant teratoma Ovary

S = Seminoma

Problem

The first and major problem with the technique is limitations of resolution on the part of the gamma camera due to the patient's tissues. At present satisfactory localisation cannot be achieved with tumour deposits less than 2cm diameter.

Sensitivity may be improved by using a better antibody and isotope. Monoclonal antibodies are not necessarily the answer, but, for defined protein antigens like AFP, may offer advantages. Further experiments will be necessary to define an isotope which is not so readily stripped off the antibody when it is interiorised within the cells. A change of isotope may also improve tissue penetration and increase the overall sensitivity of the technique.

REFERENCES

1. Goldenberg, D.M., Deland, F., Kim, E., Bennett, K., Primus, F.J., van Nagell, J.R., Estes, N., De Simone, P., and Rayburn, P., 1978. Use of radio-labelled antibodies to CEA for the detection and localisation of diverse cancers by external photoscanning. New England Journal of Medicine. 298, 1384-1388.

2. Mach, J.P., Forni, M., Ritschard, J., Carrel, S., Donath, A., and Alberto, P., 1979. Recent results concerning tumour localisation and detection by photoscanning of radio-labelled anti-CEA antibodies injected into carcinoma patients. Protides of the Biologic Fluids. 27, 205-209.

RADIOIMMUNOLOCALISATION OF MALIGNANT TERATOMA USING RADIOLABELLED ANTIBODY DIRECTED AGAINST HUMAN CHORIONIC GONADATROPHIN

R.H.J. Begent, F. Searle, G. Stanway, R.F. Jewkes,
B.E. Jones, P. Vernon and K.D. Bagshawe
Department of Medical Oncology,
Charing Cross Hospital,
Fulham Palace Road,
London W6 8RF

Human chorionic gonadatrophin (HCG) is frequently produced by malignant teratoma. It is present on cell membrane and has proved a useful serum tumour marker. The method of radioimmunolocalisation described by Goldenberg (2) for localisation of tumours producing carcino-embryonic antigen has now been applied to localisation of HCG producing tumours , including malignant teratoma (1).

Antisera to intact HCG were raised in sheep and rabbit. The antibodies were affinity-purified on a column of HCG linked to Sepharose 4B. The antibody was labelled with ^{131}Iodine by the Chloramine T method and injected intravenously after gel filtration and pyrogenicity testing. Gamma camera images of antibody distribution were obtained at twenty-four hours. Background radioactivity distribution was subtracted by the method of Goldenberg (2) and remaining distributions of ^{131}Iodine were regarded as potential sites of HCG producing tumours. Antibody was administered on fourteen occasions to eleven patients with malignant teratoma. Tumours were definitely localised by this method in seven patients. Equivocal results were obtained in four and scans were negative in two. Tumours were localised successfully when HCG exceeded 500 IU/l, with one exception. The results were negative in both patients in whom HCG was undectable in the serum. A positive result probably implies the presence of viable HCG-producing tumour and promises to be valuable in discriminating between necrotic deposits and viable tumour before surgery.

REFERENCES

1. Begent, R.H.J., Searle, F., Stanway, G., Jewkes, R.F., Jones, B.E., Vernon, P., and Bagshawe, K.D., 1980. Radioimmunolocalisation of tumours by external scintigraphy after administration of ^{131}I antibody to human chorionic gonadatrophin. Journal of the Royal Society of Medicine. 73, 624-630.

2. Goldenberg, D.M., Deland, F., Kim, E., Bennett, S., Primus, F.J., van Nagell, J.R., Estes, N., de Simone, P., and Rayburn, P., 1978. Use of radiolabelled antibodies to CEA for the detection and localisation of diverse cancers by external photoscanning. New England Journal of Medicine. 298, 1384-1388.

DISCUSSION

Dr. McIlhinney

Before one can discuss the merits and usefulness of radiolocalisation, there are one or two theoretical questions that must be answered. We started with an animal model because I cannot understand how an antibody to a soluble antigen secreted by tumour cells can achieve localisation. The phenomenon of capping is well documented for cells in culture and for lymphocytes but I am not aware that it has been demonstrated in cells in solid tumours. The use of monclonal antibodies, provides an ideal probe for radiolocalisation, particularly of membrane antigens. The animal model allows basic investigation of cell and antibody kinetics which are not possible in the patient.

Dr. Begent

A significant proportion of the injected antibody does complex with circulating marker antigen. The subtraction scanning technique allows for this antibody entrappment in the circulation. Monoclonal antibodies are not ideal for localisation in the clinical situation because one cannot get sufficient activity to the cells. A blend of monoclonal antibodies may produce the desired effect.

Dr. McIlhinney

The advantage of using monoclonal antibodies in the animal system is that one can define the parameters necessary to achieve localisation.

Dr. Grigor

Dr. Heyderman and Dr. Jacobson showed us that AFP and HCG were actually inside cells. How is localisation achieved with an antibody to intracellular antigen?

Dr. Bradwell

All tumours contain a proportion of dead and dying cells

with antigen release. Active secretion of HCG and AFP produce concentration gradients across the cell membrane. CEA is much easier to explain, the antigen being located on the cell surface.

Dr. Heyderman

Immunocytochemical data must be interpreted with caution in this context, and you must remember that the aim is to achieve cellular localisation. Conditions are adjusted to give maximum discrimination between tumour cells and the surrounding tissues and extravascular fluid.

Professor Fox

HCG is expressed as a surface antigen.

Dr. Pizzocaro

My personal series of seminoma shows 17/51 with elevated HCG in the pre-operative serum. I was surprised to find this high proportion of 35% marker positive but am comforted by the similar frequency in the Yorkshire patients.

Dr. Grigor

Will Dr. Hancock speculate on the origin of the ferritin: is it a tumour product?

Dr. Hancock

Ferritin is undoubtedly produced by some tumours, but it is also an acute phase reactant protein produced by host tissues. The origin in testicular teratoma is not clear.

Dr. Jacobsen

We have demonstrated ferritin in 80% of tumour cells in seminomas and teratomas by immunoperoxidase techniques. Ferritin may be synthesised by tumour cells and secreted as a tumour product.

Dr. Norgaard-Pedersen

Ferritin is a very sticky molecule and may cause artefacts in the serum assay and in immunocytochemistry. The nonspecific adhesiveness can be eliminated by including EDTA in the incubation buffer.

Dr. Jacobsen

Ferritin is seen in tumour cells but not in normal epithelial cells.

RADIOTHERAPY IN EARLY DISEASE STATES

Chairman of Session
Professor J. Blandy

THE CASE FOR RADIOTHERAPY IN EARLY CASES OF MALIGNANT TERATOMA

H.F. Hope-Stone
Consultant Radiotherapist,
The London Hospital,
Whitechapel,
London E1 1BB.

Twenty years ago, the question of whether to give irradiation after orchidectomy in early cases of malignant teratoma would not have been discussed, as it was assumed that all such cases would routinely be treated in that way (2). Today it is open to doubt, and this paper will endeavour to show why such treatment should continue to be used.

Early teratoma can be defined as stage I, where the tumour is confined to the testis, or stage IIa, where, if nodes are involved, they should be less than 2 cms. as measured by the lymphogram (6). However, the latter form of diagnosis is by no means accurate, and when in doubt it would be better to assume that the tumour is more advanced, and should be treated as such.

Whereas staging in the 1960's was by simple radiological tests combined with clinical skill, in the 1970's and onwards much more sophisticated methods have been used, starting with lymphangiography and progressing to whole lung tomograms, CAT scanning, ultrasound, and the use of tumour markers. Even with these methods, and particularly with lymphangiography, there are some cases who present with false positives, and up to 25% are false negatives (6).

When staging these patients, care must be taken to examine carefully the sites of extra-nodal metastases, particularly in the lungs, liver and bone.

Early results have shown a slow but steady improvement in all groups during the period 1960 to 1978. It should be noted that the stage II M.T.I. results were not particularly favourable, probably because of inadequate staging during the early years of the study. It is not until the cases between 1970 and 1978 are looked at do we begin to see any real improvement in the M.T.U. group (3), and even then the stage II and III cases have a very poor survival.

The patients were nearly all treated with irradiation as the primary method, though several did have chemotherapy on relapse. The technique of irradiation has remained unchanged over the past two decades. The target area is localised with an IVP, (and since 1968) combined with lymphangiogram. The treatment is planned with the use of a simulator. A pair of parallel opposed megavoltage T-shaped fields are used to treat the para-aortic and pelvic lymph nodes, from the level of the 10th thoracic vertebra down to the symphysis pubis. The kidneys are shielded from irradiation, and the scrotum and opposite testis are not included in the field, unless the scrotal sac was opened initially by the surgeon. Shielding of the testis is carried out indirectly by lead blocks, and the scrotal dose is routinely measured with thermo-luminescent dosimeters. Both fields are treated daily, giving a total tumour dose of 40 Gy (4,000 rads) in 20 treatments over four weeks.

Immediate morbidity is not a problem; minimal nausea, occasional diarrhoea and mild marrow depression rarely interfere with treatment, which is always given on an out-patient basis, and many of the younger patients continue to work.

Long term effects are not in the author's opinion a great problem:

Radiation myelitis

This should not occur if the tumour dose is less than 40 Gy (4,000 rads), even if the fields extend up to the level of T.10. Not a single case has been seen in The London Hospital series of 280 patients, and even transient radiation myelopathy is a rarity.

Gastro-intestinal upset

Neither gastric perforation or duodenal ulceration have been seen in The London Hospital series, although are reported from the United States by Lewis, 1948 (4). His cases all received a dose greater than 45 Gy (4,500 rads), and usually to a larger volume of tissue than is used at The London Hospital. Small bowel upset has not been a problem, nor has protein loosing enteropathy occurred. Again, this is more likely to be seen with higher doses and larger treatment volumes used by some radiotherapists.

Lung

One would not expect to see any fibrosis at all, since the width of the field at the level of the tenth thoracic vertebra will not be greater than 8 cms, and only a minute portion of the lung will receive a full tumour dose.

Kidney

Radiation nephritis does not occur if the kidneys are adequately shielded. Even if part of the kidney has to be irradiated to a dose above the known tolerance of 22.5 Gy (2,250 rads), damage does not necessarily follow. One such patient received a dose to half the kidney of 40 Gy (4,000 rads), and he is alive and well some five years later (Figure 1).

Marrow

Transient marrow aplasia will certainly occur, but there is no indication that the incidence of leukaemia will increase. One case only occurred in The London Hospital series, and this was only 2 years after treatment, the time lag being probably too short to incriminate the irradiation definitely as the cause.

Carcinogenesis

There have been no definite cases seen in The London Hospital series that could be directly attributed to the irradiation, but it should be remembered that the incidence of a second primary carcinoma is in the order of 18-20% in most post mortem series, even when no previous irradiation has been given.

Sexual function

Potency will not of course be affected by the irradiation. Infertility is a potential problem, but it is known that many patients are sub-fertile on presentation of the disease, as Dr. Barrett details later in this volume. The remaining testis, if appropriately shielded, will receive a dose in the order of 0.25 to 0.8 Gy (25-80 rads). This will produce temporary infertility, but many patients will

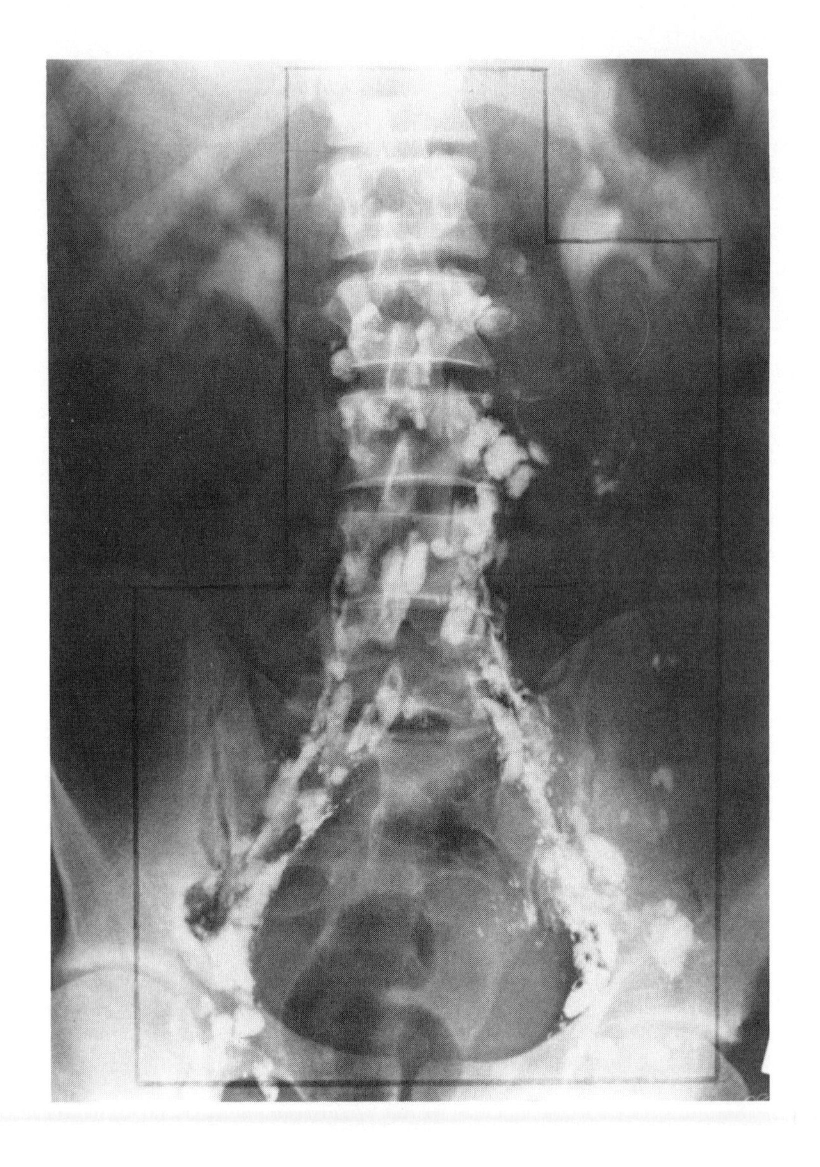

FIGURE 1 - Treatment fields, showing irradiation of part of the kidney to 40 Gy (4000 rads).

subsequently produce normal children (1). In the author's opinion, there is absolutely no indication for irradiating the remaining testis, as is practiced at the Christie Hospital and Holt Radium Institute in Manchester. The incidence of a second primary in the other testis is less that 1%, and if it does occur, then orchidectomy and further irradiation can be given. This occurred in two cases in The London Hospital series, both of whom are alive and well 5 and 7 years after the second tumour presented. The storage of sperm should be offered, providing of course that the sperm itself is of sufficiently good quality to warrant such a procedure, and sperm counts should be carried out routinely before starting treatment.

Adjuvant Surgery

The combination of radical lymph node dissection after orchidectomy as well as irradiation is to be deplored, since it will almost certainly increase the morbidity as well as producing an almost 100% impotency rate. Maier, 1977, has shown in a randomised trial that such radical surgery does nothing to improve the results (5).

Prophylactic irradiation

It is doubtful whether it is wise to use such treatment for a stage II teratoma, since if the glands are involved the chance of extra-nodal spread is fairly high, and radical chemotherapy will subsequently be required. The added irradiation to the mediastinum and neck nodes will further depress the marrow, and make such chemotherapy difficult if not impossible to give.

Recent results

Looking at the results for M.T.I. and M.T.U. staged only by clinical and simple radiological measures, it is seen that the M.T.U. stage I are particularly poor (Figure 2).

Modern radiology and tumour markers have altered the picture considerably, and this is most marked in the M.T.U. cases, which of course are more prone to metastases (Table 1).

Looking at the results over the past 20 years, it will be seen that if modern staging techniques are used, the

TABLE 1

THE LONDON HOSPITAL

Impact of Radiology and Tumour Markers on staging Teratoma of the Testis

	Year	Number	Clinical Stage I	Upstaging (Radiology and Tumour Markers
M.T.I.	1960 - 77	51	84%	26%
	1978 - 80	11	55%	33%
M.T.U.	1960 - 77	64	76%	42%
	1978 - 80	22	45%	80%

TABLE 2

THE LONDON HOSPITAL

Influence of year of diagnosis on crude survival of teratoma of the testis

Post Radiology and Tumour Markers

All Teratoma			M.T.I.		M.T.U.	
[illegible]	Number	2 year survival	Number	2 year survival	Number	2 year survival
1960	28	75%	16	94%	12	50%
1976 - 77	33	82%	15	87%	18	78%

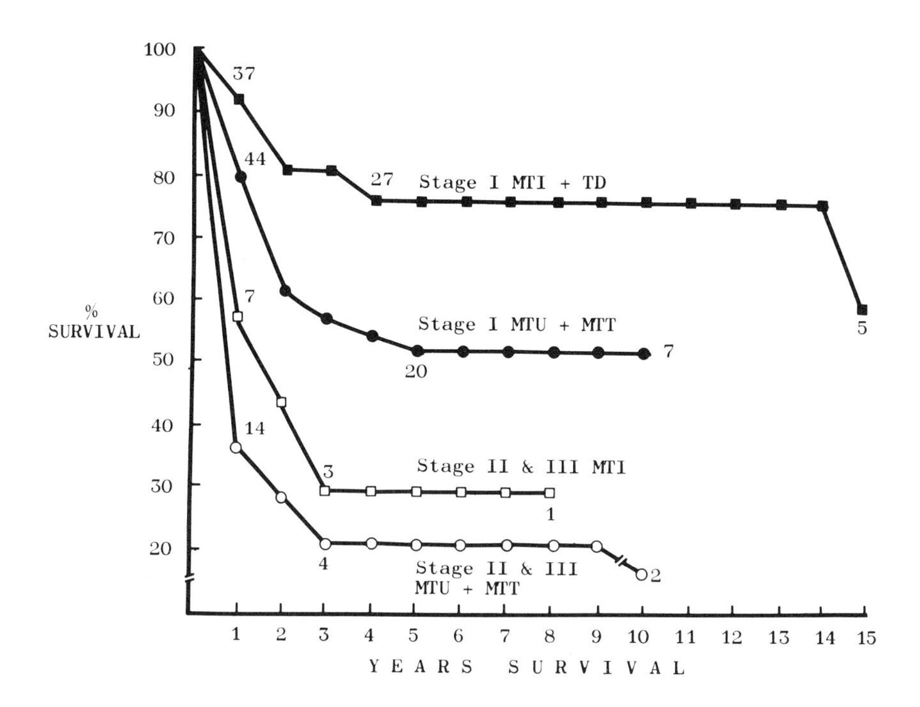

FIGURE 2 - Teratoma of the testis, MTI and MTU. Staged clinically, and by simple radiological tests.

survival figures have improved considerably for M.T.I. stage I - they are in the region of 90%, and M.T.U. 80% (Table 2).

Conclusions

In stage I teratoma, since 25% have positive glands and an 80-90% survival is achieved, then eradication of the tumour by irradiation must occur. It is admitted, however, that if irradiation is given to all stage I tumours, then about 80% will receive unnecessary treatment. Nevertheless, this form of treatment, which can be given over a relatively short period of time with minimal short and long term morbidity, and which produces such excellent results, should not, in the author's experience, be withheld from any patient with stage I teratoma.

The alternative of no irradiation but a watching policy only, will mean that a number of patients will have to be given radical chemotherapy at a later date. Although this may produce equally good results, the unpleasantness of this form of treatment given over a long period of time and associated with morbidity possibly including infertility, and the induction of leukaemia or carcinoma, is not in the author's opinion really justified. Since it is not known if the results will be as good with chemotherapy, we should not take part in the wholesale abandonment of radiotherapy, since if we do so, there is always a chance that the general surgeons who often perform the orchidectomy will not in future send the patient to the Radiotherapy and Oncology Unit. The chance of carrying out a very careful follow up (which is absolutely essential if a watching policy is to be pursued) might well be lost. Thus, all the effort over the past 26 years of urologists and radiotherapists to ensure that their patients are referred to effective uro-oncological clinics will be wasted.

There is, however, a place for a pilot study to be carried out in reputable large centres where the management of stage I testicular tumours without irradiation but with a carefully monitored watching policy can be carried out.

There is a dilemma with stage IIa teratoma; since with The London Hospital series the stage IIa M.T.I. group show a 75% survival rate, it could be argued that irradiation should be given to all these patients. However, as stated previously, the diagnosis of the IIa group is difficult, and even if accurate, some 25% at least will still need subsequent chemotherapy. It would be better therefore, to start with the latter form of treatment in the first instance. As far as stage IIa M.T.U. are concerned, the results are so poor that chemotherapy should always be used first, possibly adding irradiation to the site of bulk disease.

REFERENCES

1. Blandy, J., Hope-Stone, H.F., Dayan, A.J., 1970. Tumours of the Testicle. Heineman, London. p. 120.

2. Hope-Stone, H.F., Blandy, J.P., Dayan, A.D., 1963. Treatment of Tumours of the Testicle. British Medical Journal. i, 984-989.

3. Hope-Stone, H.F., 1979. Recent Results of Cancer Research. (Ed.) Bonadonna, G., Mathe G., Salmon, S.E., Springer-Verlag, Berlin, Heidleberg, New York. pp. 178-185.

4. Lewis, L.G., 1948. Testes Tumours - report on 250 cases. Journal of Urology (Baltimore). 59, 763-772.

5. Maier, J.G., 1977. Management of Testicular Carcinoma. Reported at 14th International Congress of Radiology, Rio de Janeiro. Abstract No. S0243.

6. Peckham, M.J., 1979. An appraisal of the role of radiation therapy in the management of non-seminomatous germ cell tumours of the testis in the era of effective chemotherapy. Cancer Treatment Reports. 63, 1653-1658.

TERATOMA OF THE TESTIS - IS RADIOTHERAPY IN EARLY DISEASE STATES DEFUNCT?

W.G. Jones
Lecturer in Radiotherapy and
Honorary Consultant Radiotherapist
University Department of Radiotherapy,
Cookridge Hospital,
Leeds LS16 6QB.

There is no doubt that radiotherapy is effective in eradicating small metastatic deposits in lymph nodes from non-seminomatous germ cell tumours of testis. The results of four series reviewed by Maier & Lee, 1977 (12) showed 3 year disease free survival figures in the order of 80-90% for the clinical Stage I groups.

The outlook for patients with advanced disease has improved tremendously with the advent of effective cytoxic chemotherapy combinations containing cis-platinum (4, 17). Even salvage therapy for refractory cases is now possible (23). We have entered an era of probable cures in the face of advanced disease.

It is now generally accepted that the definition of advanced disease is anything worse than moderate enlargement of para-aortic lymph nodes. Current staging systems define Stage I as tumour confined to the testis. Better definition of the Stage I group of patients should now be possible with the advent of sophisticated diagnostic tools such as computerised tomographic scanning and ultrasonography. These methods are undoubtedly more sensitive in detecting previously undetectable disease such as sub-pleural nodules at the lung bases. These, together with other modern methods, complement the staging procedure and should lead to the relegation of some Stage I patients to the Stage II and III categories.

Let us examine the present situation in some detail. Bilateral lower limb lymphography has been a very powerful tool in oncology to date. It has its limitations in not demonstrating nodes which are badly diseased and all the node groups of interest to the clinician are not routinely opacificed, for example the internal iliac nodes or those nodes above the level of the second lumbar vertebra. Even in the best series the accuracy of lymphography when checked by surgical resection only approaches 80-90% (2). De Wys

(3) suggests a greater range of false results in his recent review. Lymphography is not without a small morbidity. Compared to lymphography, CT scanning appears to be marginally better at assessing para-aortic nodes. In a series from the Royal Marsden Hospital (9) 5 out of 48 patients were found to have disease when the lymphogram was negative. The scanner also seemed to help solve the problem of the equivocal lymphogram. From these results it would seem therefore that CT scanning is at least 10% more accurate than lymphography. The importance of pre-operative and sequential specimens for estimation of tumour markers has been discussed elsewhere in this volume. One should note, however, that about 10% of tumours do not produce markers (10) and occasionally although the primary tumour is marker producing the metastases are not. On the other hand approximately 20% of patients with no clinical disease will have raised markers. Javadpour, 1980, (11) has shown how the clinical staging error can be reduced considerably when markers are taken into consideration.

We should also be aware that there are signs which hint at poor prognosis even in the Stage I situation, pointing to potential rapid evolution of generalised disease e.g. the 'p' stage (TNM Classification) (21), the initial marker levels, and histological type. As with other mixed tumours one must always assume the prognosis for the worst element present in this situation.

In most reported series approximately 50% of patients are in the Stage I category, which implies that what follows is of importance to a large proportion of patients. It is also true that a proportion of these patients staged by older techniques belonged to a higher stage because of later evolution of disease. Therefore, it is suggested that it is now possible to be more accurate with staging with relegation of about 20% of Stage I patients to higher stages (Figure 1).

New techniques can be repeated at intervals and thus with very close follow up very early detection of relapse should be possible.

Thus with better staging and the availability of effective salvage chemotherapy we need to ask these questions:

1. Is radiotherapy necessary as an adjunct to orchidectomy for Stage I?
2. Is adjuvant radiotherapy improving survival rates overall at the expense of patients already cured by orchidectomy alone?

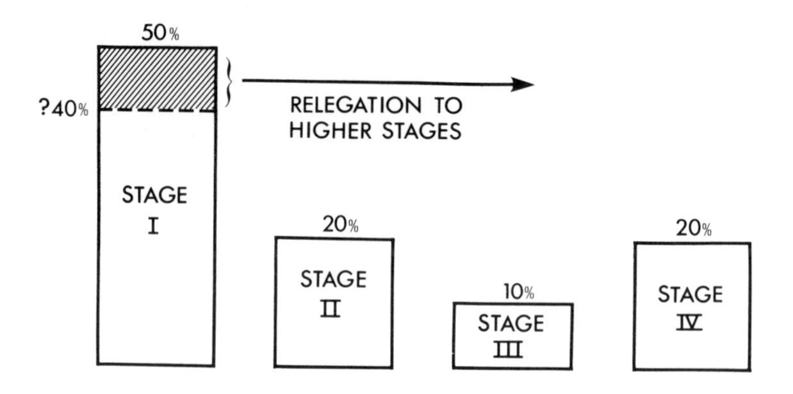

FIGURE 1 - NSGCT of testis: Effect of more accurate staging

The concept of cure by orchidectomy alone is not new. Data presented by Whitmore, 1970, (22) but based on patients treated perhaps ten or more years before, i.e. when staging was less accurate, showed that a proportion of Stage I patients survived 5 years. The burning question now is what proportion of patients would fall into this category today? In trying to find an answer from the literature, this is found to be complicated because of the natural history of untreated disease is not documented. These are rare tumours and so numbers in reported series are small and observations not statistically valid. There are a great variety of pathologies and mixtures of pathologies, and a great heterogeneity of treatments applied. Until the advent of better methods, there was difficulty in accurate determination of the extent of disease. There has been a great lack of controlled clinical trials. Add to this the confusion of pathological and anatomical staging systems and it is no wonder why it is so difficult to get a meaningful answer. However, it would appear from the days before the use of CT scanners and tumour markers that the incidence of retroperitoneal micrometastases in Stage I disease was estimated to be in the order of 25-30% according to Peckham & McElwain, 1975 (16) and 15-20% was reported in surgical series reported by Maier & Lee, 1977 (12). Another confounding factor is that disease can spread presumably by the haematogenous route, since pulmonary metastases occur in approximately 15% of patients with negative lymphograms (12). This is important when one considers the later evolution of disease and whether para-aortic/pelvic node irradiation would influence the outcome. Gilbert et al, 1976 (7), reported on the site and timing or recurrences related to the initial state of the para-aortic nodes. In the node negative group only 1 out of 8 patients relapsed within the abdomen, but 7 developed visceral disease within one year. This situation would not have been avoided by adjuvant radiotherapy. Thus, in summary, taking the negative lymphogram as our 100% standard, and acknowledging the false negative and false positive rates, it should now be possible to improve on our definition of truely Stage I patients. CT scanning alone will reduce the error by approximately 10%, radiography of the lungs alone by approximately 15%, markers alone by approximately 20% and so using a combination of all these methods it should be possible to approach the same accuracy of staging as the percentage shown in previous surgical series.

It is suggested therefore that radiotherapy is completely avoidable in the majority of Stage I patients. The argument is not that radiotherapy has a low morbidity but that it is unnecessary for the majority of patients provided that adequate follow up is maintained.

Radiotherapy treatment is not without morbidity or even mortality. There are many tissues potentially at risk, although the radiotherapist plans his treatment carefully to minimise these risks. The main tissues at risk are the bone marrow, the gastro-intestinal tract (particularly if the bowel is partly fixed by previous abdominal surgery), perhaps leading to malabsorption later (13), the spinal cord (particularly in the region of T10/11), lung, liver and the kidneys. The late effects may take years to manifest themselves, but this is likely because we are dealing with a young patient population. One fear is that the risk of second malignancy seems to be greatly increased if salvage chemotherapy is required at a later date since combination of the two therapies greatly increases the risk. Nefzger & Mostofi's report in 1972 (14) detailed 24 deaths as attributable to radiation injury and 15 deaths in which such injury was though to figure in a 17 year follow up of 786 cases and a high incidence of second malignancies (three times that expected). To be fair, however, these patients were treated by what we would now regard as antiquated techniques, often with orthovoltage therapy machines. A more recent report from Donohue et al, 1978 (4) claims four patients out of a series of 17 to have died from the complications of radiotherapy.

Post-operative radiotherapy has been reported by many authors as reducing bone marrow tolerance possibly jeopardising salvage chemotherapy (1, 4, 5, 17). We do have evidence that radiotherapy certainly causes bone marrow depletion in the area treated and at the dose levels of 40 to 45 Gy used to treat testicular teratoma permanent aplasia is likely (Sykes et al, 1964 (19, 20); Strickland, 1980 (18)). Hill et al, 1980 (8) have shown that this effect may be to all intents and purposes permanent for the majority of sites irradiated, but that it is to a degree dose dependent. How much marrow is involved in the treatment area in patients with testicular tumours? Using the data of Ellis, 1961 (6), it may be as much as 30 to 40% of the normal marrow complement.

Thus it is argued that radiotherapy is avoidable in the majority of stage I patients when modern staging methods are used. Also in this group radiotherapy should be avoided not only for the acute and long term morbidities but mainly because of the effect on the bone marrow in view of the potential need for salvage chemotherapy for the few who show signs of disease at a later date, probably in extra-nodal sites.

Twenty-five patients are now on the surveillance study at the Royal Marsden Hospital (15) receiving no post-operative radiotherapy but being very closely monitored for signs of disease (Table 1).

Four of the 25 patients have relapsed but all four have been salvaged and are now disease free. Dr. Read presents the results of the Manchester series in the following paper.

In conclusion the watch and wait policy for Stage I disease is a radical departure from previous experience. The prognosis for Stage I patients is excellent and even those who relapse are potentially curable. This requires very close follow up and recognition of relapse at an early stage, and therefore this should be done only by centres capable of doing this, obviously with easy access to the

TABLE 1

Histology	Total patients	Follow-up time (months)		Number relapsing	Time to relapse (months)	Total patients disease-free
		Range	Median			
MTI	12	3-24	8	0		12
MTU	8	7-20	9.5	4 (50%)	2,5,5,5	8
MTT	2	5,14		0		2
YS/SEM	2	7,15		0		2
MTD	1	36				1
Total	25	3-36	9	4 (16%)		25 (100%)

Surveillance Study in Stage I Non-seminoma of Testis
(Royal Marsden Hospital, 1977-1980)

sophistical techniques required. Patients treated, or rather not treated, in this way should form part of a multicentre study, and the MRC are setting up such a study at the present time.

I am grateful to Professor Peckham for allowing me to present the data shown in Table 1.

REFERENCES

1. Anon, 1979. Improved management of testicular tumours. British Medical Journal. i, 840.

2. Burney, B.T., and Klatte, E.C., 1979. Ultrasound and computed tomography of the abdomen in the staging and management of testicular carcinoma. Radiology. 132, 415-419.

3. De Wys, W., Muggia, F.M., and Jacobs, E.M., 1980. Staging of testicular cancer: a proposed clinical-surgical schema. Cancer Treatment Reports. 64, 669-674.

4. Donohue, J.P., Einhorn, L.H., and Perey, J.M., 1978. Improved management of non-seminomatous testis tumours. Cancer. 42, 2903-2908.

5. Donohue, J.P., and Skinner, D.G., 1979. Re: Why retroperitoneal lymphadenectomy for testicular tumours? Journal of Urology. 122, 140.

6. Ellis, R.E., 1961. the distribution of active bone marrow in the adult. Physics Medicine and Biology. 5, 255-258.

7. Gilbert, H.A., Shapiro, R., Kagan, A.R., Cooper, J.F., Jacobs, M.L., and Nussbaum, H., 1976. Recurrence patterns in the non-seminomatous germinal testicular tumours. International Journal Radiation Oncology, Biology and Physics. 1, 249-256.

8. Hill, D.R., Benak, S.B., Phillips, T.L., and Price, D.C., 1980. Bone marrow regeneration following fractionated radiation therapy. International Journal of Radiation Oncology, Biology and Physics. 6, 1149-1155.

9. Husband, J.E., Peckham, M.J., and MacDonald, J.S., 1980. The role of abdominal computed tomography in the management of testicular tumours. Computerised Tomography. 4, 1-16.

10. Javadpour, N., 1980. The role of biologic tumour markers in testicular cancer. Cancer. 45, 1755-1761.

11. Javadpour, N., 1980. Improved staging for testicular cancer using biologic tumour markers: a prospective study. Journal of Urology. 124, 58-59.

12. Maier, J.G., and Lee, S.N., 1977. Radiation therapy for non-seminomatous germ cell testicular cancer in adults. Urological Clinics of North America. 4, 477-493.

13. McBrien, M.P., 1973. Vitamin B12 malabsorption after cobalt teletherapy for carcinoma of the bladder. British Medical Journal. i, 648-650.

14. Nefzger, M.D., and Mostofi, F.K., 1972. Survival after surgery for germinal malignancies of the testis - II - effects of surgery and radiation therapy. Cancer. 30, 1233-1240.

15. Peckham, M.J., 1981. Personal communication.

16. Peckham, M.J., and McElwain, T.J., 1975. Testicular tumours. Clinics in Endocrinology and Metabolism. 4, 665-692.

17. Stoter, G., Sleijfer, D.Th., Vendrik, C.P.J., Schraffordt Koops, H., Struyvenberg, A., Van Oosterom, A.T., Brouwers, Th.M., and Pinedo, H.M., 1979. Combination chemotherapy with cis-diammine-dichloro-platinum, vinblastine and bleomycin in advanced testicular non-seminoma. Lancet. i, 941-945.

18. Strickland, P., 1980. Complications of radiotherapy. British Journal of Hospital Medicine. 23, 552-565.

19. Sykes, M.P., Savel, H., Chu, F.C., Bonadonna, G., Farrow, J., and Mathis, H., 1964. Long term effects of therapeutic irradiation upon bone marrow. Cancer. 17, 1144-1148.

20. Sykes, M.P., Chu, F.C., Savel, H., Bonadonna, G., and Mathis, H., 1964. The effects of varying dosages of irradiation upon sternal marrow regeneration. Radiology. 83, 1084-1088.

21. T.N.M. Classification of Malignant Tumours, Third Edition, U.I.C.C. Geneva, 1978. pp. 122-125.

22. Whitmore, W.F., 1970. Germinal tumours of the testes. Proceedings of the 6th National Cancer Conference. Lippincot, Philadelphia, pp. 219-245.

23. Williams, S.D., Einhorn, L.H., Greco, F.A., Oldham, R., and Fletcher, R., 1980. VP16-213 salvage therapy for refractory germinal neoplasms. Cancer. 46, 2154-2158.

FOLLOW UP POLICY IN STAGE I TERATOMA OF TESTIS - FIRST YEAR'S EXPERIENCE

G. Read
Consultant Radiotherapist,
Department of Radiotherapy,
Christie Hospital and Holt Radium Institute,
Wilmslow Road,
Manchester M20 9BX

Recent advances have enabled more precise identification of metastatic disease in patients with early malignant teratoma of the testis. The need for routine post-operative radiotherapy has therefore been questioned. A prospective study of follow up only in stage I teratoma patients has been carried out at the Christie Hospital since 1st November 1979. Of thirty-eight patients eighteen were found to have no evidence of residual disease (stage I) following detailed investigation including lymphography, computer assisted tomography and biochemical tumour marker estimation.

The histology was reviewed at the Christie Hospital and revealed 9 cases of MTI, 4 cases of MTU, 4 cases of combined MTI + seminoma and one case of bilateral disease, MTI on one side and teratoma differentiated + seminoma on the other.

Each patient has been seen at follow up at intervals not greater than one month, often less, particularly in the immediate post-operative period. Tumour markers, chest X-ray and any other investigations thought to be appropriate, including repeat CT scan, were performed at each attendance. The median follow up is 9 months. Three patients have recurred at 6, 8 and 12 weeks respectively after the initial assessment. The sites of recurrence were the lung, retroperitoneum and inguinal canal. In retrospect the initial investigations in two of these patients were highly suspicious of recurrence. All three patients have been successfully treated by chemotherapy followed by radiotherapy to nodal disease and remain in complete remission. These findings suggest that an intensive follow up policy is justified in stage I malignant teratoma provided the patients are assessed in a specialist centre so that treatment can be promptly instituted for the patients who relapse.

IS TREATMENT OF REGIONAL LYMPH NODES NECESSARY IN MICROINVASIVE TESTICULAR GERM CELL TUMOURS?

F.E. von Eyben[1], P. Milulowski[2], S. Krabbe[3] and C. Busch[4]
Department of Oncology[1] and Department of Pathology[2],
Malmo General Hospital, Sweden.
Children's Hospital[3],
Fuglebakken, Denmark, and
Department of Pathology[4],
University of Uppsala, Sweden.

We propose that atypical germ cells of the testis spreading from the tubules into the testicular stroma without macroscopic tumour, be called microinvasive germ cell tumour. We describe four such cases who had no dissemination of the tumour at staging procedures.

A 31 year old man with an undescended left testis underwent orchidectomy . He had a malignant teratoma-like tumour but received no treatment to the regional lymph nodes. He has been followed for 59+ months without relapse occurring.

Three men with previous undescended or maldescended testes (for which they had undergone treatment in childhood) underwent orchidectomy (2). They had seminoma-like tumours (Figure 1) and received post-operative high voltage irradiation to the para-aortic and iliac lymph nodes. They have been followed for 10+, 28+, and 29+ months without relapse.

We propose that microinvasive testicular germ cell tumour of the testis is a stage of testicular neoplasia with a very small tumour bulk, which should be treated conservatively by orchidectomy alone without treatment of the regional lymph nodes.

REFERENCES

1. von Eyben, F.E., Mikulowski, P., and Busch, C., 1981. Microinvasive germ cell tumour of the testis. Journal of Urology. In press.

2. Krabbe, S., Skakkebaek, N.E., Berthelsen, J.G., von Eyben, F., Volstedt, P., Mauritzen, K., Eldrup, J., and Nielsen, A.H., 1979. High incidence of undetected neoplasia in maldescended testes. Lancet, ii, 999-1000.

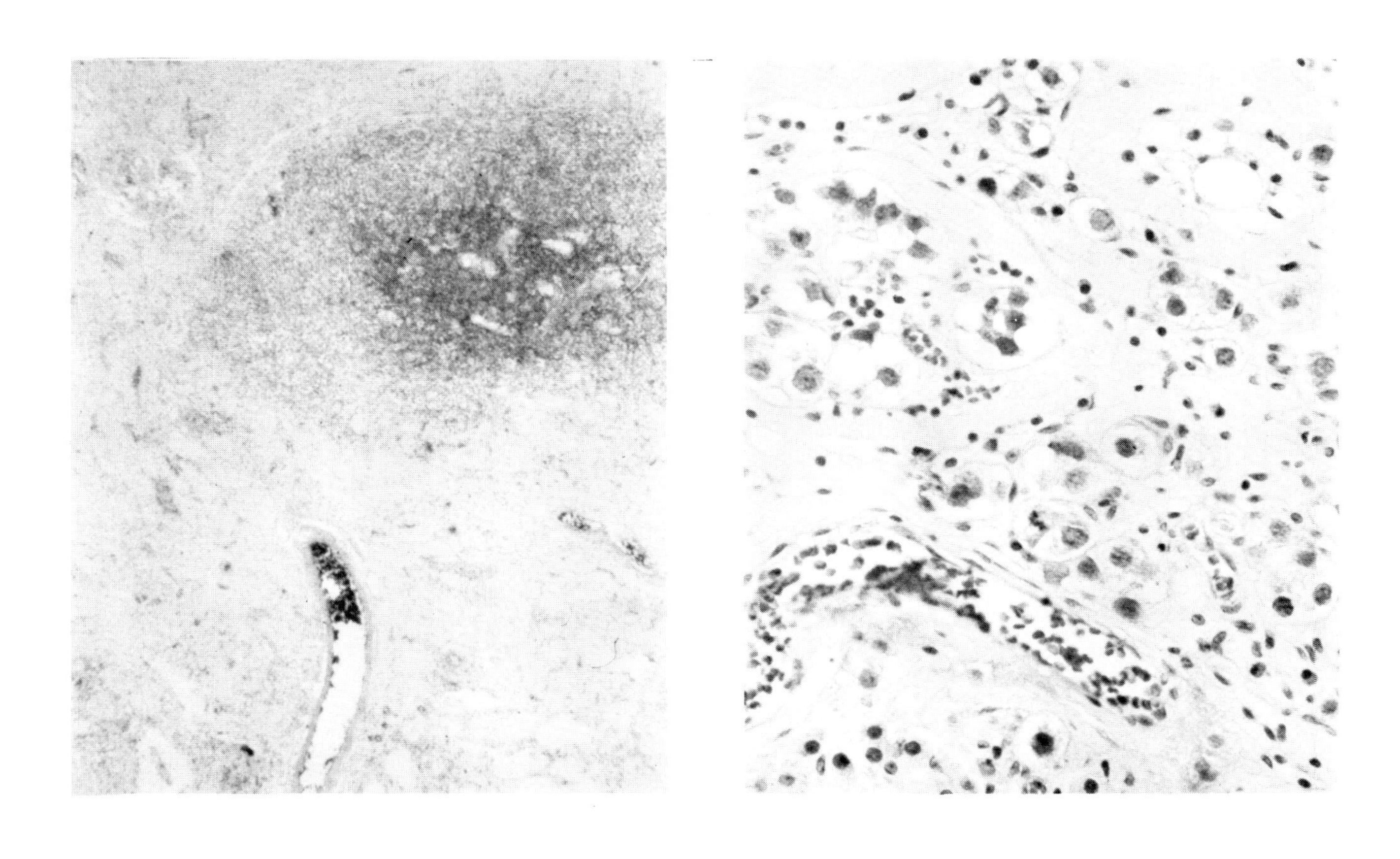

FIGURE 1a (x 8) FIGURE 1b (x 81)
Seminoma-like tumour (haemotoxylin and eosin stain)

DISCUSSION

Dr. Barrett

In my opinion it would be rather fruitless to embark on a discussion as to whether radiotherapy is or is not indicated. The aim of the Royal Marsden Stage I study is to try to determine which patients, if any, would benefit from radiotherapy. What is needed is a very careful study of the relapse patterns, thus determining those patients who might benefit from radiotherapy because they relapse only with small volume disease in the retroperitoneum, and those patients who relapse in the lung and therefore would have perhaps been prejudiced by previous radiotherapy. I think that those centres who are using this approach are aware that this is what we are looking for. It could be detrimental if it were seen as a trial of the value of radiotherapy when we know that the majority of patients are not going to benefit because of the numbers that have been presented.

Another point I would like to make is that we have found the watching policy particularly helpful in patients where an unequivocal diagnosis of Stage I disease cannot be made. One would imagine perhaps from some of the data presented that staging is always straightforward, but it is not. There are patients about whom one argues who subsequently relapse in extra-nodal sites and one sees that they were not truly in Stage I. Undoubtedly these patients have benefited from the watch policy.

Dr. Pizzocaro

I also think that the relapse pattern is very important in taking any decision on the management of retroperitoneal nodes.

Dr. James

May I put the cat amongst the pigeons and say it is only because we do not have markers that we do not omit radiotherapy in Stage I seminoma. Also the salvage rates for Stage II are very good with radiation therapy. What do the radiotherapists think about that?

Professor Blandy

I think it may be immoral not to treat Stage I seminoma.

Dr. Barrett

The cure rate for Stage I seminoma with radiotherapy is 100% in most large series, so it is a dangerous policy to consider that the radiotherapy might not be necessary.

Professor Blandy

But that is just the sort of argument that Dr. Hope-Stone makes for his MTI group.

Dr. Barrett

Yes, but the difference is the numbers. There may be a place for a study in this situation too, but we are not yet brave enough.

Professor Blandy

I think we will have to be brave enough one day.

Dr. Oliver

Does anyone have information from lymphadenectomy series of seminoma Stage I as to the incidence of metastases?

Professor Blandy

There is plenty of very dreadful information in the literature if you go back over 50 years concerning what happens to men with apparently clinically Stage I seminoma who are not treated - they die. It is not a benign disease.

Dr. Laing

I would suggest that both main speakers, of necessity, occasionally overstated their case particularly with regard to the morbidity and mortality figures.

I would be interested to know the additional cost both to the patient and to the NHS of such very careful follow up that is clearly necessary in the patients you are simply observing?

Finally I think it must be stressed from a meeting such as this is that the whole situation is not proven. I would be terrified that because of what has been said at this meeting that people assumed that radiotherapy was not necessary since I am sure that a significant number of these patients would get inadequate follow up.

Dr. Jones

It costs about £100 for a CT scan, and about £5 to do a marker estimation. Against that the patient on follow up only does not have to have a one month (or longer) course of radiotherapy, perhaps requiring an ambulance every day, in order to save him from coming into a hospital bed. He is seen regularly in the outpatients. From a mental point of view, if you tell a patient that he does not need treatment but that careful follow up is required because of the high risk of relapse, he will come regularly, and if you forget to take blood, he will remind you. There is even more mental anguish when relapse occurs, say, six months after radiotherapy, with disease, say, in the chest when they have to struggle through at least four courses of Einhorn chemotherapy.

MARKERS IN THE MANAGEMENT OF SEMINOMA

Chairman of Session
Professor J. Blandy

THE ROLE OF MULTIPLE TUMOUR MARKERS IN THE DIAGNOSIS AND MANAGEMENT OF SEMINOMA

N. Javadpour
Urologist-in-charge and
Senior Investigator,
Surgery Branch, National Cancer Institute,
Bethesda, Maryland 20205,
U.S.A.

The diagnosis and treatment of testicular seminoma relies on morphologic features of tumour cells obtained from limited sections of surgical specimens from primary or metastatic lesions. Because these tumours may contain multiple components with variable growth patterns and metastatic potential, and despite the apparent precision of the morphologic features, the diagnosis of pure seminoma, based on cellular features obtained from such limited histologic specimens, is not really satisfactory. Over the past several years, the development of specific and sensitive immunodiagnostic techniques has made an important contribution to the diagnosis and, to a certain extent, to the management of some of these tumours. Data was obtained from 160 patients with the diagnosis of pure seminoma. Serial determinations of serum AFP, and HCG were measured by radioimmunoassay and immunocytochemical techniques previously described (1, 4). For the majority of these patients, staging was determined by clinical modalities only. Because of the conventional radiotherapy administered after orchidectomy in a majority of these patients surgical and pathological staging was not feasible. Since the majority of patients were referred after orchidectomy, these patients did not have determinations of serum AFP and HCG prior to the orchidectomy. However, all the patients had histopathologic material available for serial section immunocytochemical studies. Serum HCG was measured by double antibody radioimmunoassay, utilising the antibody raised against the beta sub-unit of HCG. AFP was also measured by double antibody radioimmunoassay. Immunoperoxidase staining utilising antibodies to AFP and beta sub-unit of HCG has been previously described (1, 2, 3, 4).

Of the 160 patients investigated, 16 patients had elevated serum levels of HCG. Regarding the level of HCG, only two patients had elevation greater than 75 ng/ml. One patient proved to have an element of choriocarcinoma on further sectioning of the testis. Syncytiotrophoblastic cells were found to be capping the cytotrophoblastic cells which were positive for HCG on histologic staining. The patient underwent a retroperitoneal lymph node dissection and metastatic choriocarcinoma was found in the lymph nodes. The other patient with an elevated HCG greater than 75 ng/ml also had an elevation of AFP. Consequently it seems that in cases of pure seminoma where further sectioning does not reveal the presence of any associated germ cell tumour that the elevation of HCG will be moderate, i.e. in the range of considerably less than 75 to 100 ng/ml. Any elevation particularly greater than 100 ng/ml should immediately raise the suspicion of an associated germ cell element, and particularly anything over 100 ng/ml should raise the question whether or not it is a pure seminoma.

In 132 stage I patients only 3.7% had an elevated HCG, while in stage II seminoma, 7 of 22 patients (32%) had an elevated HCG. It appears that the more bulky the seminoma, the higher the likelihood of an elevated HCG.

Regarding elevated AFP in pure seminoma, in 160 patients with seminoma AFP levels were measured, and two patients had an elevated level. One of these patients proved to have an element of embryonal carcinoma on further histologic sectioning. Serum AFP dropped to normal after retroperitoneal lymphadenectomy, the lymph nodes containing embryonal carcinoma. The other patient with a persistently elevated serum AFP underwent debulking surgery of the retroperitoneal tumour for diagnostic and "therapeutic" purposes. On serial sections of the multiple blocks of the specimen we were unable to find any elements of non-seminomatous testicular cancer. However, on further follow up he was found to have liver metastases. Biopsy of the liver metastases revealed pure seminoma. The deposit was negative for AFP on histologic staining. However, the liver tissue from the vicinity of this tumour was positive for AFP by immunoperoxidase staining. These regenerative changes in the liver explained the modest but serially persistent elevated levels of serum AFP. This patient refused chemotherapy and died of liver metastases.

A summary of the major points regarding serum AFP in patients with seminomas currently held by myself and colleagues at the NCI is as follows:

1. The presence of liver metastases may result in moderately elevated AFP and although rare, the possibility should be considered.

2. The presence of an elevated AFP in a patient with seminoma implies the presence of non-seminomatous testicular cancer. It has been reported that there are certain cases that on multiple sectioning the tumour is of pure seminoma but there is an elevated AFP present, however we have not seen such cases.

3. Mostofi and Price, 1973, (5), have reported seminoma metastasises as seminoma in 65% of cases. In 26% of their patients the metastases consisted of embryonal carcinoma and in 4% there was teratoma. It is likely that a number of patients with seminoma with non-seminomatous metastases in these author's series have elements of embryonal carcinoma and teratoma in the primary tumour that could have been detected by determinations of serum AFP or by histochemical staining.

4. The therapy of a seminoma with an elevated AFP should be based on the treatment for that stage of non-seminomatous testicular cancer, i.e. for clinical stage II: retroperitoneal lymph node dissection, and for stage III: platinum combination chemotherapy. The prognosis of such patients with elevated levels of AFP will depend on the same parameters that predict for successful therapy of non-seminomatous testicular cancer. The major factor being the bulk of the tumour.

What then is the optimum management of patients with seminoma and an elevated HCG? Certainly it must be recognised that an increase in HCG can occur in pure seminoma. A higher frequency has been reported in the literature. We have found a lower overall incidence in the order of 10% (2). However, in 130 patients with seminoma (Table 1), that have previously been reported, 7 of 18 (40%) of the patients with clinically detectable retroperitoneal tumour (stage II) by lymphangiography had elevated serum HCG. One must search for elements of choriocarcinoma particularly if there is a marked elevation in HCG greater than 100 ng/ml. The elevation of serum HCG can be used to monitor therapy as it can in non-seminomatous testicular cancer. Whether elevation of HCG in seminoma has a prognostic significance remains to be defined. Patients with pure seminomas with moderate elevations of HCG treated by the appropriate modality, certainly radiotherapy in early

TABLE 1

CORRELATION OF CLINICAL STAGING OF SEMINOMA AND HCG[1]

Tumour	Patients (n)	Elevated Serum HCG
Stage I	109	4 (4%)
Stage II	18	7 (40%)
Stage III[2]	3	0 (0%)

[1] Cancer 42:2768, 1978

[2]Minimal disease

stages, appear not to have done any worse than patients with seminoma who have not had elevated levels of HCG. The high frequency of elevated levels of serum HCG in some reported series in the literature is a matter for concern because of the lack of serial sections and histochemical staining which substantiates the absence of choriocarcinoma or the presence of tumour giant cells responsible for HCG synthesis. In the present series of 160 patients who were comprehensively studied by serial histology sections and histochemical staining, 15 cases of pure seminoma with elevated serum HCG were detected.

There has been a reassessment by some therapists regarding the role of radiotherapy in all stages of seminoma. It has been suggested that patients with seminoma and elevated HCG should perhaps undergo lymphadenectomy routinely to rule out an element of choriocarcinoma. The therapy for seminoma with choriocarcinoma would be a lymphadenectomy for Stage I and II disease as previously discussed. This type of approach certainly needs confirmation in a clinical trial and should be compared to standard radiotherapy for that particular type of patient. If a patient develops elevated HCG levels shortly after radiotherapy he should be re-staged. If the tumours is initially bulky and persists following radiotherapy and then if after resection of the residual tumour the serum HCG is still elevated I think it is wise to institute chemotherapy at that time. The survival of Stage III patients with seminoma with or without elevated serum HCG when treated by orchidectomy and radiotherapy alone is quite dismal. Smith et al, 1979 (6), in their collected series, and from their own experience at U.C.L.A. report 5 year survival for Stage III seminomas, ranging from 14-22%. These results are certainly significantly worse than those now obtainable with treatment for Stage III non-seminomatous testicular cancer. Therefore, it is suggested that for Stage III patients, chemotherapy should be added to the armamentarium for treatment. Perhaps bulky Stage III seminoma should in fact be treated with combination chemotherapy first. Several series have now demonstrated that the same platinum containing combinations which are effective in non-seminomatous testicular cancer namely PVB or the VAB-6 type regime are equally effective in disseminated seminoma. In a collected series, again from Smith et al (6), even in Stage II the survival for these patients was only 70% compared to 80 to 90% now being reported for non-seminomatous testicular cancer patients treated with retroperitoneal dissection and chemotherapy. Smith has actually argued that therapeutic approaches which have markedly improved survival in non-seminomatous testicular cancer be now applied to seminoma.

I should like to briefly discuss the other markers that we have been investigating in patients with either seminoma or non-seminomatous testicular cancer. A common marker that may be useful in the management of testicular cancer is lactic dehydrogenase (7). Serum lactic dehydrogenase is a non-specific enzyme made up of five heterogenous isoenzymes in man that can be measured electrophorectically. Cancer cells have increased glycolysis leading to an increased synthesis of lactate, and LDH may be utilised as a non-specific tumour marker in serveral cancers. In seminoma LDH may be particularly useful because of several factors:

1. The lower frequency of serum HCG elevations in seminoma versus non-seminoma.

2. The availability and simplicity of measuring LDH compared to radioimmunoassay studies.

3. The majority of patients with bulky stage II and III seminomas seen at the NCI had an elevated serum levels of LDH which were useful in monitoring their therapy. The preliminary results suggest an elevation of LDH1 may be somewhat specific in testicular cancer compared with other neoplasms. In a series of 19 patients with testicular cancer, LDH1 was found to be normal in 4 patients with no tumour present, that is there were no false positives. LDH1 was also not elevated in 6 patients with microscopic disease only. In 9 patients with gross disease 5 had elevated LDH1 and 4 had a normal LDH1. However, all patients with an elevated LDH1 also had an elevation of total LDH. Of 4 patients with seminoma 3 had an elevated LDH1. None of these 4 patients with an elevated LDH had an elevated HCG. It appears that an elevation of LDH1 provides a marker for seminoma although many more patients will need to be studied and correlations will have to be made with stage of disease and certainly with prognosis and utility.

In the initial studies, we have found that it does not appear that LDH1 is particularly useful in patients with teratocarcinoma. All 5 patients studied had a normal LDH1 even though three of these had gross tumour present. Two of these five patients also had an elevated HCG or AFP.

The total number of patients with embryonal carcinoma currently studied with isoenzymes is too small to make any comments about potential utility of the determination. It appears that LDH is useful in monitoring patients with testicular tumour (Figure 1).

Multiple Tumour Markers in Patients with Seminoma

The object of this study was to determine the role of gamma glutamyl transpeptidase (GGT), placental alkaline phosphatase (PLAP), and human chorionic gonadotrophin (HCG) in testicular seminoma. In 89 seminoma patients with negative alpha-fetoprotein, total serum GGT was measured utilising technique of Strome and values about 30 I U/L were considered abonormal. Serum PLAP was measured by enzyme linked immunoabsorbent assay and values >1.85 mg/ml were considered abnormal. Serum HCG and AFP were measured by double antibody radioimmunoassays (normals <1 ng/ml and <20 ng/ml respectively). At the time of this study, 30 patients had detectable seminoma, 10 were histologically

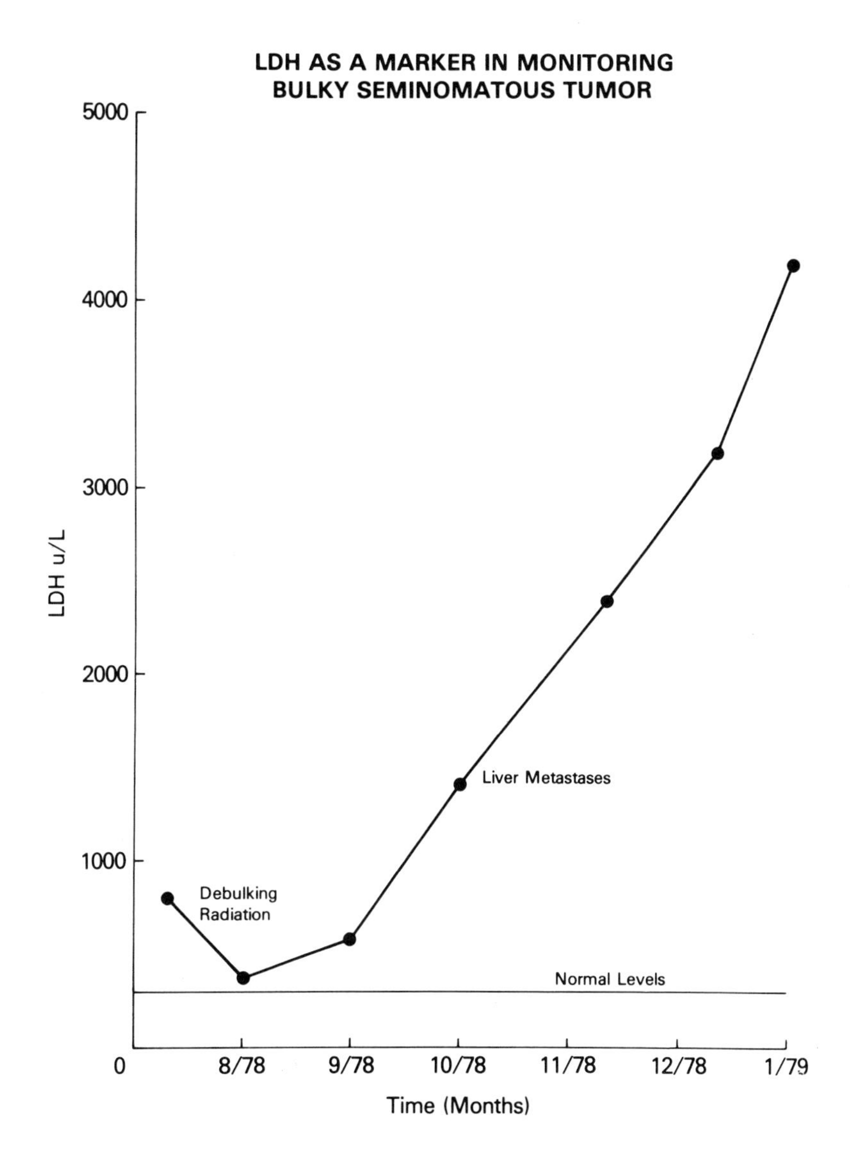
LDH AS A MARKER IN MONITORING
BULKY SEMINOMATOUS TUMOR
5000
4000
3000
2000
1000
0
LDH u/L
Liver Metastases
Debulking
Radiation
Normal Levels
8/78
9/78
10/78
11/78
12/78
1/79
Time (Months)

FIGURE 1

unconfirmed, and the remaining 49 had no evidence of tumour. Only 6 of 30 patients (20%) with active tumour had elevated levels of serum HCG. Twelve of 30 patients with active tumour (40%) had elevated serum PLAP and 10 of 30 (33%) of these patients had elevated serum levels of GGT. When these 3 serum markers were considered together over 80% of the patients with clinically active tumours had detectable serum levels of one or more of these biochemical serum markers. Since the survival of patients with stage III seminoma treated by radiation is only 28%, we advocate serial measurements of these serum markers along with early utilisation of newer chemotherapeutic regimens in these patients. However, it should be emphasised that the false positive, false negative rates of these markers, especially false positive rate for GGT, due to occasional concomittent liver diseases and the biologic half lives of these markers should be taken into consideration (Table 2).

TABLE 2

FALSE POSITIVE AND FALSE NEGATIVE OF PLACENTAL ALKALINE PHOPHATASE (PLAP), GAMMA-GLUTAMYL TRANSPEPTIDASE (GGT), AND HUMAN CHORIONIC GONADOTROPIN (HGG) IN 89 PATIENTS WITH SEMINOMA*

Status of 79 patients	Pts.(n)	PLAP(%)	GGT(%)	HGG(%)	PLAP, GGT and/or HGG
Detectable tumour	30	40	33	20	80
Nondetectable tumour	49	12	4	0	11

* Ten patients who had suspected tumour but not confirmed histologically were excluded for this analysis.

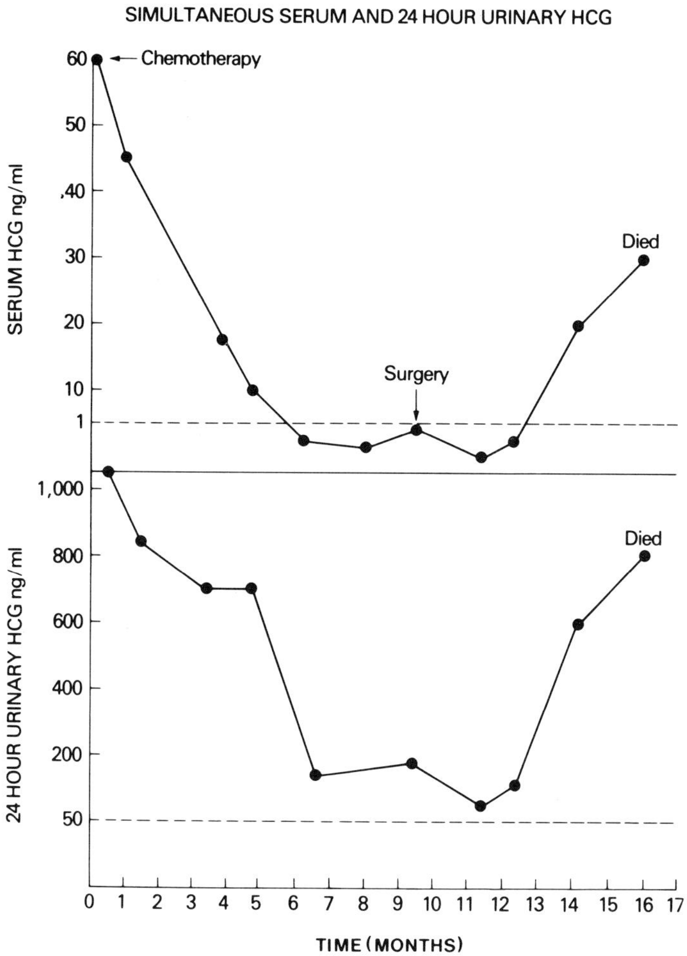

SIMULTANEOUS SERUM AND 24 HOUR URINARY HCG

SERUM HCG ng/ml

60 ← Chemotherapy

50

,40

30

20

10

1

Surgery

Died

24 HOUR URINARY HCG ng/ml

1,000

800

600

400

200

50

Died

0 1 2 3 4 5 6 7 8 9 10 11 12 13 14 15 16 17

TIME (MONTHS)

FIGURE 2

Concluding remarks

HCG is elevated in about 9% of patients with seminoma in this series. There is an increased frequency of elevation with stage, and particularly in those patients with a high volume of tumour. PLAP is elevated in 40% of patients with seminoma. In those patients with seminoma who did not have any residual tumour after therapy there was no elevation of PLAP. GGT is elevated in 33% of patients with evidence of seminoma. PLAP, GGT or HCG is likely to be elevated in about 80% of all patients with seminoma. Lastly LDH is also frequently elevated in bulky tumours and tends to reflect the bulk of disease. The isoenzyme LDH1 may be particularly useful as it may have some specificity in seminoma.

Finally, utilising a highly specific radioimmunoassay system which recognises the unique carboxyl-terminal peptide of the beta subunit of human chorionic gonadotrophin on a 24-hour urinary concentrate may prove to be more sensitive in patients when serum human chorionic gonadotrophin is not detectable by conventional radioimmunoassay (Figure 2), and merits further study.

(This paper was read by Dr. R.F. Ozols for Dr. Javadpour.)

REFERENCES

1. Javadpour, N., 1979. Serum and cellular biologic tumour markers in patient with urologic cancer. Human Pathology, 10, 557-568.

2. Javadpour, N., 1980. The role of biological tumour markers in testicular cancer. Cancer, 45, 1755-1761.

3. Javadpour, N., 1980. Management of seminoma based on tumour markers. Urological Clnics of North America, 7, 773-780.

4. Javadpour, N., 1980. Radioimmunoassay and immunoperoxidase of pregnancy specific beta-1-glycoprotein in sera and tumour cells of patients with certain testicular germ cell tumours. Journal of Urology, 123, 514-515.

5. Mostofi, F.K., and Price, E.B., 1973. Tumours of the male genital system. Atlas of Tumour Pathology, Washington D.C. Armed Forces Institute of Pathology, second series, fasc. 8.

6. Smith, R.B., Dekernion, J.B., and Skinner, D.G., 1979. Management of advanced testicular seminoma. Journal of Urology, 121, 429-431.

7. Lippert, M., and Javadpour, N., 1981. Lactic Dehydrogenase in Seminoma Urology. In press.

DISCUSSION

Professor Blandy

Are there any questions about the use of markers in early seminoma? We are grateful to Dr. Ozols for reading Dr. Javadpour's paper.

Dr. Oliver

Dr. Ozols was saying the 33% were positive for GGT and 40% for PLAP. Was that in Stage I pre-orchidectomy patients, or was this only in Stage II and III with clearly defined metastases?

Dr. Ozols

The results refer only to those patients who were found to have residual disease at some time after surgery, not before, i.e. Stages II and III.

Dr. Pizzocaro

Dr. Javadpour reports only 3% positive HCG in Stage I, 32% in Stage II and 50% in Stage III. In a smaller series, I have 35% positive in Stage I patients who have the determination of HCG before orchidectomy. Did all patients in stage I have a determination before orchidectomy or were some referred after orchidectomy?

Dr. Ozols

They were all referred after orchidectomy. None of these patients had their enzymes looked at before orchidectomy.

Dr. Pizzocaro

So the percentage of marker positive patients before orchidectomy is not known. Because we perform markers before operation we should see an increased number who are marker positive.

Dr. Milford Ward

This of course follows. Because of the short half-life the HCG marker positivity will disappear in days.

Professor Blandy

Dr. Ward, are your cases of seminoma pre-orchidectomy or not?

Dr. Milford Ward

I refer to a highly selective group of pre-orchidectomy samples.

Dr. von Eyben

I must make a protest that LDH should be regarded as specific for seminoma. In our study of the isoenzymes of LDH in patients both with seminoma and non-seminomas, we found one patient with a yolk-sac tumour and another with a mixed tumour both of whom had raised LDH as the only raised marker. This suggests that you cannot say that as a result of the serum tests this patient has metastases from a seminoma.

Dr. Ozols

I think Dr. Javadpour would be the first to agree that the data on the use of isoenzymes he allowed me to present is prelimary, although the initial patients tended to suggest that it may be specific, it would be very surprising if it was.

Dr. Jacobsen

I have to comment on Dr. von Eyben's paper. We have examined 105 cases with germ cell tumours of the testis. We looked at the adjacent tissue for carcinoma-in-situ which was found in about 90%. Nearly all had early invasion in addition, which in most cases looked like seminoma. The general view is that it is very difficult to classify the tumours in this early stage so I would hesitate to classify what Dr. von Eyben calls embryonal carcinoma as embryonal carcinoma. It is an interesting observation that the germ cell tumour has a phase or a stage where it does not show its true face.

Professor Wahlqvist

I should like to comment that pre-operatively the urologist should take a blood sample, I think that is essential and I believe the last moment is before the clamping of the testicular vein. I understand the value of having a sample from the vein in the arm but I suggest that if you take the sample from the testicular vein you are much closer to the truth. On the other hand I believe that in practice blood is not being taken routinely for HCG and AFP. It is possible that we need to take blood for the seminoma markers that we are discussing. Have you any ideas about the best way you could give urologists advice about when the sampling must be performed and why?

Professor Blandy

Well with me this is rather a sore point. 80% or more of the orchidectomies done in Great Britain are done by what some of us would regard as amateurs - otherwise well trained general surgeons who do not restrict themselves to the minor practice of urology. There is a trade union dispute here, but we are still struggling to educate this group of surgeons to not take a testicle out through the scrotum, and not poke it about or poke about inside it. The taking of pre-operative markers will mean more education.

Professor Wahlqvist

I wanted to hear from the experts on markers if we should say to the general surgeons or urologists whether we need a blood sample from the testicular vein?

Dr. Milford Ward

It would be nice I think, but I have no data that suggests that it would be of any more value than much simpler anticubital vein samples.

Dr. Norgaard-Pedersen

We have looked at a large number of samples from the testicular vein and usually concentration is 4-5 times higher than in the peripheral venous circulation. With regard to the marker positive seminoma, it must be stated that there is some "noise" in the assay requiring repeat samples as well as pre-operative specimens.

THE RETROPERITONEUM

Chairman of Session
Professor J. Blandy

THE CASE FOR RADICAL SURGERY AND COMBINED THERAPY IN TESTICULAR NON-SEMINOMA

G. Pizzocaro
Urological Surgeon,
Istituto Nazionale per lo Studie e la Cura dei Tumori,
Via Venezian 1, 20133 Milano, Italy.

There has long been controversy whether surgery or radiotherapy is the treatment of choice for retroperitoneal nodes in testicular non-seminoma. It is impossible to compare results obtained by these two treatment modalities, so it is unlikely to be proved whether one method is superior to the other. Only surgery, however, allows a careful pathological staging of retroperitoneal disease, and it combines very well with intensive chemotherapy. These are two great advantages in today's combined treatment modalities.

The aim of this paper is to determine indications for radical surgery in the management of testicular non-seminoma, alone, or in combination with intensive chemotherapy.

Pathology staging

No single clinical investigation is sufficiently exact to stage retroperitoneal nodes (12): lymphangiography (LAG) may reveal small intranodal deposits, but nodes completely replaced by tumour may be missed. CT scan and echography easily demonstrate retroperitoneal nodes larger than 1.5-2 cm, but they cannot differentiate between tumour invasion and benign hyperplasia. Post-orchiedectomy serum markers (HCG and AFP) may be positive in the presence of metastases, but not all tumour cells are marker producing. Ancilliary procedures such as urography (IVP) and inferior venocavography are useful in studying the extent of large retroperitoneal nodes and in evaluating resectability. Also, when CT scan and echography are not available, they may demonstrate metastases respectively to the left and right para-aortic testicular lymph node drainage centres which may be missed by bipedal LAG. It is concluded that several complementary procedures are thought to be necessary

TABLE 1

CONVERSION RATE OF 43 CATEGORY NO-2, MO, NON-SEMINOMAS STAGED BY LAG, IVP, ECHOGRAPHY & SERUM MARKERS (1980)

CLINICAL CATEGORIES	No. CASES	PATHOLOGICAL CATEGORIES		
		N-	N+	pN3
NO, MO	20	19	1	-
N1-2, MO	23	3	20	-
TOTAL	43	22	21	-

False negative : 1/20 (5.0%)
False positive : 3/23 (13.0%)

to clinically stage retroperitoneal nodes in testicular tumours.

Recently (Table 1), by combining bipedal LAG, IVP, echography or CT scan and post-orchidectomy serum HCG and AFP, we have succeeded in demonstrating retroperitoneal metastases in 20/21 patients with positive nodes at pathological staging (95.2%). No patient was found to bear unresectable retroperitoneal disease, but 3/22 patients (13.6%) with pathologically negative nodes had, however, a false positive clinical staging. These patients would have been overtreated if surgical staging had not been undertaken. Thus it is demonstrated that by combining several complementary investigations the false negative rate can be reduced to a very low rate (no more than 5%), but the problem of false positive clinical staging remains an open question.

Distribution of retroperitoneal metastases

A good knowledge of this topic is essential in planning a radical retroperitoneal lymph node dissection. Of course, all the retroperitoneal nodes can be surgically removed (7) and even the suprahilar nodes can be dissected (2), but one wonders whether such supraradical dissections are always necessary.

Ray et al., (13) reviewed the distribution of retroperitoneal metastases in 122 patients with positive nodes operated on at the Sloan Kettering Memorial Hospital. Contralateral deposits from a tumour in the left testis to the right side of the inferior vena cava were never found. Contralateral metastases from a tumour in the right testis were usually discovered in the left superior para-aortic nodes. Therefore, Ray et al, suggested modifications to the standard bilateral retroperitoneal lymphadenectomy in the hope of sparing ejaculatory function. We have also studied the distribution of retroperitoneal metastases in 91 patients with resectable retroperitoneal disease clinically stage NO-2, MO (Figure 1). Contralateral metastases were present in only 16.5% of cases and they were never seen in the absence of homolateral involvement. Only in two cases was a metastatic node found at the aortic bifurcation and in both cases this was accompanied by contralateral metastases. In two other cases metastases were extending to the contralateral common iliac nodes. In two patients with contralateral metastases from a left testicular tumour, right paracaval nodes were also involved. Homolateral hilar and suprahilar deposits were found in 6 patients with extensive retroperitoneal disease. It is worthy of note that suprahilar metastases were found behind the renal vessels, lying between or beneath the diaphragmatic crura.

It is therefore our policy to perform homolateral retroperitoneal lymph node dissection when there is no surgical evidence of retroperitoneal disease. By also sparing the tissue at the aortic bifurcation, post-operative loss of ejaculation is avoided. In only 1/20 patients were microscopic deposits found in spite of negative surgical findings and negative frozen section. It was decided that a second look contralateral lymphadenectomy should not be performed in this particular patient.

In our series patients with surgical evidence of retroperitoneal disease undergo a standard bilateral retroperitoneal lymphadenectomy with extension up to the diaphragmatic crura.

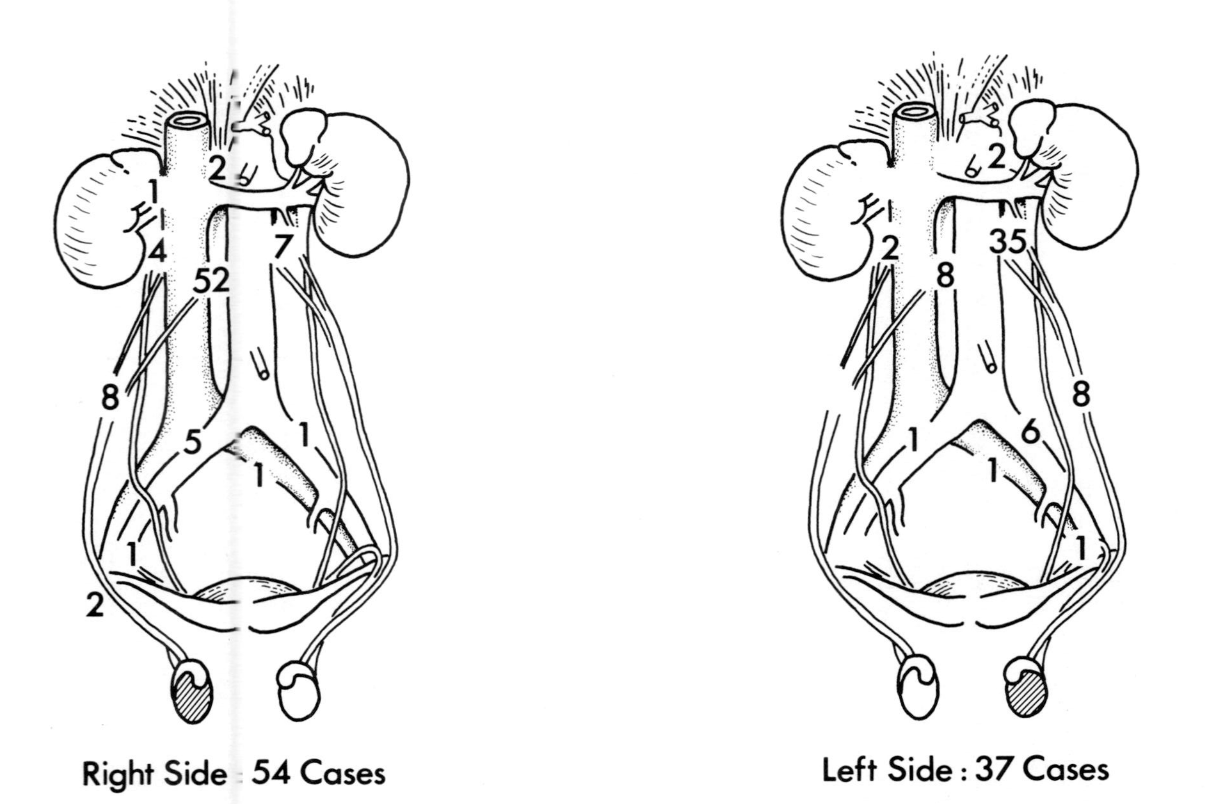

FIGURE 1 - Distribution of retroperitoneal metastases in 91 patients with resected Stage II disease (1970-80).

Retroperitoneal lymphadenectomy in stage I disease

The overall cure rate for stage I non-seminoma with retroperitoneal lymphadenectomy alone is about 90% (6, 12).

In our series of 89 patients operated on between April 1968 to December 1979, we had only 8 relapses (9.0%), occurring within 4 to 12 months after surgery (median 8 months). Two regional relapses were observed at the start of the series and they might have been avoided by more rational surgery. Four patients achieved a maintained complete remission with further therapy and only 4 (4.5%) died of disease.

We are unable to discover any difference in prognosis according to the extent of the retroperitoneal lymphadenectomy, the histology of the primary tumour and serum markers (HCG and AFP). Even adjunctive therapy did not change the prognosis: one relapse occurred in a group of 12 patients treated with pre or post-operative irradiation, and relapse occurred in one of 5 patients who received chemotherapy. On the other hand, morbidity was increased in patients receiving adjunctive therapy (4/17 experienced complications compared with 5/75 patients treated with surgery alone). The only difference of prognostic importance (Table 2) was in patients who had scrotal surgery for their testicular tumour, 5/24 relapses in this subset of patients (20.6%) compared with 3/65 relapses in patients who had primary inguinal surgery (4.5%). This difference is even more significant as far as distant metastases are concerned: 16.5% versus 3% respectively ($p < 0.05$).

Retroperitoneal lymphadenectomy in resectable stage II non-seminoma

Radical retroperitoneal lymphadenectomy alone can cure between 46 and 70% of resectable cases of stage II non-seminomas (12, 16, 18). Unfortunately, patients treated with surgery alone are few in number and have not been sufficiently studied to allow any correlation between the degree of retroperitoneal involvement and prognosis. Post-operative irradiation has not improved the results of surgery alone and usually succeeds in increasing morbidity (9, 12). Post-operative chemotherapy, on the other hand, has improved results in stage II disease and both PVB and

TABLE 2

RELAPSES AND SCROTAL SURGERY IN PATHOLOGIC STAGE I NON-SEMINOMA

PRIMARY SURGERY	No. CASES	REGIONAL REC.		DISTANT REC.	
		no.	(%)	no.	(%)
SCROTAL	24	1	(4.1)	4	(16.5)
INGUINAL	65	1	(1.5)	2	(3.0)
TOTAL	89	2	(2.3)	6	(6.7)

VAB III protocols have succeeded in avoiding post operative relapse in two series of patients (4, 17). However, delayed therapy, given at recurrence of the disease, seems to achieve equivalent results (4).

From February 1970 to December 1979, radical retroperitoneal lymph node dissection was performed in 70 consecutive patients with resectable retroperitoneal metastases. The overall relapse rate was 22.8% (16 cases). There were two regional relapses, both occurring outside the surgical field. The disease free interval ranged from 3 to 25 months (median 5 months). Eight patients died of cancer (11.4%), the others being rescued by further therapy. No correlation could be found between serum markers and prognosis. Histology of the retroperitoneal metastases did not correlate with prognosis (81.2% pure embryonal carcinoma in the metastases of the 16 patients who relapsed versus 61.1% in the 54 cases who did not relapse). On the other hand, prognosis was related to adjunctive therapy and the extent of retroperitoneal disease. Relapse occurred in

8 of the 16 patients not receiving post-operative chemotherapy (50.5%) compared with 8 of the remaining 54 (14.8%) who had been treated with adjunctive chemotherapy (Table 3). In particular, relapses occurred in only 1 of the 27 patients treated with PVB x 5 cycles (3.7%), versus 3 of the 16 treated with VB x 3 (18.7%) and 4 of the 11 treated with older chemotherapy schedules (36.3%). Post-operative irradiation, given to 17 patients, with or without chemotherapy in addition, resulted in increased morbidity; 8/17 major complications and 3 related deaths.

As to the extent of retroperitoneal disease, no relapse was observed in 9 patients with 5 or less intracapsular node metastases smaller than 2 cm (stage II A). Seven relapses occurred in 38 patients (18.4%) with extracapsular growth, nodes larger than 2 cm, or more than 5 metastases (stage II B) and 9/23 (39.1%) with metastases larger than 5 cm, or when there was fascial or vascular invasion (stage II C). As far as those patients who did not receive chemotherapy are concerned, relapses occurred in 2 of the 9 stage II A-B patients (22.2%), versus 6/7 (85.7%) in stage II C (Table 4). It is obvious that stage II C should be considered advanced disease and that these patients need intensive chemotherapy. Stages II A and B may be cured by surgery in a high percentage of cases, and further investigation is needed in order to differentiate high risk from low risk patients. We are presently treating these patients by surgery alone. Delayed chemotherapy with PVB is given to relapsing patients. Several biologic features, including cell kinetics, are being carefully studied. In the United States, on the other hand, a randomised prospective clinical trial is on-going comparing adjuvant versus delayed chemotherapy with PVB or VAB IV protocols (1).

Surgery in advanced non-seminoma

Since the initial reports of Skinner (14), Merrin (10) and Johnson (5) in 1976, many other publications have confirmed the advantages of combining surgery and chemotherapy in advanced non-seminoma. Primary cytoreductive surgery could improve results of subsequent chemotherapy by reducing the tumour burden. Nearly two thirds of partial responders to remission induction chemotherapy could be made tumour free by secondary surgery. Lange, 1980 (8), however, warns of possible accelerated growth of testicular cancer after cytoreductive surgery and advocates care in planning treatment for patients with advanced disease. Present

TABLE 3

RESECTED STAGE II DISEASE

RELAPSES AND POST OPERATIVE CHEMOTHERAPY

CHEMOTHERAPY	No. CASES	RELAPSES	
		no.	(%)
NO CHEMO	16	8	(50.0)
OLD CHEMO	11	4	(36.3)
VB x 3	16	3	(18.7)
PVB x 5	27	1	(3.7)
TOTAL	70	16	(22.8)

TABLE 4

RESECTED STAGE II DISEASE

RELAPSES IN PATIENTS NOT RECEIVING CHEMOTHERAPY

PATH. STAGE	No. CASES	RELAPSES	
		no	(%)
II A	2	-	-
II B	7	2	(28.5)
II C	7	6	(85.7)
TOTAL	16	8	(50.0)

experience suggests that surgery should be undertaken in partial responders to remission induction chemotherapy in order to remove residual disease (3, 11, 15). Patients with residual disease are prone to subsequent relapse and they should be treated with further induction chemotherapy. It is recommended that surgery is not performed at a later stage, in order to avoid the problem of severe fibrosis which ususally develops after prolonged chemotherapy and makes dissection very difficult or even impossible (12).

From June 1974 to December 1979 59 patients with bulky abdominal (stage II D) or widespread (stage III) tumours were submitted to combined surgery and chemotherapy. Forty-five patients had retroperitoneal surgery, 8 had lung resections and 6 had both. Surgical complications occurred in 5 patients (8.4%). One fatality occurred 3 months after surgery from progressive skin necrosis, haemorrhage and sepsis in a patient who had radical radiotherapy before undergoing VB chemotherapy and surgery (with contemporary resection of a double segment of bowel which was damaged by radiotherapy).

Induction chemotherapy consisted mainly of a combination of Adriamycin, Vincristine and Methotrexate (AVM) in 1974 and 1975, Vinblastine plus continuous intravenous Bleomycin (VB) in 1976 and 1977, and of a combination of cisplatinum, Vinblastine (0.3 mg/Kg) and Bleomycin (PVB) in 1978 and 1979. As the results of chemotherapy are beyond the scope of this paper, no further details of the chemotherapy will be given, except that consolidation therapy consisted of 6 months therapy with a combination of Adriamycin and Vinblastine or AVM after VB and of a combination of Adriamycin, Vincristine, Actinomycin D and Cytoxan after PVB.

Surgery was undertaken before chemotherapy in 11 patients and 9 (81.8%) were tumour free at the end of induction chemotherapy. Complete responses (CR) are maintained in 7 cases (63.6%) from between 1 and 5 years.

Forty-eight patients underwent surgery after remission induction chemotherapy (Table 5). Only 3 complete responders underwent surgery to prove their CR status histologically. This policy was soon abandoned. Eight patients were poor responders to induction chemotherapy and surgery was undertaken in the hope of enhancing further induction chemotherapy. Despite 4 patients (50%) being rendered tumour free by surgery, the disease rapidly progressed in all except one. This patient is still in CR seven years after lung resection for a solitary metastasis which had been progressing during induction chemotherapy.

TABLE 5

SURGERY AFTER REMISSION INDUCTION CHEMOTHERAPY IN ADVANCED NON-SEMINOMA (BULKY STAGE II AND STAGE III)

RESPONSE TO CHEMOTHERAPY	No. CASES	C.R. AFTER SURGERY	MAINTAINED C.R.	
			no.	(%)
C.R.	3	3	2	(66.6)
P.R.	37	23	22*	(59.4)
N.R.	8	4	1	(12.5)
TOTAL	48	30	25	(52.1)

* C.R. was achieved and maintained by further chemotherapy in 3/14 patients who did not achieve C.R. with surgery.

TABLE 6

HISTOLOGY AND OUTCOME IN 23 PATIENTS IN PR FOLLOWING REMISSION INDUCTION CHEMOTHERAPY WHO ACHIEVED SURGICAL C.R.

HISTOLOGY	No CASES	MAINTAINED C.R.	
		no.	(%)
NON-SEMINOMA	10	6	(60.0)
SEMINOMA	1	1	(100.0)
MATURE TERATOMA	8	8	(100.0)
FIBROSIS & NECROSIS*	4	4	(100.0)
TOTAL	23	19	(82.6)

* Fibrosis & necrosis was also found in 6/14 patients who did not have radical surgery, but 4 of these relapsed

This particular patient received no further therapy.

The remaining 37 patients who were operated on were partial responders (PR) to remission induction chemotherapy. Twenty-three (62.2%) were rendered tumour free by surgery and CR is maintained in 19 from between 1 to 6 years. Additionally 3 patients in the group of 14 who were not rendered tumour free by surgery achieved a maintained CR after further chemotherapy. The total number of patients operated on who achieved a maintained CR and who were partial responders to induction chemotherapy is 22/37 (59.4%).

It is interesting to examine the figures relating to the histology of the pathology specimen which was removed in partial responders to remission induction chemotherapy. In the 23 patients who had the complete removal of all apparently involved tissue, 8 (34.8%) had only residual mature teratoma and 4 (17.4%) had no residual tumour at all, but only fibrosis and tumour necrosis. These patients have all maintained their CR status (Table 6). On the other hand, the disease progressed in 4 of the 6 patients who had no demonstrable tumour in the specimen after incomplete surgery. The resected tissue could not have been representative of the pathological tissue left behind in this subset of patients.

Conclusions

Lymphadenectomy remains the best staging tool for retroperitoneal nodes in testicular non-seminoma and it may also be curative in selected cases with limited retroperitoneal disease (stages IIA and B).

When surgery demonstrates pathological stage II C disease, remission induction chemotherapy is mandatory, even in spite of radical surgery, because this is very advanced disease.

Surgery combines very well with both induction and salvage chemotherapy in advanced disease patients (bulky stage II and stage III). It carries a low morbidity and does not interfere with bone marrow reserve, which needs to be preserved for intensive chemotherapy. The major indication for surgery in advanced disease patients is the complete removal of all residual pathological tissue in partial responders to induction chemotherapy. This is to satisfy a dual purpose; to obtain a complete pathological verification of the results of therapy, and to make partial responders tumour free. In our Institution, surgery is also part of a salvage treatment programme in non-responders or relapsing patients. The tumour burden is reduced by surgery and in vitro chemosensitivity tests are carried out to determine a personalised new induction therapy regimen.

REFERENCES

1. De Wys, W.D., 1979. Basis for adjuvant chemotherapy for stage II testicular cancer. Cancer Treatment Reports, 63, 1693-1695.

2. Donohue, J.P., 1977. Retroperitoneal lymphadenectomy: the anterior approach including bilateral supra-hilar dissection. Urological Clinics of North America, 4, 509-521.

3. Donohue, J.P., Einhorn, L.H., Williams, S.D., 1980. Cytoreductive surgery for metastatic testis cancer: considerations of timing and extent. Journal of Urology, 123, 876-880.

4. Donohue, J.P., Einhorn, L.H., Williams, S.D., 1980. Is adjuvant chemotherapy following retroperitoneal lymph node dissection for non-seminomatous testis cancer necessary? Urological Clinics of North America, 7, 747-756.

5. Johnson, D.E., Bracken, R.B., Ayaca, A.G., Samuels, M.L., 1976. Retroperitoneal lymphadenectomy as adjunctive therapy in selected cases of advanced testicular carcinoma. Journal of Urology, 116, 66-68.

6. Johnson, D.E., Bracken, R.B., Blight, E.M., 1976. Prognosis for pathologic stage I non-seminomatous germ cell tumours of the testis managed by retroperitoneal lymphadenectomy. Journal of Urology, 116, 63-65.

7. Kaswick, J.A., Bloomberg, S.D., Skinner, D.G., 1976. Radical retroperitoneal lymph node dissection: how effective is removal of all retroperitoneal nodes? Journal of Urology, 115, 70-72.

8. Lange, P.M., Hekmat, K., Bosl, G., Kennedy, B.J., Fraley, E.E., 1980. Accelerated growth of testicular cancer after cytoreductive surgery. Cancer, 45, 1498-1506.

9. Lynch, D.F., McCord, L.P., Nicholson, T.C., Richie, J.P., Sargents, C.R., 1978. Sandwich terapy in testis tumour: current experience. Journal of Urology, 119, 612-613.

10. Merrin, C., Takita, H., Weber, R., Wajsman, Z., Baumgartner, G., Murphy, P., 1976. Combination of radical surgery and multiple sequential chemotherapy for the treatment of advanced carcinoma of the testis (stage III). Cancer, 37, 20-29.

11. Peckham, M.J., Barrett, A., McElwain, T.J., Hendry, W.F., 1979. Combined management of malignant teratoma of the testis. Lancet, 2, 267-270.

12. Pizzocaro, G., Durand, J.C., Fuchs, W.A., Merrin, C.E., Musumeci, R., Schmucki, O., Vahlensieck, W., Whitmore, W.F. Jr., Zvara, V.L., 1981. Staging and surgery in testicular cancer. European Urology, 7, 1-10.

13. Ray, B., Hajdu, S.I., Whitmore, W.F. Jr., 1974. Distribution of retroperitoneal lymph node metastases in testicular germinal tumours. Cancer, 33, 340-348.

14. Skinner, D.G., 1976. Non-seminomatous testis tumours: a plan of management based on 96 patients to improve survival in all stages by combined therapeutic modalities. Journal of Urology, 115, 65-69.

15. Sogani, P.C., Vugrin, D., Whitmore, W.F., Bains, M., Herr, H., Golbey, R., 1980. Experience with combination chemotheray and surgery in the management of advanced germ cell tumours. Proceedings of the American Association of Cancer Research, 21, 401.

16. Staubitz, W.J., Early, K.S., Magoss, I.V., Murphy, G.P., 1973. Surgical management of testis tumour. Journal of Urology, 65, 113-117.

17. Vougrin, D., Cvitkovic, E., Whitmore, W.F., Golbey, R.B., 1979. Adjuvant chemotherapy in resected non-seminomatous germ cell tumours of testis Stages I and II. Seminars in Oncology, 6, 94-98.

18. Whitmore, W.F. Jr., 1970. Germinal tumours of the testis. Proceedings of the 6th National Cancer Conference. Lippincot, Philadelphia, pp. 219-245.

ELECTIVE DELAYED EXCISION OF BULKY PARA-AORTIC LYMPH NODE METASTASES FROM TESTICULAR TERATOMAS

W.F. Hendry
Consultant Urologist
Royal Marsden Hospital,
Fulham Road,
London SW3 6JJ

More than 10 years ago at the Royal Marsden Hospital, Sir David Smithers noticed that large para-aortic lymph nodes in some patients with testicular teratomas did not resolve with radiotherapy and he asked my predecessor Mr. D. Wallace to excise the lymph nodes. In the era before effective chemotherapy, 13 patients were treated in this way and undifferentiated malignant teratomatous tissue or embryonal carcinoma was present in 8 (61%) of these patients. Nevertheless, 9 of the 13 are still alive today and it was concluded from this that removal of bulky para-aortic lymph node metastases could be beneficial. However, we felt this was best done after preliminary treatment, originally with radiotherapy and now more recently with chemotherapy. The advantage of electively delaying surgery, we believe, is that radiotherapy can still be included as a pre-operative treatment option since the abdomen should be free of adhesions at this stage.

And so my colleagues at the Royal Marsden Hospital evolved a protocol for patients with bulky para-aortic metastases, bulky meaning more than 2 cm in diameter as defined by lymphography, ultrasonagraphy or CT scanning, even if there were initially small volume metastases elsewhere, for example in the neck or lungs. The protocol starts with chemotherapy followed by radiotherapy and finally the surgical removal of persistent masses. A typical patient has four courses of chemotherapy, at the usual intervals of time, then 45 Gy (4500 rads) to the para-aortic lymph nodes, then a gap of about 4 weeks before going for surgical excision of the residual masses. These are removed completely, but only with the aim of removing the enlarged lymph nodes, not attempting to do a complete clearance of all the para-aortic nodes by any means.

I should like to give examples of the sorts of problems which arise with this form of surgery and indicate the experiences we have had with them. A rather typical example is a residual solitary node placed high in the upper para-arotic node chain, with no evidence of involved nodes elsewhere. It is straightforward to define this by ultrasonography to show the extent and position, and to remove it without undue difficulty by peeling it off the front of the great vessels. With a larger mass in a typical postion just below the left kidney, surgery here starts by stripping down the front of the vena cava and moving across to the left. One or two points about technique - it is very helpful to divide the inferior mesenteric artery at its origin to be able to get the dissection going across in that way. The most difficult point is probably where the testicular vein comes into the inferior aspect of the left renal vein which tends to be surrounded by lymph nodes. The insertion of that vein tends to be obliterated, and there is a tendency to tear this out if one is not careful. The other big problem is the lumbar vessels arising from the posterior of the aorta which can be defined by lifting it forward with slings. An additional problem may be encountered if there are some lower polar renal vessels running across.

For the purposes of choosing the best surgical approach, the exact position of the tumour is defined pre-operatively by radiological means. The difficulty for the surgeon is knowing whether an anterior approach or a thoraco-abdominal incision is required so that he can get above the kidney if necessary. The way this is done is to perform repeated cuts on the CT scanner going higher and higher until the tumour mass disappears, resulting in a decision about the most appropriate surgical approach. Sometimes the masses are very large indeed, for example occupying the large part of the upper left abdomen. Again, the CT scan is useful in demonstrating the extent and size of the mass. Following the chemotherapy, the mass shrinks to some extent but usually not completely. Many of these massive lesions do not have the radiotherapy part of the protocol but go to surgery at this point. Occasionally the masses involve the renal vessels, and under these circumstances a nephrectomy may well be necessary as well. This sort of case undoubtedly benefits from a thoraco-abdominal approach to obtain adequate exposure of the mass. Some of these tumours tend to grow into the psoas muscle behind and large parts of this have to be removed along with the tumour mass. The scanogram facility of some CT scanners is extremely helpful in planning the approach, showing the relation of the tumour mass to the renal vessels and to the kidney, and also outlining additional masses behind the vena cava. The services of a thoracic surgeon may be required to help remove masses higher up behind the crura of the diaphragm. We have seen this on two or three occasions and this additional deposit hidden high up on the right side does seem to be one to bear in mind. Something also to be aware of is possible tumour involvement of, or indeed as in one of our cases, tumour inside, the vena cava. Again the CT scan detected this, and we were able to make all the necessary preparations for dealing with this problem, including having a vascular surgeon standing by. In this case, a vena-cavogram and a right atriogram were done to show the extent of the tumour and the approach could be modified accordingly.

Now, from a surgical point of view, the long midline abdominal incision from xyphi - sternum down to symphysis pubis has been extremely satsifactory. They always heal well and have given no trouble whatever, despite the prior chemotherapy and radiotheray. The thoraco-abdominal approach is also very straightforward if there is thoracic disease as well. Loin and other approaches have proved unsatisfactory because they tend to come in at the back of the mass where there is no plane of cleavage. We have had one post-operative death, a boy with massive right sided tumour who presented with a paralysed right leg. He had an enormous mass that shrank nicely with chemotherapy. The problem was that the tumour was growing back through the psoas muscle and into the para spinal vessels, invading the roots of the femoral nerve. It was not possible to remove the mass completely. He died of a secondary haemorrhage at about 10 days post-operatively.

A point about morbidity in the form of loss of ejaculatory function. There is some non-descript tissue in front of the aorta between the bifurcation of the iliac vessels which I have shown by electrostimulation to be necessary for ejaculation. This is very often divided in the course of this particular operation. I have since tried to avoid this tissue. It is very difficult, however, when doing a radical dissection, to avoid cutting through this particular piece of tissue.

Now to the results of the series of patients treated at the Royal Marsden Hospital. We have performed 40 operations on patients who have had chemotherapy alone or chemotherapy and radiotherapy. Active malignancy was found in 5 out of the 40 cases (12.5%). One of these has died and another has evidence of advancing disease, so clearly residual active disease is a bad prognostic feature. Among the 35 patients with differentiated teratoma or no evidence of malignancy in the excised nodes, only one has developed disseminated disease. There was one post-operative death. Overall, 36 (90%) of the 40 patients are alive and apparently disease free at the present time with a maximum follow up of four years, although many of them have a much shorter follow up.

REFERENCES

1. Hendry, W.F., Goldstraw, P., Barrett, A., Husband, J.E. and Peckham, M.J., 1981. Elective delayed excision of bulky para-aortic metastases from non-seminomatous germ cell tumours of the testis. British Journal of Urology, (In press).

2. Peckham, M.J., Barrett, A., McElwain, T.J., Hendry, W.F. and Raghavan, D., 1981. Non-seminoma germ cell tumours (malignant teratoma) of the testis. British Journal of Urology, 53, 162-172.

CYTOREDUCTIVE SURGERY IN ADVANCED TESTICULAR CANCER

R. Ozols and N. Javadpour
Medicine and Surgery Branches,
Division of Cancer Treatment, National Cancer Institute,
Bethesda, Maryland, 20205,
U.S.A.

The exact role of cytoreductive surgery in the treatment of disseminated cancer has been an area of frequent debate among chemotherapists, radiotherapists, and surgeons. The clinical observation that small volumes of tumour respond better to either radiotherapy or chemotherapy, together with experimental studies which demonstrated improved survival in tumour-bearing animals treated with chemotherapy, have provided the impetus for surgeons to remove as much bulk of tumour as possible in preparation for post-operative chemotherapy or radiotherapy (1, 4). Cytoreductive surgery has been utilised in the management of advanced disseminated non-seminomatous testicular cancer in three distinct clinical settings: 1) prior to any systemic chemotherapy; 2) after 3-4 cycles of chemotherapy when there is known persistent carcinoma indicated by elevated levels of either human chorionic gonadotrophin (HCG) or alpha-fetoprotein (AFP); and 3) after 3-4 cycles of chemotherapy if there is a persistent mass but with normal levels of AFP and HCG. However, only in the first clinical setting has there been a clinical trial in which the role of cytoreductive surgery has been examined in a prospective randomised manner.

Surgical Cytoreduction Prior to Chemotherapy

Cytoreductive surgery has been advocated for patients with advanced testicular cancer in an effort to improve the response rate to subsequent chemotherapy. Merrin et al demonstrated that a simultaneous excision of abdominal and thoracic metastases could be performed in patients with Stage III bulky testicular cancer (6, 7). However, they reported no difference in survival in a group of 11 patients who had cytoreductive surgery prior to combination chemotherapy compared to a group of 8 patients who were treated with the same chemotherapy but without initial cytoreductive surgery (6). The role of cytoreductive surgery cannot be assessed in this non-randomised study since patients were assigned to the two groups on the basis of whether the metastatic disease was amenable to surgical cytoreduction. Patients who had multiple surgically inaccessible metastases were treated with chemotherapy.

The Medicine and Surgery Branches of the Division of Cancer Treatment, NCI have recently reported the results of their prospective randomised trial of cytoreductive surgery followed by chemotherapy versus chemotherapy alone in patients with advanced (bulky) Stage III testicular cancer (5). Patients selected for this trial all had one or more of the following poor prognostic features: a metastasis in the thorax greater than 2.0 cm in diameter, a palpable abdominal mass, liver involvement, obstructive uropathy or involvement of the inferior vena cava. The cytoreductive surgery consisted of removal of as much disease as possible from the thorax or abdomen. Chemotherapy consisted of 3 intensive 21-day induction cycles with cyclophosphamide (1000 mg/m^2), vinblastine (4 mg/m^2), actinomycin D (1 mg/m^2) on day 1, and cis-platinum (100 mg/m^2) on day 7. Bleomycin was administered as a continuous infusion (15 u/m^2) on days 1-5. Patients also received 2 non-intensive induction cycles with lower doses of cyclophosphamide (100 mg/m^2 p.o. days 1-8) and cis-platinum (25 mg/m^2) and bleomycin (15 u/m^2) on day 1. Patients achieving a complete response were started on 13 cycles of maintenance therapy without bleomycin, with cis-platinum administered on alternate cycles. The results of this trial are summarised in Table 1.

The patients randomised to initial surgery had an estimated 70-90% reduction in tumour burden by the debulking procedure. Following chemotherapy, the response rates for the two groups were similar. Patients treated with chemotherapy as initial treatment had a complete remission rate of 37%, partial response rate of 42%, and 21% had no response to therapy compared to 44%, 39% and 16% for those patients treated with surgery prior to chemotherapy. Three patients in the chemotherapy alone arm were rendered complete responders by excision of a mature teratoma. Life table analysis did not reveal statistically significant differences in survival between the two groups. There has been only one relapse in a patient who attained a complete response to therapy.

To our knowledge, this is the first reported prospective randomised trial of cytoreductive surgery in any malignancy.

TABLE 1

NCI Trial of Cytoreductive Surgery in Advanced Testicular Cancer (5)

Characteristic	Surgery + Chemotherapy	Chemotherapy
Number of patients:	20	19
Stage III	20	19
Advanced disease (*)	20	19
Percent tumour reduction by surgery:	70-90%	N/A
Response following chemotherapy:		
Complete response	8/18(**)(44%)	7/19(37%)
Partial response	7/18(39%)	8/19(42%)
No response	3/16(16%)	4/19(21%)

(*) Including any of the following: thoracic metastases>2cm, palpable abdominal mass, obstructive uropathy, liver metastases, inferior vena cava involvement

(**) One patient was not evaluable and there was one surgical death

TABLE 2

CYTOREDUCTIVE SURGERY FOLLOWING CHEMOTHERAPY (2)

Type of Surgery	Number	Histology	Number NED
Thoracotomy	20	Teratoma	8/8
		Necrosis/Fibrosis	2/4
		Carcinoma	1/8
Laparotomy	40	Teratoma	10/12
		Necrosis/Fibrosis	15/15
		Carcinoma	3/13

It has demonstrated what while cytoreductive surgery is technically feasible in untreated bulky Stage III patients with testicular cancer, there is no improvement in response to subsequent chemotherapy or in survival compared to initial treatment with chemotherapy alone.

Cytoreduction for Residual Carcinoma Following Chemotherapy

Cytoreductive surgery has also been used in patients with known persistent disease following 3-4 cycles of chemotherapy (1, 8). Donohue reported 5 patients with positive pre-operative serum markers who underwent cytoreductive surgery following platinum, vinblastine, and bleomycin induction chemotherapy. These patients all had residual cancer in the resected specimen and most of these patients have subsequently relapsed. In contrast, Vurgin et al have recently reported on the use of VAB-6 intensive induction chemotherapy followed by surgery (8). Cytoreduction is performed after 3 cycles of chemotherapy and if the resected specimen contains cancer, 2 more cycles of induction chemotherapy are given. A complete response rate of 90% has been achieved, and only one of 19 patients attaining a complete response has relapsed. The number of patients who had residual carcinoma after 3 cycles of initial induction therapy was not specified in this preliminary report.

Cytoreduction for a Residual Mass Following Chemotherapy

Frequently, following chemotherapy, a residual mass persists, the nature of which cannot be determined by non-invasive studies. Einhorn has recently summarised the Indiana University experience with cytoreductive surgery in this clinical setting (Table 2) (2). In this series, it appears that the surgery is more of a diagnostic and prognostic procedure than an actual therapeutic intervention. Eighteen of 20 patients who either had a mature teratoma, necrotic tissue, or a fibrous mass, had a prolonged disease-free survival. In contrast, of the 21 patients in whom carcinoma was identified in the surgical specimen, only 4 have remained free of disease, even though post-operative chemotherapy was administered. These results appear to be in contrast to the previously cited Memorial Hospital experience with VAB-6 (8).

Discussion

A review of the reported studies of cytoreductive surgery in advanced testicular cancer supports the view that combination chemotherapy should be the initial therapeutic modality used. While surgery is frequently necessary to define the exact nature of a residual mass following chemotherapy, it is unclear whether the survival of patients rendered a complete response by surgical removal of a carcinoma is equivalent to that of patients who achieve a complete response with chemotherapy alone.

The NCI trial has demonstrated that while cytoreductive surgery is technically feasible in previously untreated patients with advanced poor prognosis testicular cancer, it does not improve the response to subsequent chemotherapy compared to treatment with chemotherapy alone. The overall response rate to chemotherapy in these groups of patients with advanced bulky disease is similar to that reported in other studies, 29-50% (3, 9). Since current platinum-containing regimens have a complete response rate of 70-90% in patients with small volume of disease, the current emphasis on patients with advanced disease is to improve upon the therapy of patients with bulky tumours and other poor prognostic features. The use of new combination chemotherapy regimens and high dose schedules with autologous bone marrow infusion are potential new approaches for this group of patients. Such studies are currently in progress at the NCI.

Whether surgery has more than a diagnostic role in advanced Stage III patients following chemotherapy remains to be determined. The data of Einhorn (2) indicate that if carcinoma persists following 3-4 cycles of chemotherapy, the vast majority of patients will not continue in a complete response following surgical resection of a residual mass (Table 2). The preliminary results with VAB-6, however, suggest that resection of residual carcinoma followed by additional intensive cycles of chemotherapy can result in prolonged complete responses (9). The apparent differences in these studies may be due, in part, to the different chemotherapeutic regimens used following the cytoreductive surgery.

REFERENCES

1. Donohue, J.P., Einhorn, L.E., and Williams, S.D., 1980. Cytoreductive surgery for metastatic testis cancer: considerations of timing and extent. Journal of Urology, 123, 876-880.

2. Einhorn, L.H., 1980. The role of surgery in disseminated testicular cancer. Proceedings of the American Society for Clinical Oncology, 21, 159.

3. Einhorn, L.H., Williams, S.D., Chemotherapy of disseminated testicular cancer. A random prospective study. Cancer, 46, 1339-1344.

4. Griffiths, C.T., and Fuller, A.T., 1978. Intensive surgical and chemotherapy management of advanced ovarian cancer. Surgical Clinics of North America, 58, 131-142.

5. Javadpour, N., Ozols, R.F., Barlock, A., et al. 1981 Cytoreductive surgery followed by chemotherapy versus chemotherapy alone in bulky Stage II (poor prognosis) testicular cancer. Proceedings of the American Society for Clinical Oncology, 22, in press.

6. Merrin, C., Takita, H., Weber, R., Wajsman, Z., Baumgartner, G., & Murphy, G.P., 1976. Combination radical surgery and multiple sequential chemotherapy for the treatment of advanced carcinoma of the testes (Stage III). Cancer, 37, 20-29.

7. Merrin, CE., and Takita, H., 1978. Cancer reductive surgery. Report on the simultaneous excision of abdominal and thoracic metastases from widespread testicular tumours. Cancer, 42, 495-501.

8. Vugrin, D., Dukeman, M., Whitmore, W., and Golbey, R., 1980. VAB-6: Progress in chemotherapy of germ cell tumours (GCT). Proceedings of the American Society for Clinical Oncology, 21, 426.

9. Vugrin, D., Cvitkovic, E., Whitmore, W.F., Cheng, E., and Golbey, R., 1981. VAB-4 combination chemotherapy in the treatment of metastatic testicular tumours. Cancer, 47, 833-839.

SELECTIVE ANGIOGRAPHY OF THE TESTICULAR ARTERY AND ITS USE IN TESTICULAR TUMOURS

L. Wahlqvist, S. Cajander, L. Nordmark and G. Nyberg
Departments of Urology, Pathology and Roentgenology,
University of Umea, S-901 85 Umea,
Sweden

Since 1974 selective angiography of the testicular artery has been performed in 151 patients. Forty-eight patients had a non-palpable testis, 28 long-standing epididymitis, 6 long-standing torsion of the testis and 16 malignant testicular tumour. Other patients had traumatic lesions, hydrocoeles or spermatocoeles. The diagnosis has usually been verified by surgical exploration. The course of the disease in patients who were not operated on confirmed the diagnosis. Angiography gives information about the branches of the artery, the parenchyma of the testis, the tumour tissue, the veins in the testis, epididymis and funicle as well as the gonadal vein. Epididymitis or torsion of the testis or a malignant testicular tumour has not been mis-diagnosed as any other lesion in our series. An infected spermatocoele was once mistaken for a tumour.

Eight non-seminomatous germ cell tumours all had contrast filled tumour vessels and accumulation of contrast medium in the tumour tissue. All 8 seminomas had accumulated contrast medium in the tumour tissue and possibly one had contrast filled tumours vessels. Normal intra-testicular arteries were not found in any malignant tumour. If arteries were seen, they were dislocated by the tumour. Elevated AFP and/or β-HCG was found in 57% and elevated LDH in 71% of the patients with malignant testicular tumour and should be of value in aiding the diagnosis.

Selective angiography of the testicular artery might thus be of value, when the urologist has to distinguish malignant testicular tumours from long-standing epididymitis. Angiography might also be of value when AFP and β-HCG estimations are normal.

THE RESULTS OF BILATERAL RETROPERITONEAL LYMPH NODE DISSECTION (RPLND) IN TURKEY

A. Adkas, D. Remzi, C. Tasar, M. Baddaloglu
Department of Urology, Faculty of Medicine,
Hacettepe University Hospital,
Ankara, Turkey.

Testicular tumours are the most common malignancies in young men. In Turkey bilateral retroperitoneal lymph node dissection (RPLND) for the treatment of testicular tumours other than pure seminoma is undertaken only in our department.

This operation has been performed on 29 patients with non-seminomatous testicular tumours in clinical stages I and II.

The accuracy of lymphangiography in the staging of 21 patients was 85.7% when compared with the pathological examination of the surgical specimen. According to this histopatholocial staging on the whole group of 29 patients, 16 (55%) were found to be in Stage I and 13 (44.8%) were in Stage II. There were 9 patients (31%) with embryonal carcinoma, 8 (27.5%) with teratocarcinoma and 12 (41.5%) had a mixed type of tumour.

Serious complications were experienced by 2 (6.9%) of our patients. One patient had a left renal vein thrombosis and the other a urinoma. Eleven (84.6%) of the 13 Stage II patients also received radiotheray and/or chemotherapy. The others were subjected to bilateral RPLND only.

Since these patients come from different parts of Turkey, follow up can be very difficult, but among the patients followed, 3/13 (23.1%) in Stage I and 6/12 (50%) in Stage II had recurrences within one year.

In conclusion, vigorous follow up is essential after bilateral RPLND in Stage I disease, and adjuvant chemotherapy should be given in Stage II disease patients in order to achieve longer survival.

DISCUSSION

Dr. Laing

What proportion of the total number of patients in the Royal Marsden Series who have been given chemotherapy and/or radiotherapy have come to surgery?

Dr. Barrett

The reason for going on to surgery is a residual node mass of greater than 2 cm. That happens in a very small proportion of patients, although I cannot give the exact numbers off hand.

Professor Blandy

Can I make sure I have understood that. You do not send them to Mr. Hendry if the tumour has gone down to something less than 2 cm.?

Dr. Barrett

Yes, because we have found that these patients have continuing regression over the next few months, with complete disappearance of the tumour, usually by about 6 months.

Dr. Oliver

What percentage of patients going to radiotherapy with a residual mass actually progressed during radiotherapy?

Dr. Barrett

I can only think of one or two patients who have had an apparent increase in the size of tumour mass during radiotherapy. We have certainly seen one patient who had a cystic mass which enlarged during radiotherapy.

Dr. Oliver

You never start radiotherapy in the face of raised serum markers?

Dr. Barrett

No.

Professor Blandy

I think it would be interesting at this point to ask Dr. Barrett, in view of what has been said about early disease, whether these patients with retroperitoneal tumour who go into complete remission subsequently relapse and if so do they get metastases elsewhere, such as in the chest, or do they relapse in the retroperitoneum?

Dr. Barrett

I am not aware of any recurrences in that group of patients. They have done well.

Dr. Newlands

There seem to be common threads from a number of presentations in that it does seem that one can predict with the biochemical markers HCG and AFP, with a pretty close correlation, what is going to be the outcome of surgery. Of those with raised markers pre-surgery, by and large nearly all the cases are found to have active tumour still present, and as other investigators have found, this carries a bad prognostic future. But if the markers are normal, and have been normal for some time prior to surgery, the residium is found to be mature teratoma or just purely necrotic tissue. Now I have two questions for Dr. Ozols. With the early elective surgery cases, were you not worried that in those patients with trophoblastic tumours, in particular in those with pulmonary metastases, you were running the risk of CNS disease developing while you are doing the surgery, because of the long period before starting chemotherapy, and whether this was a problem? Secondly you said you had to stop platinum therapy in a number of patients because of reactions. Could you specify what the reactions were?

Dr. Ozols

The first question - there was a 16 day delay between cytoreductive surgery and chemotherapy. That was a median time, between initial surgery and when chemotherapy was started in that group, randomised to cytoreductive surgery. Patients with pure choriocarcinoma were excluded from this trial, so we did not have a group that had higher risk of CNS disease. Overall we did not see any CNS metastases in this group of patients. Second question - the type of platinum reactions we saw were anaphylactic in nature but did not involve serious life threatening severe hypertension or orolanyngeal obstruction. They consisted basically of urticaria, some mild wheezing, which was very discomforting to the patients, leading to refusal of subsequent chemotherapy, even though the reactions could be moderated with benedril and steroids.

Dr. Oliver

May I·take Dr. Newlands point up again because I think it is rather crucial vis a vis whether or not radiotherapy is likely to be doing anything in the retroperitoneum in this situation. Could I ask if Dr. Pizzocaro has information from his series about the percentage incidence of viable cancer in tumour marker negative patients who have excision. From the Einhorn data, and I do not know about your own data, I understand this surgery is being done on marker negative patients. If Dr. Newlands is correct in stating that surgery on marker negative patients is not giving a 33% malignancy rate then the presence of 12% malignancy in the radiotherapy series is not as significant.

Dr. Pizzocaro

I have reviewed the data on markers in patients who underwent surgery for Stage II disease. 50% of them were marker negative before lymphadenectomy in spite of retroperitoneal metastases. I do not know the exact figures regarding markers in patients who had advanced disease who had chemotherapy and then surgery, but on many occasions we have seen markers become negative in partial responders after only one course of chemotherapy. So I think markers alone are not sufficient in predicting if there is residual malignant disease or not, because not all malignant cells are marker producing.

Dr. Oliver

Dr. Newlands, I believe, was trying to make the very important point that, from the N.C.I. experience, the incidence of viable malignancy was in the 12% range when you do excision on a marker negative individual. Is that correct?

Dr. Ozols

I could not put a percentage on it, but it is very low.

Dr. Oliver

It is in that range, not in the order of 33% ?

Dr. Newlands

Yes, if the marker has been persistently negative, not if it has just come down. It is also a matter of what you give the pathologist. Ours has been caught out once when what the surgeon gave him was purely necrotic tissue and the patient did relapse later. Obviously we missed the active focus that was there.

Professor Blandy

I think the question that the surgeon would like to be asking of the pundits is: are we really doing any good by taking out dead cheese?

Dr. Ozols

I think Dr. Einhorn's data speaks to that issue in a sense, in that basically the surgery was of prognostic value, if they in fact had disease, but most of the patients were marker negative.

Dr. Oliver

But Einhorn does not give the actual information specifically on that point and that is what is missing.

Professor Blandy

That is not really the question that surgeons want to ask, which is: are we doing any good? It is a tremendous incision just for a biopsy to give you a better guide to prognosis.

Dr. Newlands

We have recently analysed our series of abdominal operations and some of the patients certainly had raised markers. In retrospect one could say with a fair degree of confidence that the surgery had helped the outcome in the six patients out of 30. But your question about whether or not the markers are completely negative is a different thing.

One is looking at a much broader biological problem. Obviously defacto we are going to have the controls in every series because they are unresectable in proportion. We really need to know what this tissue is so that we can get some idea of predicitability so that the pathologist can analyse what is the tissue that is safe to leave and what is not.

Professor Blandy

Marker negative dead cheese is alright.

Dr. von Eyben

What was the interval from randomisation to chemotherapy in the surgical group compared to the interval from randomisation to chemotherapy in the other group? Furthermore did you look into a possible effect of the delay in chemotherapy in the patients who had surgery prior to chemotherapy? Did this interval have any influence on the results of the chemotherapy?

Dr. Ozols

There is no way in this type of study to totally make the arms equal. We did have a delay in starting chemotherapy, a median delay of 16 days in those patients who were randomised to undergo surgery. The markers were lower at the time chemotherapy was started in 7 out of 8 patients with markedly elevated markers who were operated on. We did not see any sort of explosion of growth after surgery that has been mentioned before, and reported by the Minnesota group. That may be one of the reasons why there was no difference between the arms of our study. We did the surgery to attempt to make many of these patients more susceptible to a good response to chemotherapy but we did not see any significant differences between the two arms. It may be that there was some growth after surgery negating the benefits of surgery. However, we did not see evidence for this on the basis of a rapid increase in markers during the period. This is the first randomised study of cytoreductive surgery in any malignancy that I am aware of. In any such study you are bound to have an imbalance. It was not felt ethical for those patients who were randomised to chemotherapy only to wait a comparable time period. We treated then immediately.

Professor Blandy

Debulking is really a very novel concept in surgery and I think Dr. Pizzocaro must have some views about this.

Dr. Pizzocaro

I try to perform a radical dissection every time it is possible. I never undertake debulking surgery unless it is impossible at operation to do a complete resection. In the few patients who have had this type of surgery I have not seen any evidence of improvement.

Professor Blandy

If I interpret the situation correctly it seems that

debulking is a waste of time and even though viable tumour is removed this does not make the patients live any longer.

Dr. Ozols

I think that is correct. Debulking surgery prior to chemotherapy is certainly not to be recommended now. The study was based on a theoretical plan to convert poor prognosis patients into a good prognosis patients, but I do not think we were able to do that by debulking surgery.

Dr. Basting

We have done about eight second look operations in the last year because of rising tumour markers after chemotherapy and four out of those eight patients had explosive growth of lung metastases. I am not quite sure if the timing of the operation was correct while their tumour markers are still raised, or whether second line chemotherapy should be given before the second look operation. I would be interested if anyone in the audience has seen the same phenomenon.

Dr. Newlands

I think the point is that in the face of rising markers if the chemotherapy is stopped for three weeks or so while the surgery is performed, you will see an explosion of disease at other sites in these rapidly growing tumours. It is not actually the surgery that is stirring things up, it is the fact that one has resistant tumour growing through chemotherapy and if the chemotherapy is stopped it just grows faster.

I want to throw out a general question to all the surgeons. Dr. Pizzocaro mentioned one patient in whom he had trouble with fibrosis following chemotherapy. This has not been our experience. After chemotherapy alone it has been relatively straightforward to get a plane of cleavage, so our surgeons say, it is difficult when the patients have had radiotherapy previously in the area. We had one patient with Stage II disease, who was given radiotherapy and then had a massive para-aortic relapse. On chemotherapy he had a perforation of the aorta and died from this. I assume that the radiotherapy had bound down the residual tissue onto the aortic wall and then it had eroded through. I wonder what other surgeons have experienced in terms of fibrosis.

Professor Blandy

Certainly the fibrosis just after chemotherapy can sometimes be very miserable.

Dr. Barrett

The problems that we have encountered surgically have been in patients with seminoma largely, and not in teratoma. Elective radiotherapy after chemotherapy has produced problems, but we have seen problems in patients who have had radiotherapy before chemotherapy, which seems to produce more fibrosis. We have had two patients with exploding disease after surgery. They were marker negative and apparently no different from any of the other patients in the series. I do not know what the explanation is for this.

Mr. Hendry

One group I would be especially cautious about are those who have been explored previously. We have had two patients who had a laparotomy to diagnose the disease. They went into the study and had chemotherapy, radiotherapy and then repeat surgery. The tissues were as stuck as anything, the usual planes of cleavage on which one relies were simply not there, and I got into terrible trouble. This is why I favour the idea of going in as one clean operation as the last manoeuvre, having previously prepared the field as carefully as possible.

Professor Blandy

Routine node dissection is still the standard treatment in North America for the non-seminomas without enormous bulky disease. What is the proportion of people who have the laparotomy and are found to have inoperable tumour? In this case the poor patient has had this great incision for nothing. Now I suppose with better investigations the failure rate is not very high. In your series Dr. Pizzocaro, how many did you find to be impossible to completely dissect?

Dr. Pizzocaro

It depends on the clinical staging. In 165 patients staged mainly by lymphangiography and IVP, 89 were node negative, 76 were node positive and of these 6 were unresectable.

Professor Blandy

That is a very small proportion and is very creditable.

Dr. Pizzocaro

In the 43 patients who were staged by lymphangiography, IVP ultrasonography or CT scan, no patient was found to be unresectable.

Professor Blandy

So with better investigations the proportion gets less.

Dr. Oliver

Could we focus on the question of surgery in the earlier stage disease, because this is something which has not been discussed in detail. What proportion of positive radiologically staged patients are found at surgery to have no disease? Which of these patients would be staged as Stage II and go onto chemotherapy? I believe this is something we have to bear in mind in basing the definition of Stage II on one single parameter which might be one cut on a CT scan or one minor filling defect on a lymphogram without having other supportive information that this patient has metastatic disease.

Mr. Hendry

We had this problem in one patient who had a horseshoe kidney and we could not be sure from the CT scan if it was the middle of the horseshoe kidney or a lymph node metastasis. At laparotomy there was nothing except the horseshoe kidney. The patient is now in the watch policy group as a Stage I.

Professor Blandy

One thing has not been discussed about which I am very concerned. If you do a laparotomy, it precludes an effective dose of radiotherapy because the adhesions tether the gut and increase the risk of morbity from the radiotherapy. If you give full dose radiotherapy, it precludes full dose chemotherapy so that option is closed. If you routinely irradiate the early non-seminomas as Dr. Hope-Stone and I do then they are not going to be given the option of full chemotherapy later on. In Dr. Pizzocaro's series if they have had that kind of radical surgery they are not going to be able to be given radiotherapy. So when you use one option you close another. Is there any solution to this and has anyone got any views about it?

Dr. Ozols

As a chemotherapist I think the major advances in this disease have been made with chemotherapy. You should not do anything that will render the giving of effective chemotherapy difficult. I think that we should be looking at early stage disease and non-bulky Stage II as having 100% survival and the only way we can achieve that is if we continue to give optimum chemotherapy in full doses. In the early stages radiotherapy should not be routinely used.

Dr. Begent

I can not give exact figures but in our series there have been a number of patients having had previous para-aortic node radiation, who have achieved complete remission with chemotherapy. I do not think radiotherapy necessarily excludes subsequent chemotherapy.

Professor Blandy

Maybe I have misunderstood what the medical oncologists have been telling us.

Dr. C. Williams

I would agree with Dr. Ozols and point out that the data shown has really demonstrated that patients who have minimal disease which is treated with local treatment, whatever it is, the relapses were nearly always at distant sites, requiring treatement with chemotherapy. So if you are going to hold any therapy in reserve I think it is right that you need to preserve the chemotherapy option.

Dr. Newlands

To enlarge on what Dr. Begent was saying, when we analysed the toxicity data, although the myelosuppression was somewhat more profound and raised serum creatinines slightly more frequent in the group given prior para-aortic node irradiation, this did not actually compromise therapy. With our chemotherapy regimes we do not give this very high dose Vinblastine which I think is the major problem if you have had that amount of bone marrow compromised by radiotherapy as indicated by Dr. Jones. One point from the discussion on radiotherapy in early disease states of relevance here, is that I do not think there is now any indication for going above the diaphragm with radiotherapy as a "prophylactic" exercise.

Dr. Hope-Stone

I think the fundamental point is that there are radiotherapists who give far too large a dose of radiotherapy and to too large a volume. This is why I said prophylactic radiation should not be given. Otherwise anything over 40 Gy (4000 rads) is unnecessary. You will run into trouble with chemotherapy if too big a dose level is used. We have heard from the Charing Cross group that you can get away with it, albeit you have to be more careful.

CLINICAL TRIALS

Chairman of Session
Mr B. Richards

THE NEED FOR CLINICAL TRIALS

P.H. Smith
Consultant Urological Surgeon,
St. James's University Hospital,
Leeds LS9 7TF

Some of the difficulties of clinical trials in patients with testicular tumour have already been highlighted at this conference. There will, however, continue to be problems for surgeons, radiotherapists and chemotherapists to resolve in this group of disease and, if progress is to be made, hypotheses have to be put forward and confirmed and perhaps re-evaluated by others. The results when published may be accepted, or questioned. Obviously clinical trials are not necessary if the treatment fails. They are also not necessary if the treatment cures. For instance one did not need a clinical trial in 1943 to prove that penicillin will cure lobar pneumonia. Trials are, however, essential in the rather more ill defined areas where treatment A may or may not be marginally better than treatment B.

In the chemotherapeutic field the first task is to look at new agents, in Phase I studies, given to patients with advanced disease. The object is to determine the dose and route by which the drug can be given and to find out whether the toxicity of the agent is acceptable. That is a task for a medical oncologist. Once it is found that an agent may be given with acceptable toxicity, its value in a given disease is then established by determining the incidence of remission in a Phase II study. Such a study requires at least one positive objective response in the first nineteen patients and at least three objective responses in 29 patients for the agent to be considered as potentially effective (1). Once a number of drugs have shown activity in Phase II studies, it becomes desirable to compare different agents or different regimes in what are called Phase III trials.

The accepted concept in a Phase III trial is that, in order to observe a 20% improvement in results of one treatment over another, and to have only a 5% chance of missing this observation, 93 evaluable patients are required in each arm of the study. In testicular cancer the problem of accumulating 200 evaluable patients with a given stage of testicular teratoma is a task of some magnitude. This difficulty also applies with adjuvant studies, which are a variant of Phase III trials, in which a comparison of A + B versus A alone is made. The field for international collaboration in these types of studies in patients with testicular tumours is open for exploration and is, in my view, ripe for exploitation as the best method to obtain the number of patients needed for such studies.

Before starting any trial, it is necessary to know the number of patients required; this has been referred to already in terms of Phase II and III studies. It is also necessary to define the percentage improvement which is thought to be acceptable or desirable. If only a 20% difference is looked for, many patients are needed (v.supra). If a 50% difference is the minimum required, the numbers fall to something between 50 and 75 in each arm. Definition of the acceptable risk of missing the difference is also necessary. After these points have been agreed, the next important criterion is the time to recruitment. Certainly in the EORTC, the American concept that interest falls off if the trial is not recruited within two years, has been found to be valid. The EORTC Urological Group has now completed several studies and the major input has come between 6 and 18 months after the start of the trial. It is necessary to establish very carefully the number of patients likely to be entered per year and the number of centres who will actively co-operate, to try to confirm that the trial will finish within two years. Lastly, the most careful statistical control is necessary from the very start of the design of any study and is also essential for any evaluation which is undertaken.

A Phase III chemotherapy study is equally concerned with disease free interval and acute toxicity, and with overall survival and chronic toxicity. Such trials often use combinations of agents, chosen from drugs which are active alone, which have differing actions, and which should have differing toxicities. The highest tolerable doses should be used.

Table 1 is a list showing the way in which EORTC protocols have to be written. It may be thought to be unnecessarily complex but all protocols have to be approved by a Protocol Review Committee whose membership is very broad. As a result, if a protocol of a study on testicular cancer is received for consideration by a man whose basic interest is in the leukaemias, he can, by looking at the introduction, see the justification for such a study and can

TABLE 1

EORTC STUDIES - CONTENTS OF THE PROTOCOLS

The following paragraphs should be included in all protocols to be reviewed:

1. Background and introduction
2. Objectives of the trial
3. Selection of patients
4. Design of the trial (including a schema)
5. Therapeutic regimens and toxicity
6. Required clinical evaluations, laboratory tests, and follow up.
7. Criteria of evaluation
8. Registration and randomisation of patients
9. Forms and procedures for collecting data
10. Statistical considerations
11. Administrative responsibilities
12. References

Appendices: Performance status scales
TNM or other classifications

assess its feasibility. The objectives obviously have to be defined, the patients selected to include or exclude certain criteria and the design of the trial determined. The treatment and its expected toxicity are clearly outlined. The investigations before entry to the study and at follow up are concisely stated. The evaluation criteria however, may prove more difficult and may at times have to be modified at the time of evaluation if the value of a predictive factor has by that time been questioned. Patients are registered and randomised by a telephone call or telex to the Data Centre and forms approved by the statistician and clinician must be filled in at the required intervals. One clinician is given the responsibility for co-ordinating the study. Finally there should be a list of references, so that a well written protocol is almost a scientific paper.

The problems of clinical trials are familiar to all medical oncologists and radiotherapists and argument rages. Can one use historical controls? Is it necessary to randomise? Why stratify? What are the other variables?

In considering historical controls, diseases and patients do change, as do the attitudes of doctors and the effectiveness of the general support, e.g. intensive care, which can be offered to the patient. In general, we become more competent, not through any intellectual superiority over previous generations of doctors, but because we have better arms with which to fight, thanks largely to the pharmaceutical industry.

Randomisation is a critical factor and I feel this must be done within centres, expecially in sizeable collaborative trials. There also has to be an even distribution of cases between therapies; in other words a set of random numbers has to repeat itself fairly frequently, perhaps every eight patients, if one is to avoid unnecessary complications. Stratification may also be vital particularly with regard to histological grade. It may also be necessary to stratify for other factors depending upon the nature of the trial.

Inevitably there will in addition be national and local variations, for example the management of bladder cancer in Spain or prostatic cancer in Germany will to some extent differ from that in this country. If multi-national studies are considered, this has to be acepted. There will also be differences in standards between institutions and even within institutions between different doctors. Surgical competence manifestly must differ and inevitably there will be certain unknown factors so that those who object to the idea of clinical trials can always legitimately question any conclusion that may be reached. However, the adoption of a prospective study, randomised within centres and suitably stratified, minimises the validity of any such objection and offers the best chance of affording a fair comparison of the effects of alternative forms of therapy.

In patients with testicular tumours the necessity for additional treatment following orchidectomy in those with Stage I disease, the superiority of retroperitoneal lymph node dissection over radiotherapy, and the role of maintenance chemotherapy after remission induction are some of the current problems which are not yet revealed and which in my view at least, demonstrate the need for clinical trials.

REFERENCES

1. Staquet, M., and Sylvester, R., 1977. A decision theory approach to Phase II clinical trials. Biomedicine, 26, 262-264.

THE DEVELOPMENT OF MODERN CHEMOTHERAPY FOR MALIGNANT TERATOMAS AND RESULTS OF SEQUENTIAL CHEMOTHERAPY AT CHARING CROSS HOSPITAL

E.S. Newlands, R.H.J. Begent, G.J.S. Rustin, K.D. Bagshawe
Department of Medical Oncology,
Charing Cross Hospital,
Fulham Palace Road, London W6 8RF

Initial responses to single agent therapy and combination chemotherapy in patients with metastatic malignant teratomas can be dramatic. Li et al in 1960 reported three out of 23 complete remissions with a drug combination of actinomycin-D, methotrexate and chlorambucil (4). By 1972 Smithers (11) was able to collect reports of 65 cases of documented complete remissions in patients with malignant teratoma but only 25 of these lasted more than two years and 11 more than five years. The problem in all centres at that stage was that while a high proportion of patients with malignant teratoma responded initially, the tumour became resistant to further cytotoxic chemotherapy, and the patients died rapidly from drug resistant disease. The next major development in cytotoxic chemotherapy were the reports by Samuels et al in 1975 and 1976, using high-dose vinblastine and bleomycin by infusion (9, 10). With this combination they obtained 22 out of 70 (31%) complete remissions. While this combination was a definite advance in therapy, it was at the cost of considerable morbidity and some mortality from the side effects of the chemotherapy. In 1974 Higby et al reported a Phase I study of cis-platinum which included several responses in patients with malignant teratoma (3). This new cytotoxic agent was rapidly introduced by Einhorn and Donohue who reported in 1977 that they had obtained 32 out of 50 (64%) complete remissions in patients with a combination of vinblastine, bleomycin and cis-platinum (1). By 1980 the same series showed that 28 (56%) of these patients were still in complete remission with a minimum follow up time of three years. Following this report, many other centres have used this combination of vinblastine, bleoymcin and cis-platinum and obtained remission rates of 24 out of 40 (60%) (12), and 64 out of 126 (51%) (8). However the very high initial response rate using vinblastine, bleomycin and cis-platinum tended to mask the fact that only a relatively low proportion of patients with large volume metastatic disease in the abdomen, lungs, liver or brain went into a sustained remission. Since the introduction of cis-platinum, a further new agent has been identified which is active in heavily pre-treated patients with metastatic teratoma, and this is VP 16-213 (2, 5). In 1977 it was felt at Charing Cross Hospital that the petential for cure in patients with large volume metastatic disease was potentially higher than had been obtained with vinblastine, bleomycin and cis-platinum. Therefore a study was set up using sequential chemotherapy, trying to optimise therapy using most of the available active cytotoxic drugs.

The counterpart of the introduction of new active cytotoxic agents has been the improved monitoring of malignant teratomas allowing accurate confirmation of complete remissions. At the Charing Cross Hospital patients with malignant teratomas have been routinely monitored with twice-weekly assays for human chorionic gonadotrophin (HCG) and alpha-fetoprotein (AFP) for many years. In a very high proportion of patients this provides the most accurate monitor of disease activity. The other monitoring procedure that has contributed considerably to the management of these tumours has been computerised tomography. The sites where computerised tomography has contributed particularly have been the brain, lungs, para-aortic region and, in some cases, the liver. This has allowed selection of patients for surgical removal of residual mass lesion(s) as a much more accurate procedure.

Between 1977 and 1st February 1981 64 male patients with metastatic malignant teratoma have been treated and 17 female patients with metastatic ovarian teratoma. The stage of the male patients at presentation was:

Stage I :	4 patients	(including Stage I(M) = 3 raised tumour markers as the only evidence of disease activity)
Stage II :	11 patients	
Stage III :	5 patients	
Stage IV :	44 patients	

The incidence of having raised tumour markers was:

Raised HCG alone :	13 patients
Raised AFP alone :	14 patients
Raised HCG and AFP :	34 patients
Non-marker producers :	3 patients

The treatment schedules and analysis of toxicity with these treatments have already been published (6) and were in part based on previous pilot studies. Patients were started on:

Treatment A

On Day 1, vincristine 1.0 mg/m^2 i.v., followed by methotrexate 100 mg/m^2 i.v. stat., and then methotrexate infusion of 200 mg/m^2 as a 12-hour infusion.

On Day 2, bleomycin 15 mg. was given as an i.v. infusion over 24 hours and folinic acid rescue started 24 hours after the start of the methotrexate as 15 mg. 12-hourly for 4 doses.

On Day 3, bleomycin infusion 15 mg. over 24 hours.

On Day 4, following forced diuresis with mannitol and hydration at the rate of 1 litre per hour for 3 hours, cis-platinum was given in a dose of 120 mg/m^2 by short i.v. infusion. The hydration was continued at the rate of 1 litre hourly for a further 3 hours.

Treatment B

VP 16-213 (etoposide) 100 mg/m^2 i.v. Days 1 to 5; actinomycin D 0.5 mg. i.v. Days 3, 4 and 5; cyclophosphamide 500 mg/m^2 i.v. Day 5.

Since the publication of our initial results (6), the combination treatment C with hydroxyurea, vinblastine and chlorambucil has been discontinued as a number of patients developed biochemical evidence of recrudescence of their disease during treatment with this combination.

Treatment D

Is identical to treatment A, with omission of the hydration and cis-platinum.

Originally patients were treated with two courses of treatment A, followed by treatment B, C, D, B, C and D etc., and if any evidence of drug resistance developed, that particular treatment schedule was omitted from subsequent management. One of the reasons for initially limiting patients to two courses of treatment A was the anxiety that patients might develop cumulative ototoxicity from the

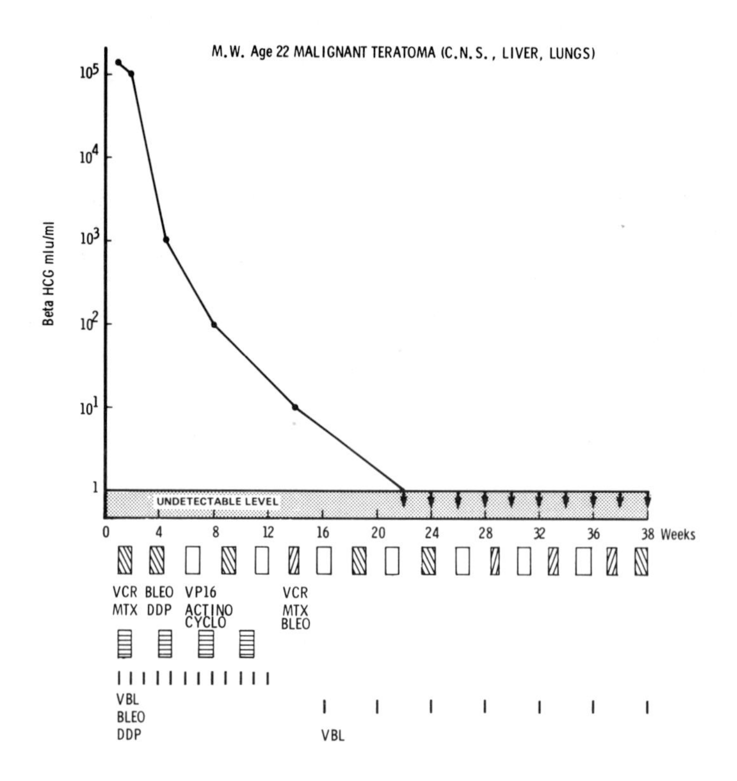

FIGURE 1 - Tumour marker response in a patient with Stage IV malignant teratoma (including brain metastases) comparing the schedule of Charing Cross Hospital and the Einhorn regime.

Abbreviations:

DDP - cis-platinum
VBL - vinblastine
VCR - vincristine
MTX - methotrexate
BLEO - bleomycin
VP16 - VP16-213 (etoposide)
AD - actinomycin-D
CYCLO - cyclophosphamide

high-dose cis-platinum. Although some patients develop high-frequency hearing loss on audiogram, this is rarely severe enough to become socially apparent. Therefore, patients with particularly large volume disease have been treated with a sequence of treatment A, A, B, A, B, A until the disease was in biochemical remission (according to HCG and AFP concentrations). Remission was maintained by alternating treatments B and D until the patient had been in complete remission for approximately 12 - 16 weeks. At this point, patients were restaged with CT scanning, and any residual mass lesion excised wherever this was possible. The pattern of chemotherapy and its duration as compared with the original schedule described by Einhorn and Donohue, 1977 (1), is shown in Figure 1. This patient had large volume liver and lung disease, together with a cerebral frontal lobe secondary. As can be seen, agressive chemotherapy is continued longer on the Charing Cross schedule until the patient is in complete biochemical remission, and if that remission remains stable no further chemotherapy is given. In this patient, the brain deposit cleared completely and the patient is now back at work and

TABLE 1

RELATIONSHIP OF INITIAL EXTENT OF DISEASE TO THE RESULTS OF CHEMOTHERAPY IN 64 MALE PATIENTS WITH MALIGNANT TERATOMA (1.2.81)

Extent of Disease	Number of Patients	Alive	Dead
Minimal pulmonary and abdominal disease (L_1, a, b)	21	19	2
Advanced pulmonary disease (L_3, a, b)	12	9	3
Advanced abdominal disease (c, L_1, L_2)	12	10	2
Advanced abdominal and pulmonary disease (c, L_3, H^+, CNS^+)	19	11	8

has been off treatment for nine months. We have given no maintenance chemotherapy, and so far this policy has been justified by the small number of patients relapsing after stopping treatment.

Table 1 shows the results by disease site and volume in the male patients. The staging classification used describes tumour extent, site(s), and volume (7) (Table 2).

TABLE 2

Stage I	Disease limited to testis. No evidence of metastases.
Stage II	Para-aortic node spread.
	A - metastases < 2 cm. diameter.
	B - metastases 2-5 cm. diameter.
	C - metastases > 5 cm. diameter.
Stage III	Supradiaphragmatic lymph node involvement.
	Abdominal status A, B and C, as above.
Stage IV	Extralymphatic metastases. L = lung.
	L1 - up to 3 metastases < 2 cm. diameter.
	L2 - > 3 metastases < 2 cm. diameter.
	L3 - metastases > 2 cm. diameter.
	H+ = liver involvement.
	CNS+ = central nervous involvement.

Where there was minimal metastatic spread, only two out of 21 patients have died, which contrasts with eight deaths out of 19 patients with very advanced pulmonary and abdominal disease and, in some cases, liver and brain disease as well. In this latter group there have been 10 cases with clear-cut metastatic liver disease, and of these, six are off treatment (range 3 - 24 months), two are on treatment, and two patients have died (one from a ruptured aorta and one from septicaemia). The causes of death in the 15 male patients who died are: resistant malignant teratoma -8, initial extent of disease -1, pulmonary embolism -1, ruptured aorta -1, septicaemia due to neutropaenia -2, pulmonary oedema while off treatment -1, encephalitis - unknown cause -1.

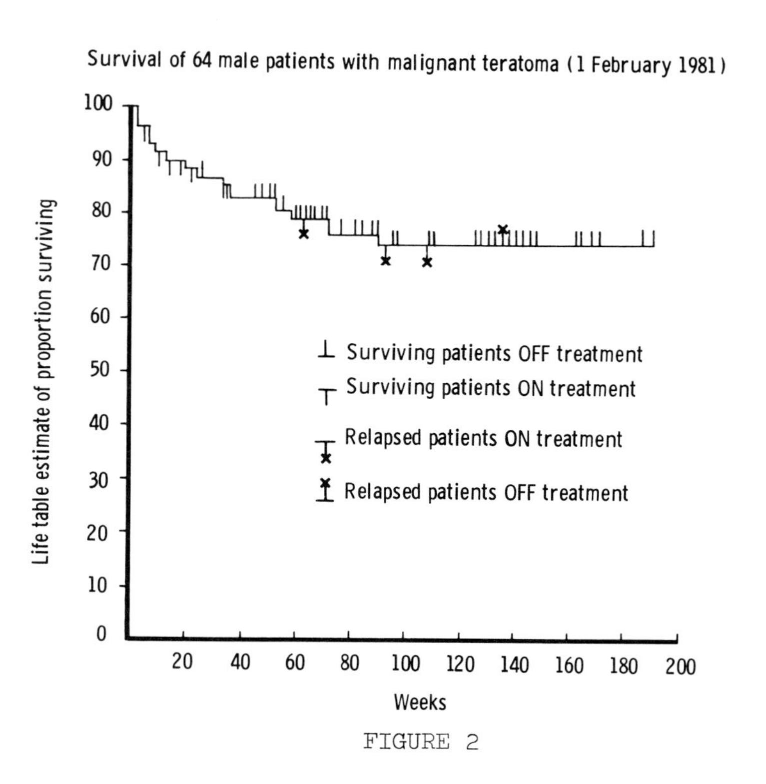
Survival of 64 male patients with malignant teratoma (1 February 1981)
Life table estimate of proportion surviving
100
90
80
70
60
50
40
30
20
10
0
20
40
60
80
100
120
140
160
180
200
Weeks
Surviving patients OFF treatment
Surviving patients ON treatment
Relapsed patients ON treatment
Relapsed patients OFF treatment

FIGURE 2

Up to 1st February 1981, life table analysis of the 64 male patients gives a survival of 74% (Figure 2). There have been five relapses off treatment: one patient has achieved a further complete remission and is off treatment; one has died; three are alive, with biochemical evidence of disease activity.

In the female patients, the results have been very comparable to the male series and, coincidentally, the life table to 1st February also indicates the survival of 74% in the 14 of 17 patients who are alive (13 are off treatment; range 2 - 40 months).

We feel there are several advantages in this sequential chemotherapy over the platinum, vinblastine and bleomycin regimes. Firstly, it would appear that a higher proportion of patients with very advanced disease obtain a complete remission. Secondly, the toxicity from the patient's point of view is less than the original high dose of vinblastine which was used in combination with bleomycin and cis-platinum. Thirdly, by combining a degree of flexibility in the chemotherapy with biochemical monitoring, some patients have obtained a complete remission who would otherwise have become resistant to a particular drug schedule. Fourthly, having obtained a complete remission, patients do not continue on maintenance chemotherapy which not only adds to the toxicity but also potentially to long-term complications of chemotherapy, such as the development of second tumours.

REFERENCES

1. Einhorn, L.H., Donohue, J., 1977. Cis-diamminedichloroplatinum, vinblastine and bleomycin combination chemotherapy in disseminated testicular cancer. Annals of Internal Medicine, 87, 293-298.

2. Fitzharris, B.M., Kaye, S.B., Saverymuttu, S., Newlands, E.S., Barrett, A., Peckham, M.J., McElwain, J.J., 1980. VP 16-213 as a single agent in advanced testicular tumours. European Journal of Cancer, 16, 1193-1197.

3. Higby, D.J., Wallace, H.J., Albert, D.J., Holland, J.F., 1974. Diamminedichloroplatinum: A phase I study showing resonse in testicular and other tumours. Cancer, 33, 1219.

4. Li, M.C., Whitmore, W.F., Golbey, R., Grabstald, H., 1960. Effect of combined drug therapy on metastatic cancer of the testis. Journal of thc Amcrican Medical Association, 174, 1291-1299.

5. Newlands, E.S., Bagshawe, K.D., 1977. Epipodophyllin derivative (VP 16-213) in malignant teratomas and choriocarcinomas. Lancet, ii, 87.

6. Newlands, E.S., Begent, R.H.J., Kaye, S.B., Rustin, G.J.S., Bagshawe, K.D., 1980. Chemotherapy in advanced malignant teratomas. British Journal of Cancer, 42, 378-384.

7. Peckham, M.J., McElwain, T.J., Barrett, A., Hendry, W.F., 1979. Combined management of malignant teratoma of the testis. Lancet, ii, 267-270.

8. Samson, M.K., Stephens, R.L., Rivkin, S., Opipari, M., Maloney, T., Groppe, C.W., and Fisher, R., 1979. Vinblastine, bleomycin and cis-diamminedichloroplatinum (II) in disseminated testicular cancer; preliminary report of a Southwest Oncology Group finding. Cancer Treatment Reports, 63, 1663-1667.

9. Samuels, M.L., Holoye, P.Y., and Johnson, D.E., 1975. Bleomycin combination chemotherapy in the management of testicular neoplasia. Cancer, 36, 318-326.

10. Samuels, M.L., Lanzotti, V.J., Holoye, P.Y., Boyle, L.E., Smith, T.L., and Johnson, D.E., 1976. Combination chemotherapy in germinal cell tumours. Cancer Treatment Reviews, 3, 185-204.

11. Smithers, D.W., 1972. Chemotherapy for metastatic teratomas of the testis. British Journal of Urology, 44, 217-228.

12. Stoter, G., Sleijfer, D.T., Bendrik, C.P.J., Schraffordt Koops, H., Struyvenberg, A., Van Oosterom, A.T., Brouwers, T.M., and Pinedo, H.M., 1980. Combination chemotherapy with cis-diamminedichloroplatinum, vinblastine and bleomycin in advanced testicular non-seminomas. Lancet, i, 941-945.

CURRENT AND FUTURE TRIALS IN TESTICULAR NON-SEMINOMATOUS CANCER

G. Stoter
Medical Oncologist,
Deparment of Oncology,
Free University of Amsterdam,
Amsterdam, The Netherlands.

Combination chemotherapy in advanced disease

Over the past few years, testicular non-seminomatous cancer has become a potentially curable disease. As a result of the development of highly active cisplatin containing remission induction regimes, complete remissions are achieved in 60-70% of the patients with a potential cure rate ranging from 45-60% (3, 4, 9, 16, 18, 19, 20, 22, 23).

However, toxicity accompanying these regimes is considerable, and 30-50% of the patients do not achieve complete remission or relapse subsequently. Therefore, the research priorities are:

1. The reduction of the toxicity of these regimes without loss of activity, and

2. the development of "salvage" chemotherapy for patients failing first-line chemotherapy or relapsing subsequently.

The reduction of the dose of vinblastine from 0.4 to 0.3 mg/kg in the PVB regime (cisplatin, vinblastine, bleomycin) by Einhorn (9) has led to a decrease from 35% to 15% of granulocytopaenic fever without decrease of activity.

In a current EORTC trial, 0.4 versus 0.3 mg/kg of vinblastine is also examined to see if the results of Einhorn can be confirmed.

Bosl (2) has further modified Einhorn's PVB regime, using cisplatin 60 mg/m^2 on day 2 instead of 20 mg/m^2, days 1-5. Of 16 patients treated, only 4 (25%) achived a complete remission. No decrease of adverse effects was noted. The low activity of this regime is due to the low dose of cisplatin rather than to the short course, since a short course is also given in the highly active VAB-III combination of cisplatin 120 mg/m^2 on day 8, vinblastine, actinomycin-D, bleomycin and cyclophosphamide.

The development of "salvage" chemotherapy depends mainly on new drugs. Conclusive data (5, 11, 17, 25) show that VP-16-213 is a very active drug in the treatment of testicular cancer, and more importantly constitutes an essential component of combination regimes for patients with advanced refractory germ cell tumours. With the use of VP-16-213: 100-120 mg/m^2 i.v. on days 1-5, every 2 to 4 weeks, remission rates of 35-46% have been achieved in a total number of 76 patients with tumours refractory to prior chemotherapy (5, 11, 17). Patients who initially respond to PVB chemotherapy but relapse subsequently, respond very favourably to cisplatin plus VP 16-213 containing re-induction regimes. At the Indiana University (5, 25) a remission rate of 82% (40% CR) was obtained in a series of 51 such patients with a median survival of 52 weeks. Haematologic and organ-system toxicity was considerable with one drug-related death. Of note, some of the complete responders are probably cured. It appears that VP 16-213 is an essential component of "salvage" chemotherapy regimes. Its role in first line combination chemotherapy of germinal neoplasms should be examined (4).

Maintenance chemotherapy in advanced disease

The value of maintenance chemotherapy is still undefined. Parallel to the situation in Hodgkin's disease, acute lymphoblastic leukaemia in children and gestational choriocarcinoma, several investigators feel that maintenance chemotherapy is unnecessary in the case of a well-documented complete remission because in such patients relapse hardly ever occurs. Patients who enter into complete remission as late as at the end of the scheduled therapy, may require longer induction treatment - e.g. 6 insteady of 4 PVB courses to prevent relapse (3). In ongoing EORTC trials, cisplatin plus vinblastine maintenance chemotherapy for one year is compared to no maintenance therapy in complete responders to PVB induction therapy.

Multimodality treatment in advanced disease

Cytoreductive surgery many considerably enhance the results of chemotherapy in patients who present with bulky metastases. However, improvements in combination chemotherapy are so impressive that surgery has become adjunctive to the primary treatment goal of curative combination chemotherapy. The timing of debulking surgery remains controversial. Large metastases at presentation may appear inoperable. Even if the volume can be reduced new metastases may appear or rapid growth may occur during the convalescent period. A "sandwich" approach of initial chemotherapy of short duration followed by surgery and chemotherapy carries the risk of a long interval between chemotherapy cycles and therefore may have the same disadvantages.

Mathisen and Javadpour (14) stress the fact that retroperitoneal tumour dissection often is extremely difficult and hazardous in patients with bulky abdominal metastases. Since these investigators are involved in an ongoing NCI study of initial chemotherapy followed by reductive surgery versus initial surgery followed by chemotherapy in patients with stage III testicular non-seminomas whose abdominal metastases are considered operable at the time of randomisation, they developed a technique of en-block resection of the inferior vena cava together with the tumour bulk to allow better access to the area behind the aorta in order to remove more of the tumour and to decrease the risk of major bleeding from injury to the vena cava or the aorta and also to decrease the risk of pulmonary embolism from clot or tumour located in the vena cava. However, analysis of the 6 cases presented, shows that macroscopic tumour remained in situ after the operation in 5 patients, that the patients suffered from transient or permanent oedema of the legs with one episode of venous thrombosis, that in one patient the aorta and right iliac artery had to be resected with replacement by a graft, and that one patient died of cachexia shortly after the operation while suffering from chemotherapy complications. These complications clearly indicate that initial surgery does not benefit the patient.

Lange (13) reports 8 cases with very rapid tumour growth after cytoreductive surgery. However, in all cases surgical intervention was carried out during the active growth phase of either chemotherapy refractory or newly diagnosed untreated tumour (5 and 3 patients, respectively). Very rapid tumour doubling is not uncommon in non-seminomatous cancer so that surgery can hardly be considered the cause of the rapid tumour growth. Besides, it is difficult to understand how an operation could alter cell-cycle kinetics.

When surgery is applied to patients after the completion of induction chemotherapy, fibrocystic elements are found in about one third of the resected specimens, one third will have mature teratomatous elements - still with malignant potential - and the remaining third will have persistent cancer including those with small malignant foci. Unless serum tumour markers are still positive at the time of surgery, there is no way to distinguish clinically between the three pathological categories. Since it is important to recognise the patients with persistent cancer and mature teratoma for the planning of further treatment, all patients with clinical or roentgenographic evidence of residual tumour at the completion of induction chemotherapy should be subjected to cytoreductive surgery.

Donohue (6) reports the Indiana experience with retroperitoneal lymph node dissection after remission induction chemotherapy in 26 patients, 23 of whom presented with bulky retroperitoneal metastases and still had evidence of residual abdominal metastases after 3-4 PVB cycles. Although in 23 patients the abdominal tumour bulk was considered inoperable at the beginning, radical resections could be carried out in 22 patients after induction chemotherapy. A second important observation was that 9 patients with mature teratoma and 8 fibrocystic patients fared well, with only 2 relapses in the mature teratoma group, in whom complete remission was again obtained after further surgery and chemotherapy. However, of 9 patients with persistent cancer, 3 died of cancer and 2 others were salvaged only after further cisplatin plus VP 16-213 chemotherapy. Thus, the most important prognostic factor appears to be the histology of the resected tissue (6, 16). Since at least half of the patients with so-called radically resected residual cancer will subsequently relapse, this subset of patients should have further remission induction treatment post-operatively. In this study 22 (85%) of 26 patients have no evidence of disease for a follow up period of 1-5 years post-operatively.

Bearing in mind the disadvantage of initial or interim cytoreductive surgery as mentioned earlier, and looking at the excellent results of the Indiana study, supported by observations of Skinner (15), the best approach to patients with well advanced disease seems to be initial remission induction chemotherapy followed by radical resection of residual tumour with additional induction chemotherapy for patients with persistent cancer in the resected specimen.

Management of stage I and II disease

Both retroperitoneal lymph node dissection and radiotherapy

have proved unsatisfactory, when used alone to cure patients with extensive stage II disease. Surgery is frequently difficult or impossible and persistent disease is common following radiotherapy (1). Besides, patients with extensive retroperitoneal disease run a high risk of developing haematogenous metastases (7, 12). Therefore, the best treatment for patients with extensive stage II disease seems to be primary remission induction chemotherapy to be followed by retroperitoneal lymph node dissection in those patients who have evidence of residual tumour after induction treatment (1, 6).

The situation is different for patients with resectable stage II disease. Einhorn (8) reports on 52 patients with resectable stage II disease, treated with orchidectomy and retroperitoneal node dissection. Eighteen of these patients (35%) developed metastases. All 18 achieved complete remission with prompt combination chemotherapy. One patient relapsed subsequently and died and one died of Klebsiella pneumonia 4 months after chemotherapy. Based on uncontrolled data, Samuels (21) and Skinner (24) advocate retroperitoneal lymph node dissection followed by adjuvant chemotherapy with vinblastine and bleomycin indicating a projected 5 year relapse-free survival over 80%. To resolve this question, a U.S. Intergroup Study (12) has been started to compare patients with resectable stage II treated with adjuvant chemotherapy to patients treated at first recurrence after retroperitoneal node dissection. Patients with pathological stage I and negative markers after orchidectomy will not be randomised, but will be registered and followed closely.

The disease-free survival in clinical stage I disease exceeds 90%, whether the modality of treatment is lymph node dissection or radiotherapy. The failure rate to detect regional lymph node metastases, using lymphangiography, CT-scan, and markers is approximately 20%. Apparently, about 80% of the patients with clinical stage I have their radiotherapy or node dissection unnecessarily. It is tempting to suggest a "wait and watch" policy after inguinal orchidectomy for these patients. However, the main problem here is the difficulty in evaluating the status of the retroperitoneal region. Ultrasonography seems the only practicable means for repeated screening.

The British Medical Research Council is in the process of developing a "wait and watch" study for clinical stage I testicular non-seminomas.

REFERENCES

1. Babaian, R.J., Johnson, D.E., 1980. Management of stages I and II non-seminomatous germ cell tumours of the testis. Cancer, 45, 1775-1781.

2. Bosl, G.J., Kwong, R., Lange, P.H., 1980. Vinblastine, intermittent bleomycin and single-dose cis-dichlorodiammine platinum (II) in the management of stage III testicular cancer. Cancer Treatment Reports, 64, 331-334.

3. Bosl, G.J., Lange, P.H., Fraley, E.E., 1980. Vinblastine, bleomycin and cis-diammine-dichloroplatinum in the treatment of advanced testicular carcinoma. Possible importance of longer induction and shorter maintenance schedules. American Journal of Medicine, 68, 492-496.

4. Cavalli, F., Monfardini, S., Pizzocaro, G., 1980. Report on the international workshop on staging and treatment of testicular cancer. European Journal of Cancer, 16, 1367-1372.

5. Dhafir, R.A., Einhorn, L.H., Rosenbaum, P., 1981. The effect of etoposide (VP-16), used alone or in combination on the survival of patients with refractory testicular cancer. Proceedings of the American Society for Clinical Oncology, 22, (in press).

6. Donohue, J.P., Einhorn, L.H., Williams, S.D., 1980. Cytoreductive surgery for metastatic testis cancer: considerations of timing and extent. Journal of Urology, 123, 876-880.

7. Edson, M., 1980. Testis cancer: the pendulum swings. Experience in 430 patients. Journal of Urology, 122, 763-765.

8. Einhorn, L.H., 1979. Proceedings of the 2nd International Conference on adjuvant therapy of cancer. (Ed.) Steven E. Jones. Grune/Stratton, New York.

9. Einhorn, L.H., Williams, S.D., 1980. Chemotherapy of disseminated testicular cancer. A random prospective study. Cancer, 46, 1339-1344.

10. Feun, L.G., Samson, M.K., Stephens, R.L., 1980. Vinblastine (VLB), bleomycin (Bleo), Cis-diamminedichloroplatinum (DDP) in disseminated extragonadal germ cell tumours. Cancer, 45, 2543-2549.

11. Fitzharris, B.M., Kaye, S.B., Saverymuttu, S., Newlands, E.S., Barrett, A., Peckham, M.J., McElwain, T.J., 1980. VP 16-213 as a single agent in advanced testicular tumours. European Journal of Cancer, 16, 1193-1197.

12. Jacobs, E.M., Muggia, F.M., 1980. Testicular Cancer: Risk factors and the role of adjuvant chemotherapy. Cancer, 45, 1782-1790.

13. Lange, P.H., Hekmat, K., Bosl, G.J., Kennedy, B.J., and Fraley, E., 1980. Accelerated growth of testicular cancer after cytoreductive surgery. Cancer, 45, 1498-1506.

14. Mathisen, D.J., Javadpour, N., 1980. En-bloc resection of inferior vena cava in cytoreductive surgery for bulky retroperitoneal metastatic testicular cancer. Urology, 16, 51-54.

15. McLorie, G.A., Skinner, D.G., 1980. Metastatic non-seminomatous testis tumours: morbidity of treatment. Journal of Urology, 124, 479-481.

16. Neidhart, J.A., Memo, R., Metz, E.N., 1980. Probable cure of metastatic testicular tumours treated with sequential therapy. Cancer Treatment Reports, 64, 553-558.

17. Newlands, E.S., Bagshawe, K.D., 1980. Anti-tumour activity of the epipodophyllin derivative VP 16-213 (etoposide: NSC-141540) in gestational choriocarcinoma. European Journal of Cancer, 16, 401-405.

18. Newlands, E.S., Begent, R.H.J., Kaye, S.B., Rustin, G.J.S., Bagshawe, K.D., 1980. Chemotherapy of advanced malignant teratomas. British Journal of Cancer, 42, 378-384.

19. Oliver, R.T.D., Ama Rohatiner, A., Wrigley, P.F.M., and Malpas, J.S., 1980. Chemotherapy of metastatic testicular tumours. British Journal of Urology, 52, 34-37.

20. Samson, M.K., Fisher, R., Stephens, R.L., 1980. Vinblastine, Bleomycin and cis-diamminedichloroplatinum in disseminated testicular cancer: Response to treatment and prognostic correlations. European Journal of Cancer, 16, 1359-1366.

21. Samuels, M.L., Johnson, D.E., 1980. Adjuvant therapy of testis cancer: The role of vinblastine and bleomycin. Journal of Urology, 124, 369-371.

22. Scheulen, M.E., Higi, M., Schilcher, R.B., 1980. Sequentiell alternierende Chemotherapie nicht seminomatoser Hodentumoren mit Velbe, Bleomycin und Adriamycin/Cisplatin. Klinische Wochenschrift, 58, 811-821.

23. Scheulen, M.E., Seeber, S., Schilcher, R.B., 1980. Sequential combination chemotherapy with vinblastine, bleomycin and doxorubicin, cis-dichlorodiammine platinum (II) in disseminated non-seminomatous testicular cancer. Cancer Treatment Reports, 64, 599-609.

24. Skinner, D.G., Scardino, P.T., 1980. Relevance of biochemical tumour markers and lymphadenectomy in management of non-seminomatous testis tumours: Current persepctive. Journal of Urology, 123, 378-382.

25. Williams, S.D., Einhorn, L.H., Greco, F.A., Oldham, R., and Fletcher, R., 1980. VP 16-213 salvage therapy for refractory germinal neoplasm. Cancer, 46, 2154-2158.

CIS-PLATINUM, VINBLASTINE AND BLEOMYCIN IN ADVANCED TERATOMA

J.A. Green, C.J. Williams and J.M.A. Whitehouse
Medical Oncology Unit,
Centre Block CF 99,
Southampton General Hospital,
Tremona Road,
Southampton S09 4XY

Twenty-one patients with advanced malignant teratoma were treated with Cis - DDP 20 mg/m^2 by intravenous injection (I.V.) on days 1 to 5, Vinblastine 0.15 mg/kg I.V. on days 1 and 2, and Bleomycin 30 mgms. I.V. on days 2, 9 and 16. The courses were repeated every three weeks to a total of four cycles. Three patients also received VP16 - 213 at a dose of 100 mgm/m^2 I.V. for 5 days.

Fifteen of the tumours were of testicular origin, 2 were primary retroperitoneal tumours, 3 from the mediastinum and one was of ovarian origin. Seven cases had MTU histology, 12 were MTI and 2 contained trophoblastic elements. Serum HCG alone was elevated in one case, AFP alone in seven, and both markers were raised in 10 cases. The mean age of the series was 28.6 years (range 13 - 47). Two patients had received prior para-aortic node irradiation and two further cases had received single agent chemotherapy. The drugs used were Vinblastine, Bleomycin and Adriamycin.

Results

Eleven (52%) patients achieved a complete remission (C.R.) (documented by CAT scan in all but 3 cases) after 4 cycles. Three patients who initially showed a partial remission (P.R.) were converted to C.R. status after two further cycles of Cis - DDP and VP16 - 213, or Cis - DDP and Adriamycin. The C.R. rate, therefore, is 14/21 (67%), or 73% if only the 15 patients with testicular tumour are considered.

The six patients who achieved P.R. and the one case of stable disease have all died (Figure 1). Three of the patients in C.R. relapsed at 2, 3 and 4 months after the completion of chemotherapy, and have subsequently died.

The prinicpal causes of treatment failure in 10 cases

were as follows:

No response to chemotherapy	1 case
Failure to achieve a C.R.	5 cases
Relapse after C.R.	3 cases
Bleomycin lung	1 case

Survival and time to progression deteriorated with advancing stage. Only two of the six patients with abdominal nodes more than 5 cm in diameter are alive. Only one of the two with large pulmonary metastases and none of those with hepatic disease are surviving. The site of relapse was that of the predominant initial disease in every case. Adverse prognostic factors identified are:- bulky chest or abdominal disease, undifferentiated histology, high serum marker levels, hepatic involvement, and prior chemotherapy.

The principal conclusion from this study is that 4 cycles of PVB, while effective in low volume disease, is inadequate therapy for those patients with bulky teratomas.

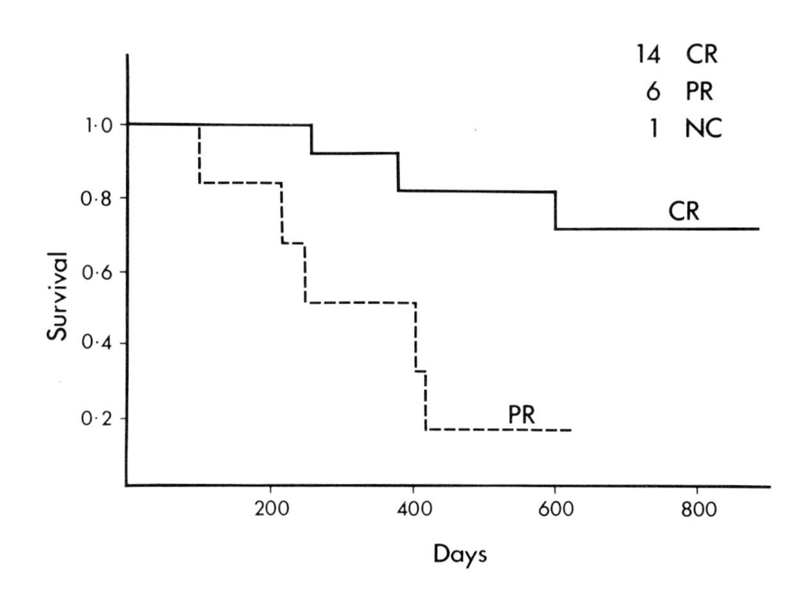

FIGURE 1 - Survival in relation to remission status

THE TREATMENT OF TESTICULAR TERATOMA BY A MODIFICATION OF EINHORN'S SCHEDULE

K.W. James, J.S. Scoble, R.A. Belcher, and D.K. Davies
Velindre Hospital,
Cardiff,
and University College,
Swansea, Wales.

From late 1978 to December 1980, 25 patients with metastatic testicular teratoma were treated as follows:-

Day 1: Vinblastine 5mg/m^2 i.v., Cis-platin 100mg/m^2 i.v. (with hydration and mannitol-diuresis).
Day 2: Vinblastine 5mg/m^2 i.v., Bleomycin 30mg. i.v.
Day 8: Bleomycin 30mg, in 1% Lignocaine, i.m.
Day 15: Bleomycin 30mg, in 1% Lignocaine, i.m.

Four complete courses, each starting on day 22 of the previous course, were given. Hydrocortisone was given with the Bleomycin to minimise any reaction. Toxicity included various degrees of constipation, skin pigmentation, hyperkeratosis, alopecia, nausea and vomiting in most patients. When a patient with a serum creatinine of 130μ mol/L was given 50% of the Cis-platin dose on his third course, the creatinine level rose to 185μ mol/L but has since returned to normal. There was one case of type-I hypersensitivity to Cis-platin. The mean WBC nadir was 2.0 x 10^9 /L and the mean vinblastine dose 10.1mg/m^2 per course (range 4.8 to 13.9; standard deviation 3.2). Over this range there was no significant correlation between Vinblastine dose and nadir. There was one case of synergistic gangrene of the ischio-rectal fossa and 3 cases of septicaemia (one fatal case being admitted to another hospital and dying within four hours, despite having been reasonably well with a WBC of 2.1 x 10^9 /L in our clinic two days previously).

The results are shown in Table 1. Fourteen of 17 evaluable cases (82%) are alive and in complete clinical and biochemical remission at a median follow up time of 14 months (range 4 to 29). Five cases have had surgery following treatment, one had bilateral thoracotomies, and there have been no anaesthetic deaths (1). The histology in two cases was reported as necrosis and fibrosis and differentiated teratoma in two others. In one case the polycystic mass showed residual tumour of intermediate

TABLE I:

STATUS OF PATIENTS ON MARCH 1st. 1981

Initial stage (all irrespective of marker levels)	No. Cases	Cases★ Eval.	CR to Chemo.	Further treatment	Relapses & deaths (months)	Survival in Months
Pulmonary disease I. Up to 5 mets. in each lung with the largest diam. of any single lesion < 2cm.	2	1	1	0	0	6
Pulmonary disease II. More than 5 masses in a lung with the largest diam. of any single lesion < 2cm.	1	1	1	0	0	16(A)
Pulmonary disease III. Mediastinal or hilar mass, pleural effusion or intrapulmonary mass>2cm.	6	6	3	2(S)	1(B)	6 12 14 17 29
Advanced abdominal disease. Palpable abdo. mass: ureteral displacement or obstructive uropathy.	5	3	0	2(R) 1(S)	0	25,29 4
Para-aortic node deposits>2cm. in diameter.	6	4	1	2(S)	2(C)	7 21
Lymphogram +ve cases with no deposit >2cm. in diameter.	5	2	1(D)	1(R)	0	12 14
All cases treated by chemotherapy.	25	17	7	8	3	median 14 Months

★ The 8 non-evaluable cases comprise: 3 still undergoing chemotherapy, 3 awaiting decisions regarding surgery or radiotherapy and 2 awaiting surgery.

(R) Post-chemotherapy radiotherapy.
(S) Post-chemotherapy surgery.
(A) Patient refused further treatment after two courses of chemotherapy.
(B) Patient did not have complete remission and died 16 months with cerebral metastases.
(C) One patient died with Klebsiella septicaemia after first course of treatment.
One patient relapsed at 9 months and is receiving further chemotherapy.
(D) Patient refused surgery after chemotherapy.

differentiation.

Computerisation of the records of these patients with a graphical display of tumour burden and marker levels against time was felt to be of value in the monitoring of treatment. Microcomputers with the ability to display graphics are now available. We have used the Motorola Exorset (kindly donated by the Aberdare Cancer Education Committee) which has two disc-drives for data storage (a total of 160 K-bytes of information). Programs have been written and tested on retrospective data. Future plans include the development of programs to allow the calculation and display of statistical data of patients in current studies and the easy selection of such displays from a menu on the screen by the clinician using a light-pen.

REFERENCES

1. Allen, S.C., Riddell, G.S., and Butchart, E.G., 1981. Bleomycin therapy and anaesthesia, Anaesthesia 36, 60-63.

BLEOMYCIN LUNG AND PEROPERATIVE OXYGEN TENSION IN TERATOMA PATIENTS

E.D. Rubery and M.J. Lindop
Department of Clinical Biochemistry,
University of Cambridge,
Addenbrooke's Hospital,
Hills Road,
Cambridge CB2 2QR

Goldiner et al (1) suggested that a high peroperative inspired oxygen concentration led to fatal post-operative pulmonary fibrosis (PPF) developing in patients previously given high-dose Bleomycin therapy.

We report two patients who underwent successful debulking surgery without any pulmonary sequelae in spite of per-operative inspired oxygen concentrations of 33%. When these patients are compared with the Goldiner series (Table 1) several other differences are apparent:- 1) Average duration of surgery in our cases was 1½ hours as opposed to 6 hours in the American series. 2) Bleomycin therapy ceased 6 months prior to surgery in Goldiner's cases but only a few weeks prior to surgery in our cases. 3) We were much less vigorous with post operative fluid replacement therapy. 4) Pre-operative respiratory function appeared much more depressed in Goldiner's patients than in our patients, in spite of an average lower total Bleomycin dose in their patients. Other reports in the literature (2, 3) also conflict with Goldiner et at (1).

Because of these inconsistencies we injected 2 groups of eighty (25gm) mice with 0.5mg or 0.05mg Bleomycin twice a week for 8 weeks. When 40 of each group were exposed to 60% oxygen for 24-42 hours there was no difference in survival or lung pathology in either group when compared to the air-breathing controls. The high Bleomycin group all exhibited weight loss and 15% died suggesting the dose of Bleomycin was approaching toxic levels.

Further experiments are in progress using longer exposures to oxygen, but these preliminary experiments do not suggest it is easy to demonstrate Bleomycin potentiation of oxygen toxicity in mice.

In conclusion, Goldiner's attractive hypothesis that oxygen is a dangerous drug for a patient who has received Bleomycin must remain unproven. The relevance of factors

TABLE ONE

SUMMARY OF CLINICAL, OPERATIVE AND ANAESTHETIC DETAILS OF PATIENTS

Patient	Age (yrs.)	Previous Lung Disease	Total Bleo. Dose (mg.)	Bleo. Surgery Interval in days	Pre-op FVC (%)	Pre-op trans-fer factor	Opera-tion *	Duration of operation (hours)	Anaesthetic Details †	Blood Loss	Perop-erative fluids	Perop-erative oxygen %	Post-operative Course
R.G.	30	Resolved lung met-astases	750	12	91	84	RND	1.5	Trichlorethy-lene	550 ml. Transfused 2 pints post-operatively	1 litre crystal-loid	33	Uneventful Home 7th postopera-tive day
B.R.C.	23	? early lung fibrosis (ref. 4)	600	50	99	92	RND	1.5	Fentanyl Droperidol	600 ml.	1 litre crystal-loid	33	Uneventful Home 5th postopera-tive day
Goldiner et al. 5 pts.	31 average	1 pt.	426 average	288 average	80.5 average	58.5 average	4 RND 1 PM	5.86 average	Opiates	Not Stated	5.86 1 crystal-loid 2.38 1 Colloid	39	All died of respiratory distress Path pulmon-ary fibrosis
Goldiner et al. 12 pts.	27.6 average	3 pts.	599 average	272 average	83.0 average	61.7 average	10 RND 2 PM	5.74 average	Opiates	Not Stated	3.54 1 crystal-loid 3.13 1 Colloid	24	Uneventful Home 12-25 days post-operation

* RND = Radical lymph node dissection
PM = Removal of pulmonary metastases

† All patients had positive pressure ventilation with Nitrous oxide and Oxygen; thiopentone and a muscle relaxant.

such as fluid overload and other anaesthetic drugs such as muscle relaxants and opiates perhaps also warrant investigation.

With a cured population of young men now completing Bleomycin therapy the factors involved in the subsequent development of pulmonary fibrosis urgently need to be clarified.

REFERENCES

1. Goldiner, P.L., Carlon, G.C., Cvitkovic, E., Schweizer, O., and Howland, W.S., 1978. Factors influencing post-operative morbidity and mortality in patients treated with bleomycin. British Medical Journal, i, 1664-1667.

2. Nugaard, K., Smith-Erichsen, N., Hatlevoll, R., Refsum, S.B., 1978. Pulmonary complications after Bleomycin, irradiation and surgery for esophageal cancer. Cancer, 41, 17-22.

3. Douglas, M.J., and Coppin, C.M.L., 1980. Bleomycin and subsequent anaesthesia. A retrospective study at Vancouver General Hospital, Canadian Anaesthetists' Society Journal, 27, 449-452.

4. Rubery, E.D., and Caokley, A.J., 1980. Early detection of lung toxicity after bleomycin therapy. Cancer Treatment Reports, 64, 732-734.

KNOWING THE INITIAL DIAGNOSIS

C.D. Collins
Consultant Radiotherapist,
Lambeth Palace Road,
St. Thomas' Hospital,
London SE1

With all the help available from various scientific tests it is not always possible to predict the prognosis from the initial diagnosis. To illustrate this, two case histories are presented.

The first patient, aged 37, was admitted with a five month history of pain in the right testis. A tumour mass was confirmed and pre-operatively a chest X-ray showed multiple metastases. A right orchidectomy was performed and the histology reported as an undifferentiated malignant teratoma. The patient refused all further treatment or investigations, having, it is believed, seen his chest X-ray. He returned to hospital six weeks later when his gynaecomastia was settling. The serum HCG was now 1880 and there was insufficient blood to do an AFP, but this had been previously raised. His chest X-ray showed the metastases to be smaller. Two weeks later his HCG was 884 and AFP 4. One month later his HCG was rising again (938). He died eight months later after palliative treatment with radiotherapy and chemotherapy. He had massive recurrences in the chest and multiple metastases. His HCG had risen to 11,400. At no time would his General Practitioner or his very dominating mother allow us to tell the patient the diagnosis.

The second patient, aged 35, presented with a lump in the testis of two weeks duration. An orchidectomy was performed. The histology was reported as seminoma with spread to the rete and epididymis together with vascular invasion. All tests including CT scanning and markers (both pre-and post-op) were normal. The patient and his relatives were told the full diagnosis and given the information that he had a reasonable chance of being cured. Routine post operative radiotherapy was given. Six months later a chest X-ray showed a probable metastasis in the right lung, but later chest X-ray showed what could have been spontaneous regression of the metastasis. A few months later a definite mass appeared behind the heart. This was biopsied and said to be consistent with a seminoma. The original histology was reviewed, confirmed as a seminoma but with cutting of further sections possibly of one or two small areas of teratoma were seen. The markers were still normal. The tumour responded to radiotherapy. The subsequent progress was that of dealing with a patient with a mixed tumour and a raised AFP. The AFP started to rise one year after surgery. Some deposits responded well to local radiotherapy and the disease was confined to the right chest for many months. He finally died on his birthday two years later as a result of cerebral metastases, with persistent disease in his right chest and and AFP of 84702.

There are points of interest from these two cases. There was apparent spontaneous regression of the lung metastases and resolution of gynaecomastia after orchidectomy in the first case. We also had great difficulty with communication with this patient. In the second case, the disease evolved although the initial diagnosis was a Stage I seminoma. It took over one year before the AFP began to rise. There was a possible spontaneous regression of a lung metastasis. The lung metastases were confined to the right side of the chest and most of them responded well to radiotherapy.

DISCUSSION

Mr. Richards

May I ask Mr. Smith to comment about the difficulty that arises out of the numbers required for clinical trials and conclusions being drawn from very small series. Do we have to get cases from a very wide area for a very long time if we are going to get the required numbers into any series with conditions as rare as non-seminomatous germ cell tumours?

Mr. Smith

There are only 43 new testicular tumour patients per 4½ million population in Yorkshire per annum, of which about half will be non-seminomas. If I then extrapolate and assume that there will be not more than 600 per year of all categories in the U.K., then to run a Phase III study in any sub-category is a task, even in the whole country, which could hardly be managed in one or two years. Now whether that means inevitably one must look beyond national borders is something for the MRC testicular tumour panel to consider.

Mr. Richards

I am sure this is going to become an important feature of discussion of any trial in this disease.

Continuing in this vein, could I ask Dr. Oliver to comment about the MRC Stage I study?

Dr. Oliver

I cannot speak for the global view of the Committee, but I can comment on some of the discussions which in a sense have also been made in previous presentations at this Conference. The current problem of the Stage I patient is that in a situation where there is 80 - 95% cure in that category, the numbers required are going to be in the region of thousands for any randomised trial of the surveillance versus elective therapy. For that reason the MRC has not gone straight to the randomised trial, but have simply suggested that a register be developed to document information.

Mr. Richards

The introduction of VP16 - 213 has obviously been a very important step forward. I was very impressed with the figures Dr. Ozols presented. You said something which to a urologist seemed a remarkable statement, that you pressed on with giving relatively high doses of PVB, despite compromised renal function. Could you comment on your toxicity figures?

Dr. Ozols

We have treated 11 patients, 5 of them were previously untreated. I think one point that should be made is that this type of treatment is toxic and it should be stressed that it should be aimed at only that group of patients with poor prognostic signs. I think PVB as standard treatment is very effective for minimal disease and even that perhaps is too toxic for patients with minimal disease. That prompts a whole new avenue of investigation to modify the toxicity for patients with small volume disease. For patients with bulky disease this type of therapy with new chemotherapy approaches is certainly warranted and there will be more toxicity, there is no question. With the way we have been giving VP16 at 100mg/m^2 for 5 days, we have seen dramatic white cell count nadirs of between 0.5 x 10^9 /l and 0.2 x 10^9 /l occurring at about 10 - 12 days. All patients have regained adequate counts for continued therapy even though there is no dose modification. In fact we are escalating the VP16 in those patients who do not drop their white count below 10 x 10^9 /l. Our policy is to treat them, send them home or to the motel and if they develop a fever they come back to hospital and are treated as any leucopaenic patient and put on antibiotics. I would suspect that most of the patients at some time during their first three courses will have some episode of leucopaenic fever.

As regards renal disease, a lot of these patients have very bulky abdominal disease. We have CAT scans and renograms demonstrating in three of them complete obstruction and non functioning kidneys. It is our philosophy for bone marrow involvement, and disease at other sites that, if the tumour is causing problems, there is no use in dose modifying because the main thing is to attack the tumour. So we continue to hydrate these patients very vigorously with 5 L saline per day, and then go ahead and treat them with full doses. In those patients who had complete obstruction of the kidney documented by flow studies, after the first course of therapy their kidneys are completely opened up, so if the obstruction is due to tumour the name of the game is to get rid of the tumour and the kidneys will start working again.

Dr. Newlands

To comment on Dr. Rubery's presentation, we have operated on over 50 patients and have never had any problems. We do not recommend that the anaesthetists change their practice at all. The Bleomycin we give is all by infusion and is not a terribly high dose whereas most American work is at fairly high pulse dose, and this may be important in terms of sensitising the lung.

Dr. Rubery

One of our patients had an infusion, but the other had high dose pulses of Bleomycin. It is a possibility that there may be a difference because of the method of administration.

Dr. Stoter

In a Dutch series of 90 patients we operated on more than 40 patients after 4 cycles of PVB. In the first 20 patients we were not aware of the paper by Goldiner et al, so we did not pay any attention to the oxygen tension. In the first half of the series we gave Bleomycin by rapid intravenous injection. Afterwards we gave an infusion over half an hour to decrease peak serum levels. We did see Bleomycin toxicity but it was not related to oxygen. The three patients who died of Bleomycin toxicity did so after their chemotherapy, not following anaesthesia.

Mr. Richards

May I raise the question of maintenance chemotherapy after complete remission? There is no evidence that I have seen that it is of value. What is the view about maintenance chemotherapy after complete remission?

Dr. Newlands

There is no evidence at all, it should be induction, complete remission and stop. One thing that does worry me is that second tumour induction does seem to be related not only to alkylating agents (which are not part of most maintenance regimes) but also to the duration of use of these drugs. Short term chemotherapy seems relatively safe, in certain Hodgkin's patients who have had no radiotherapy and in gestational choriocarcinoma, second tumours are extremely rare after fairly short lasting chemotherapy.

Dr. Ozols

There has been a controlled trial of maintenance therapy by the South West Oncology Group. Patients were treated with PVB and those patients in complete remission were randomised to a year of Vinblastine maintenance or no maintenance. There was absolutely no difference in survival or relapse rates. So I agree that there is no role for maintenance chemotherpay in this disease.

Dr. Oliver

I would welcome a discussion with respect to the Charing Cross approach of many different drugs. They are using a number of less effective drugs in addition to the what would probably be considered the most effective, i.e. platinum and VP16, and the duration of treatment on average goes out to about 6 months. With the escalating dose of platinum and VP which has been mentioned by Dr. Ozols one may well see advantage, from the patient's point of view, of treatment that can be over and done with in three months as opposed to the more continuous treatment.

Dr. Newlands

In reply to your comments about the drugs we are primarily using, Vincristine and Methotrexate have never been adequately evaluated although there is certainly a lot of anecdotal evidence. We have certainly seen responses to single agents and we are in no doubt about the activity of Methotrexate although it has not been adequately published. All the other agents are known to be active, Bleomycin, Actinomycin-D, Cyclophosphamide, VP16 and Platinum. I think that we probably could shorten the chemotherapy in some patients, and certainly the patients with small volume disease get less chemotherapy.

We are aiming for what we regard as a stable remission. If you examine the survival curve from the South West Oncology Group for treatment with PVB, it is a waste of time for those with partial remissions. This is one of the reasons why people using fairly short PVB therapy in patients with fairly large volume disease are seeing relapses. The problem is not one of getting a response, the response rate on our chemotherapy is 100%, it is a matter of complete remissions.

Dr. Oliver

I agree the duration of continuation of treatment beyond the point of disappearance of disease is critical, rather than simply having a straight four courses of Einhorn. Your combination has VP16 in it and I am not sure that we will be able to see any difference between just adding VP16 to Einhorn and your treatment because of all the other drugs which you have in your combination.

Mr. Richards

That raises a very fundamental point since you say you will not be able to see the difference. The difficulty is that different research groups start off with different raw material, and they produce results obtained on that raw material with that particular method of treatment. Unless one does a controlled study you will never find out which is better. For example how are you going to demonstrate that your treatment is better than the Einhorn regime plus VP16?

Dr. Newlands

This of course does need a controlled trial, but I think there are two sides to this as Philip Smith was arguing earlier about controlled trials. This is fine if you have a clearly defined condition, but we are dealing with a very heterogeneous group of patients, with rare tumours, and stratification of patients will be necessary in any trial. While we have been treating our patients, our strategies have changed because of the introduction of newer techniques such as CT scanning which has prompted us to do more surgery with a greater confidence in asking the surgeon to go to the right place. But I agree with you that there comes a time once it has been defined what a single institution study can do with a bigger combination and they have refined it, then it is relevant to look at that combination on a wider scale to see whether the results can be confirmed on a wider basis. It does seem that other institutions can get complete remissions using this chemotherapy, the two ovarian teratomas in the group were treated at the Royal Marsden Hospital with complete remission and there are five complete remitters not included in the series who were treated at Guy's.

Dr. C. Williams

To be realistic, I am a great supporter of clinical trials and think they are very important but I worry about the number of patients that are available. We are going to break the patients down into:-

(a) Stage I, which we may simply want to follow

(b) Minimal disease which in some centres may go to radiotherapy or to surgery with or without chemotherapy afterwards

(c) Advanced sub-group I, which may have a moderate chemotherapy and

(d) Advanced sub-group II, which may have more aggressive chemotherapy regimes.

We have also seen some studies from the United States which are looking at adjuvant chemotherapy in earlier disease states of which only perhaps 15 - 20% of the patients will relapse. If we are talking about relatively small numbers nationwide or even in continents, are we actually going to be able to break down all these various sub-groups and take into account that we already have survival which even in advanced disease is 50%? Are we going to be able to detect changes which may be of the order of 10% at the most between treatments? It seems to me that we may need trials with 500 - 1000 patients in each arm, and I cannot see that to be possible. Would anyone like to respond as to whether we can actually run studies that we want to and will we have enough patients for all the various sub-groups?

Mr. Richards

That has been the EORTC experience in these diseases. There are some groups in which one can get sufficient cases and some which one cannot.

Dr. Stoter

This is a matter for concern of course. Nevertheless, we have succeeded in the EORTC to run an induction chemotherapy study comparing the 0.4 and 0.3 mg /Kg Vinblastine dose, which has recruited about 100 patients in 1½ years. These patients are predominantly coming from Yorkshire, Glasgow and Holland. So I think there are opportunities to do such studies. Another consideration is that if you start off such a study hoping to find a 20% difference in the results, and you collect 300 evaluable cases in which you find less than that proportion of difference, I think you are morally justified to close the study and say that there is no significant difference and go on to something of more interest in this field of research.

Dr. Von Eyben

I have a question about the EORTC study. You said that Einhorn had randomised and found no difference between the two drug levels of Vinblastine in the treatment with PVB. Why did you choose to do the same study in Europe when there are many possible studies in this area?

Dr. Stoter

I find the presentations from Einhorn are always very understandable. He claims to find a significant decrease of toxicity but on this one occasion I had the feeling and I still have reservations about his proof that is as effective. That is why we wanted to confirm or deny it. As things are, 3 years after the first steps in developing these studies, I think there may now be more interesting areas of research, for example I would much more like to compare PVB versus VP16/Bleomycin, and that is what you mean I think.

FERTILITY OF TREATED PATIENTS

Chairman of Session
Professor J. Blandy

FERTILITY OF PATIENTS WITH TESTICULAR TUMOURS AND THE EFFECTS OF TREATMENT A PRELIMINARY REPORT

A. Barrett, J. Stedronska*, W.F. Hendry, M.J. Peckham
The Institute of Cancer Research and
The Royal Marsden Hospital,
London and Surrey.

*Chelsea Hospital for Women,
Dovehouse Street,
London SW3

All patients presenting to the Teratoma Unit at the Royal Marsden Hospital from December 1978 to December 1980 following orchidectomy for a malignant teratoma have been offered seminal analysis with the possibility of sperm conservation if appropriate. A profile of hormonal measurements have also been performed in all patients the results of which will be reported in detail elsewhere.

Of 166 patients seen, 75 only requested a sperm count. A considerable proportion of patients not wishing to be tested had completed their families.

16% of patients were found to be azo-ospermic after orchidectomy, 53% were oligospermic (<20 million/ml) and only 31% had counts in the normal range. The distribution of counts is shown in Figure 1. Impaired motility was also a frequent finding (Figure 2). Because of low counts and impaired motility only 25% of patients had sperm considered suitable for banking.

Post-orchidectomy studies in patients who had had children before orchidectomy showed azo-ospermia in 1, oligo-spermia in 8 and normal counts in 6.

Details of sperm counts in patients referred for treatment of relapse after radiotherapy are shown in Table 1 and indicate that radiotherapy does not inevitably lead to impaired fertility.

So far only a limited amount of information is available for men treated with chemotherapy with either vinblastine and bleomycin, or vinblastine, bleomycin and cis-platinum. Fourteen patients have been studied at varying intervals up to 2 years after completion of treatment.

Five whose counts were normal before treatment maintained normal counts, but 9 with previously normal counts were azo-ospermic when tested after chemotherapy. This

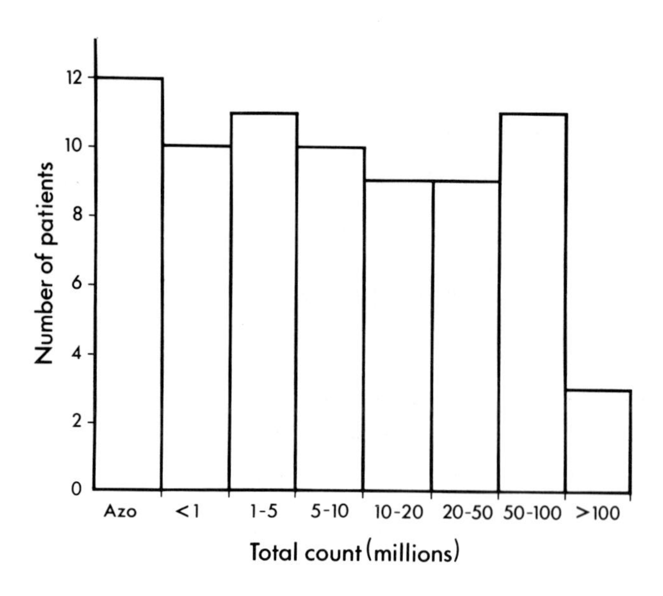

FIGURE 1

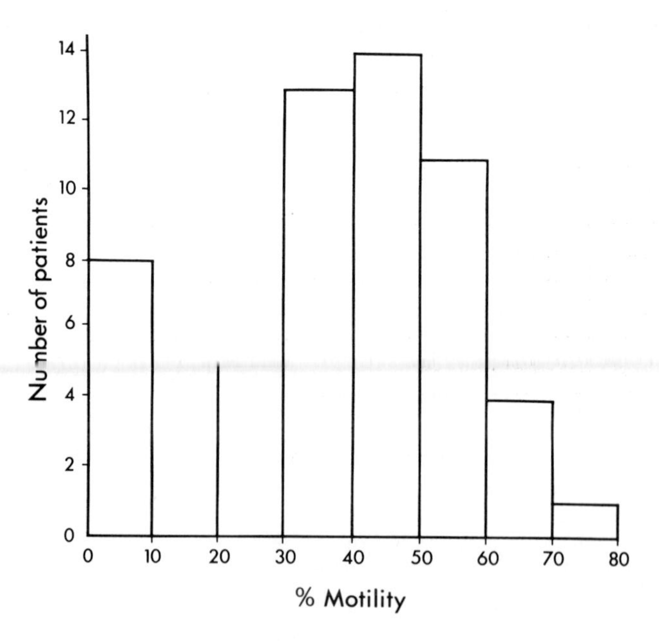

FIGURE 2

TABLE 1

PATIENTS STUDIED ON RELAPSE

STAGE	SPERM COUNT	CHILDREN	TIME FROM R/T	ABNORMAL	MOTILITY
IV_oL_2	49.5	0	2/12	+	5%
IV_cL_1	42	0	6½ yrs	-	30%
III_{N2}	11.2	0	2⅓ yrs	-	50%
IIa	10	(1 died)	10/12 post chemo 1 yr	-	40%
Marker only	<3	1	6¼ yrs	-	-
IVL_1	Azospermic	0	1 yr	-	60%
III_{N2}	Azospermic	2	4/12	-	-
IV_cL_1	Azospermic	0	4/12	-	-
IV_cL_1	Azospermic	0	2/12	-	-

observation may be a function of the time at which the analysis was made because recovery may be delayed for up to two years. No relationship between impairment of sperm counts and histology of the tumour has been found. Neither elevation of HCG or AFP appears to influence fertility, nor does stage of disease, although patients with very advanced disease were not offered sperm testing in view of the urgency of instituting treatment.

No obviously significant hormonal abnormalities were observed in patients with normal sperm counts but in those with azo-ospermia 67% and 62% respectively have raised FSH and LH and 50% have low testosterone levels.

There have been 13 pregnancies in the partners of patients treated with chemotherapy. Two of these men were known to have been azo-ospermic before treatment but in the one patient tested at the time of his wife's pregnancy, the count had returned to normal. Eight pregnancies ended in the delivery of a normal child and in four cases sperm counts were known to be normal.

There were three early miscarriages (at < 3 months). There was one stillbirth in a child born 11 weeks prematurely and one child had minor congenital abnormalities. The significance of these problems cannot be evaluated in such a small group of patients and further studies are needed.

It appears that there may be an inherent defect of sperm production in some patients with testicular tumours but in others, changes seem to be related to orchidectomy either because of stress, general ill health or anaesthesia. Further work is necessary to elucidate the reasons for infertility in association with testicular tumours but it appears from this preliminary information that radiotherapy and chemotherapy did not inevitably lead to sterility and if there is impairment it may be reversible. If these findings are confirmed, there are important implications when considering new drug combinations in the treatment of seminoma and teratoma as the use of, for example, alkylating agents might mean the loss of this advantage.

GERM CELL FUNCTION IN PATIENTS WITH TESTICULAR TUMOURS

R.T.D. Oliver
Medical Oncologist,
Institute of Urology,
Shaftesbury Hospital,
London WC2H 8JE

A review of the question of fertility of patients with germ cell tumours is presented on the basis of an analysis I made of the patients whose clinical follow up Dr. Hope-Stone has recently reviewed. In this series of patients (238 overall) who were treated primarily by post-orchidectomy radiotherapy, there was a large group of 149 about whom no information on fertility was available, due in the majority of instances to the fact that they were unmarried and their fertility had not been tested, though there was a minority of patients whose notes contained no information as to whether or not they had fathered children. There were 12 patients who had no children despite at least five years of marriage. No sperm count was performed on these patients though some of them commented on their lack of fertility but had gone no further in investigating it. There were three unequivocal azo-ospermias and 74 had proven fertility in terms of siring children, although one did in addition have a normal sperm count and not had a child.

Following treatment there is the same proportional distribution i.e. 14 proven fertility, 5 azo-ospermia, 4 no children despite at least 5 years of marriage. There were many more in the 'no information' category mainly because they already had children. This clear cut incidence of azo-ospermia is well in excess of that expected in a normal population and confirms the more substantial data of Dr. Barrett.

I should also like to present a brief review of more limited information from my own experience of treating patients with Bleomycin, Vinblastine and Cis-Platinum chemotherapy. Three of eleven patients prior to treatment had azo-ospermia and there was no relation between sperm count and serum HCG level as reflected by the observation that the patient with the highest HCG level (4,600) had a low normal count while the three aso-ospermic patients had no detectable HCG. Follow up on these patients is limited, but within 6^5 months of treatment all had counts below 1 x 10 per ejaculate. By 18 months following completion of treatment 4 of 5 patients had more than 10×10^6 sperm per ejaculate. The only patient without any sperm had bilateral atrophic testes and spontaneously regressed primary tumour at presentation, despite previously having conceived 4 children prior to becoming ill.

From these observations it is clear that:

1. Patients with testicular tumours have an unequivocal but low incidence of azo-ospermia in the region of 8 - 20% which is substantially in excess of that in the normal population.

2. Recovery of sperm count does occur after Bleomycin, Vincristine and Cis-Platinum, though insufficient cases have been followed to establish that this recovery is in the same range as that in patients who have received radiotherapy to the para-aortic lymph nodes.

FINAL DISCUSSION SESSION

POINTS FROM FINAL DISCUSSION SESSION

Embryogenesis and Histopathology

Dr. K. Anderson

These sections I found particularly interesting and informative. Drs. Graham and Evans described the various modes of origin of murine teratomas and counselled against a too-close analogy between teratogenesis in the animal model and human pathology. The so-called embryonal carcinoma cell in murine tumours is readily recognised by such workers. Histopathologists, however, are not able to identify comparable cells in the human. The embryonal carcinoma cell is true "stem cell" capable of endless, monotonous reduplication or of differentiation to the highest level. This is a very interesting concept.

In the field of human histopathology the chief interest lay, as I had expected, in refining a histopathological classification of germ cell tumours which would positively identify the various constituent elements in the growth and apportion to these a prognostic value which would be of real clinical significance. Considerable advances have been made here, but many problems remain unanswered and some appear, at the moment, insoluble.

Apart from a commencing search for new and potentially useful immunocytochemical markers within the tumours, the principal interest has centred first on the positive identification of yolk-sac tumour. This has to be a positive diagnosis and not one by exclusion. Secondly, there is the increasing recognition of the importance of a certain type of giant cell, which is neither an effete tumour cell nor a macrophage, but an important component of the tumour.

Yolk-sac tumours are composed of distinctive cells, although these may appear in polymorphous forms (Figures 1, 2 and 3). Such tissue stains capriciously for AFP but this staining correlates well with the level of this marker in the blood. The inclusion of such tissue within a germ cell tumour may decide prognosis and survival.

Giant cells of trophoblastic type, but without the formation of villi, occur in some germ cell tumours (Figure 4) and occasionally in seminomas. This last finding is very important because it may be a link between embryonic and extraembryonic structures and the fixed-cell type testicular tumour.

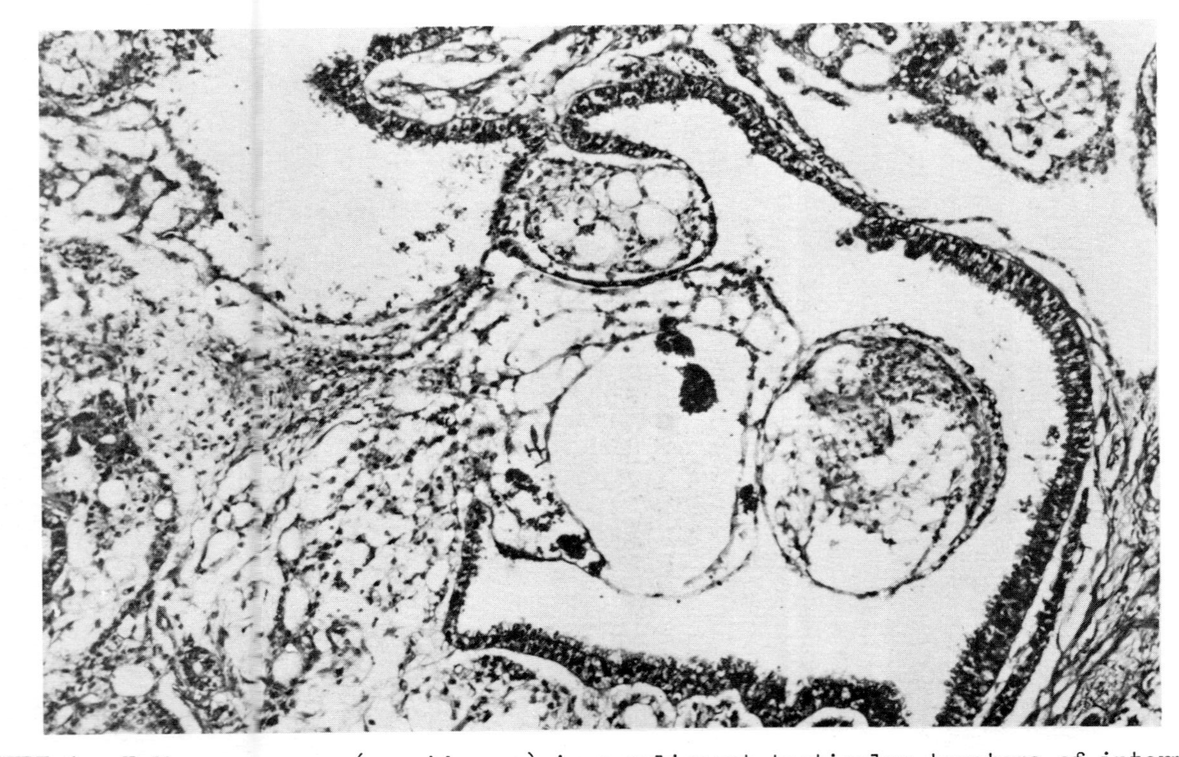

FIGURE 1 - Yolk sac tumour (mucoid area) in a malignant testicular teratoma of intermediate type (MTI) (Embryonal carcinoma with yolk sac tumour) PAS x 105.

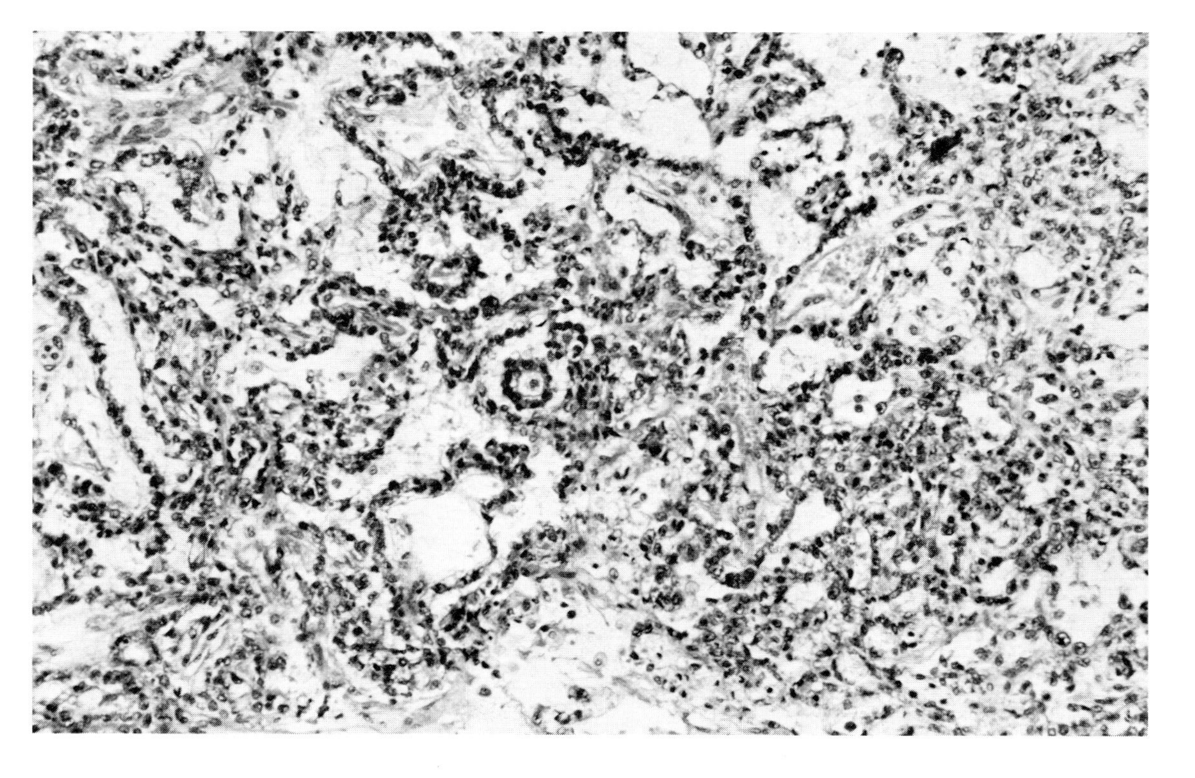

FIGURE 2 - Yolk sac tumour (microcystic pattern) in a malignant testicular teratoma of intermediate type (MTI) (Embryonal carcinoma with yolk sac tumour) H and E x 150.

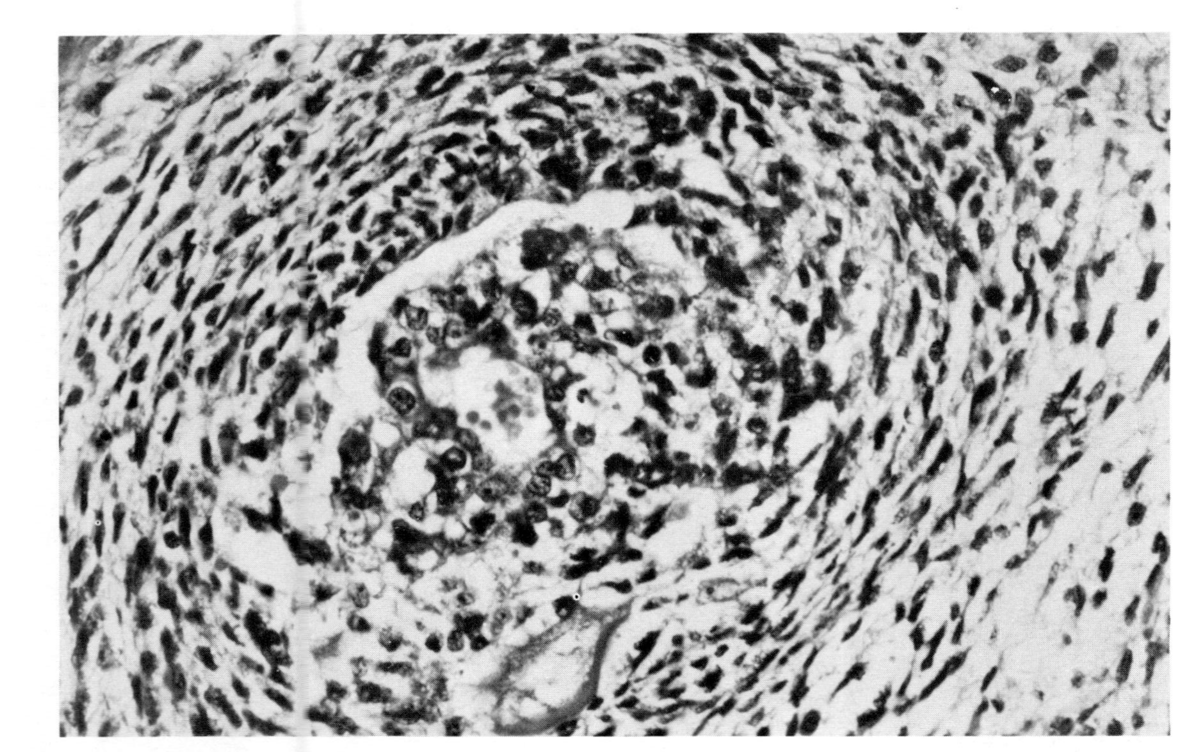

FIGURE 3 - Duval-Schiller body from an area of yolk sac tumour in a malignant testicular teratoma of intermediate type (MTI) (Embryonal carcinoma with yolk sac tumour) H and E x 390

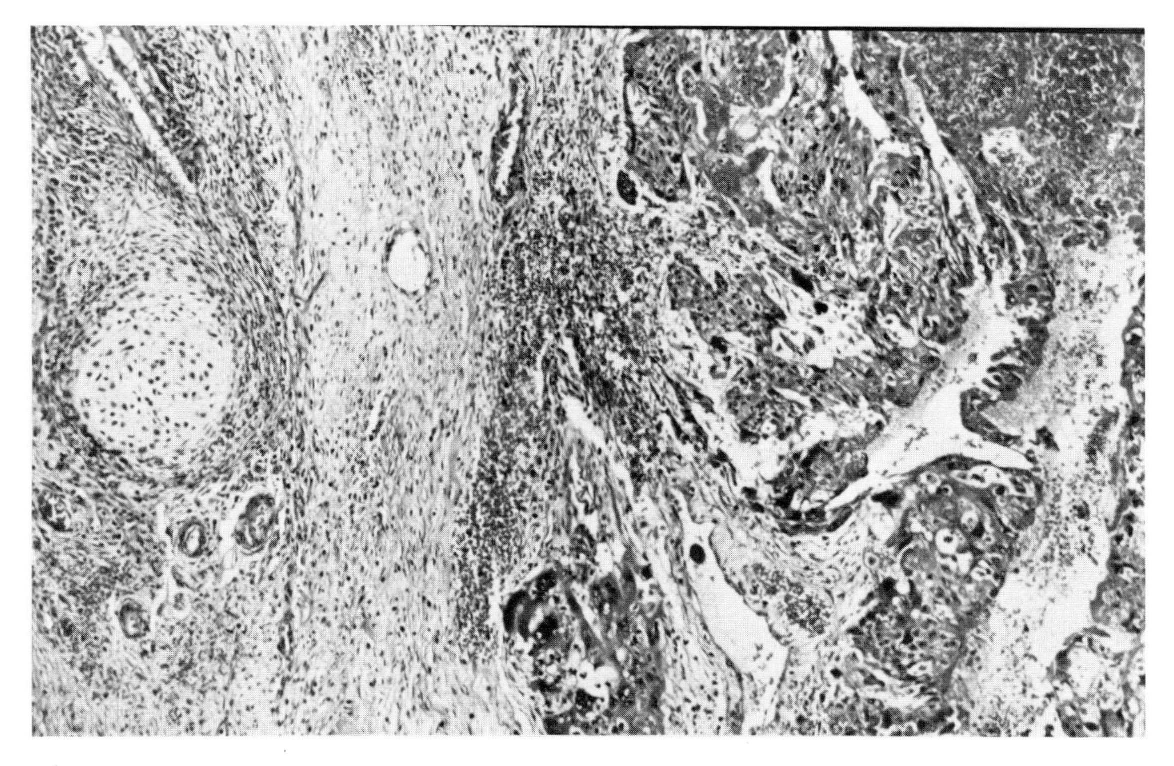

FIGURE 4 - Trophoblastic elements in a malignant testicular teratoma of intermediate type (MTI) (Embryonal carcinoma with trophoblastic elements) H and E x 105

These giant cells often stain for HGG and one is tempted to regard them as being a form of syncytiotrophoblast. A certain validity is given to this concept when we consider that some of these tumours are capable of metabolising dehydroandrosterone sulphate, which is adrenal derived and turning it into oestrogen. These tumour giant cells, therefore, are truly behaving like first trimester trophoblast. The significance of such giant cells may be considerable in terms of ontogeny, but when it comes to prognosis the position is much less clear and their presence may not worsen the outlook beyond that already dictated by the tumour type and its stage.

Now I should like to ask two questions. First of all from Dr. Martin Evans as to whether he thinks embryonal carcinoma cell systems have any significance in human pathology and it they have, what sort of significance is it? Then to ask Dr. Grigor and Dr. Parkinson if they had any further thoughts on the classification of germ cell tumours?

Dr. Evans

I obviously think that we are in the same ball game with mouse teratocarcinoma and the human, or at least I think it is a Ocham's razor approach. I would think that in there somewhere are the potential stem cells which are giving rise both to the yolk-sac and to the giant cells and to the other differentiated elements that you can recognise. But I think we academics can learn that you have a strictly practical job to do. I do think we can tell you that we do not think that this is something you are going to sort out merely by looking at the histology of tumours. I think it can only be sorted out experimentally, and, therefore, I think there are really only two ways it can be done. The most rigorous would be to clone these cells either in vivo in xenografts, as Professor Peckham has suggested, or by cloning in vitro and possibly putting back into an animal model system so that you can look at the histology, specific cell type markers and possibly by looking rather more widely for antisera which we know are specific in the mouse system for the stem cells, which we shall continue to call embryonal carcinoma cells I'm afraid. I think there are possibly a number of antisera that have not been looked at yet, and possibly rather closer collaboration is needed between the histopathologists, who are used to looking at the human tumours, and people like myself, who have expended a lot of ill-spent time looking at the mouse tumours. One might possibly be able to elucidate on these cells purely on morphological and staining grounds, but I would not be so hopeful about that.

Dr. Grigor

Thank you for giving me this opportunity to express my views on the biology and classification of testicular tumours.

I agree with the widely accepted concept that most testicular tumours arise from germ cells. However, the pathways of subsequent differentiation are less well defined. Many classifications follow the Pierce and Abell (Pathology Annual, 5, 27, 1970) theory that malignant germ cells may develop into either seminomas or embryonal carcinomas, and the latter may remain undifferentiated or may differentiate along embryonic or extraembryonic lines. Professor Fox pointed out that "embryonal carcinoma" is a bad term because we have no real evidence that it is either embryonal or carcinoma. In addition, I find it difficult to accept that all teratomas and extraembryonic tumours were originally embryonal carcinomas.

A practical disadvantage of maintaining the term "embryonal carcinoma" is that it necessitates the use of the phrase "non-seminomatous germ cell tumour" when referring to the group of neoplasms incorporating the embryonal carcinoma and the other tumours purported to be derived from it. This long-winded and cumbersome terminology is a negative approach to a well-defined group of tumours which should be grouped together in a positive way.

At this stage I would like to postulate a possible alternative histogenesis of some germ cell tumours. Many seminomas contain giant cells with the morphological and immunocytochemical staining properties of syncytiotrophoblast. Such cells may also resemble morulae, as demonstrated by Dr. Heyderman, or even blastocysts, and both the morula and preimplantation blastocysts are known to produce gonadotrophin (Dickmann et al., Vit. Horm. 34, 215, 1976). Therefore the seminoma giant cells may be analogous to an early stage of embryonic development destined to become an embryoid body from which embryonic and extraembryonic structures may develop. This could explain why apparently pure seminomas have been known to metastases as non-seminomatous tumours, and would indicate that the flow diagram of Pierce and Abell is inappropriate.

It would therefore appear that the classification of Pugh (Pathology of the Testis, Blackwell, 1976) is more applicable for common usage, non-seminomatous germ cell tumours being positively grouped together under the heading of "teratoma" which may show embryonic or extraembryonic differentiation. The embryonic tumours may be graded as differentiated (teratoma differentiated : TD), intermediate (malignant teratoma intermediate : MTI), or undifferentiated (malignant teratoma undifferentiated : MTU), and the extraembryonic elements as either yolk-sac or trophoblastic tumours. This classification is simple to use, is readily understood by pathologists and clinicians, and is the classification giving the best guide to prognosis. It is often criticised for not including all the component parts of a tumour, but identification and quantification of individual componenets are greatly influenced by sampling error and probably have no significant prognostic relevance. Moreover, a pathology report should include a full description of the different elements identified before the tumour is classified into a specific category.

The main defects of the Pugh classification are in the recognition of extra-embryonic differentiation and the failure to incorporate functional studies. It is primarily a prognostic classification and any proposed modifications have not yet been shown to be of significant prognostic relevance, therefore any addition should be incorporated into the classification as it stands without changing its basic structure.

Pugh's strict criteria for diagnosing trophoblastic differentiation (malignant teratoma trophoblastic : MTT) have prognostic significance but not necessarily functional or histogenetic relevance. A tumour failing to meet the diagnostic requirements for MTT may nevertheless show areas of definite chorionic differentiation which are often the basis of raised serum gonadotrophin.

The Pugh classification totally underestimates the frequency of yolk-sac differentiation in testicular tumours although he now admits that yolk-sac elements may occur in adults (Pugh, Int. J. Andrology Suppl. 4, 50, 1981). It is in the field of yolk-sac tumour (YST) that we must acknowledge the significant contribution of the late Professor Teilum to the classification of germ cell tumours. He gives a very full account of the various histological patterns identifiable as yolk-sac (endodermal sinus) tumour and his diagnostic criteria should be followed very closely. YST must not be used as a terminological "dumping ground" for undifferentiated areas, and if Teilum's criteria are followed, about two thirds of testicular tumours from patients of all ages will be seen to contain YST foci

(Grigor, Int. J. Andrology Suppl. 4, 35, 1981).

The prognosis of YST has not been fully documented although preliminary data suggest it is similar to MTU (Grigor, 1981). It would be interesting and rewarding for Mr. Corbett to review the histology of his series of 217 teratomas to examine the prognostic significance of YST.

Increasingly more workers are examining the functional aspects of testicular tumours by immunocytochemistry which has played a major role in the understanding of the biology of germ cell neoplasms. There is strong, but not absolute, correlation between tissue localisation of tumour antigens and serum markers, however, the clinical relevance of immunocytochemistry remains to be proven. All testicular tumour patients should have serum levels of AFP and HCG monitored regardless of marker localisation within tumours and in the search for new tumour markers, serum assays are probably more applicable than immunoperoxidase.

From a practical point of view Dr. Parkinson pointed out that it is not feasible for every laboratory to examine all testicular tumours by immunoperoxidase for a wide range of marker substances which have doubtful prognostic significance. Therefore, the inclusion of immunocytochemical results in the basic classification should not be advocated. Moreover, localisation of tumour markers should not be considered diagnostic of specific tumour types, in particular AFP localisation does not necessarily signify YST, as demonstrated by Dr. Jacobsen in her immunoperoxidase studies and by Dr. Buamah who described AFP production by ovarian cystadenocarcinomas.

In the discussion of the biology and classification of germ cell tumours, consideration of animal tumour models should be included. Dr. Graham and Dr. Evans described the origin of murine teratomas and the nature of the tumour cells. Although there are many similarities between human and mouse teratomas, there are many points of divergence including differences in cell surface markers as described by Dr. McIlhinney.

Our experimental colleagues have clearly defined the embryonal carcinoma cell which is recognisable in murine tumours and is the stem cell for further differentiation: such a cell is not recognised in human tumours. They also mention a "teratocarcinoma cell" but I find it difficult to believe that a single cell can demonstrate the features of both differentiation and non differentiation! This apart, it would appear that the terms "embryonal carcinoma", "teratocarcinoma" and "teratoma" are suitable for the classification of murine germ cell tumours. These terms are less applicable to human tumours and therefore the classification proposed by Pugh seems more sensible for clinical practice.

Dr. Parkinson

I think it has really all been put very well indeed. I was talking to Dr. Heyderman about classification, and we were trying to create various artistic ways of getting away from a tree shaped diagram that ended up with 4 or 5 rigid groups. I would rather see it in some sort of circle, because, as I said at the beginning of my talk, I think in any classification we are trying to convert testicular tumours which, to me are a circle, into a hectagon or a pentagon, and we are always going to get difficult grey areas like the finding of giant cells in seminomas. Such a tumour to me is something like the lamprey; the link between the vertebrates and the invertebrates, and that is why I think such rare tumours are so important. I do not want to be negative about it but you are never going to get a perfect classification. I think the other point is that while I am happy to accept one classification, and I would accept Dr. Pugh's with its prognostic significance and with its deficiences, we should look at and compare others. It we are going to look at large numbers of tumours in something like the MRC trial, we should classify them in different ways, and liaise far more closely with experimentalists. I think we are coming pretty near the end of the road as to how far we can go on morphology. There has been tremendous work done which we acknowledge, but I think we must look outside to experimentalists, to biochemists, to immunocytochemists, and I think these should be active members on any panels of pathologists that are constructed in the future.

Conceptual basis of therapy

Dr. Jones

This session was interesting because the two speakers approached the problem from different angles, Professor Peckham from the clinical and para-clinical situation and

Professor van Putten from the theoretical point of view. They came to remarkably similar conclusions, that there are a large number of treatment options, that we have to be aware of the toxicities of the therapies and apply the treatment correctly for different clinical situations.

One point that has been repeatedly stressed throughout this conference is that volume of disease is more important as a prognostic factor than anatomical staging.

Professor van Putten made the very valid point that we need to know a lot more about the pharmacokinetics of cytotoxic drugs.

Epidemiology

Dr. Jones

I feel the presentation give by Mr Corbett will prove to be of historical value, because the survey indicates the natural history of disease to some extent since the patients were subjected to little more than orchidectomy and perhaps radiotherapy.

Dr. Waterhouse's presentation contains a wealth of information on epidemiology of ovarian as well as testicular germ cell tumours, mainly in graphical form.

Animal Tumour Models, Kinetics etc.

Mr. Corbett

The papers by Dr. Brigid Hogan and Miss Robertson, both on mouse teratocarcinomas in culture, emphasised that different types of cells are found in these tumours - something we often do not take sufficient cognisance of clinically, I suspect. We tend to think in terms of single cell types responding in a particular way to chemotherapy. The presence of both pluripotential and nullipotential cells was stressed and the completely different responses these had to different agents.

From the histological point of view I think it will become important for the pathologist to attempt to quantify the amounts of the different tissues involved.

Dr. Silvestrini presented results of ^{3}H-thymidine labelling and illustrated a cut-off between Stage I and

Stage IIa and Stage IIb and beyond in terms of labelling characteristics.

Dr. Ash presented data on regression and re-growth of human tumours in patients, and showed that the regression lines were the same for the same tumour and independent of the agent(s) used.

Germ Cell Tumours in Extragonadal sites

Dr. Anderson

The main theme that did come out very clearly was the extraordinary difficulty in diagnosis. These tumours do strikingly badly, possibly because they are rapidly growing, or because they reach a large size before clinical presentation. One needs a high index of suspicion and here serum markers may be of great help.

Accurate diagnosis and staging are imperative and some patients seem to die because they have received inappropriate therapy, or too little therapy too late or the treatment options may have been compromised by the time a final diagnosis is made. The worse thing that can happen is for the patient to have a "therapeutic trial" of some regime or be treated as though for Hodgkin's disease or lymphoma.

Tumour Markers

Dr. Jones

This field is relatively new and yet such a tremendous amount of work has been done in a short period of time, a very exciting era. Markers are of tremendous usefulness to the clinician, but since not all germ cell tumours produce markers the search for new substances goes on.

The techniques of radioimmuno-localisation described by Drs. Begent and Bradwell are really exciting, and I feel that nuclear medicine will take a large step forward in this direction in the next few years.

Clinical Trials

Mr. Richards

There is a need for rather more controlled, careful, studies and trials in order that further improvements can take place and be proved to have occurred, since the results of treatment are now so much improved. A very valid aim, also is to reduce toxicity by modifying approaches or therapies.

Clinical Sessions

Professor Blandy

The problem of carcinoma in situ of the testis presented by Dr. von Eyben is very worrying. However, I think the general consensus is that these patients do not need immediate orchidectomy, but they do need to be followed carefully in case the disease evolves.

The next point was whether radiotherapy was necessary in the early stage cases of testicular teratoma or not. Many of us will now be more willing to participate in properly supervised clinical trials in which there is a no treatment arm in this situation. However, proper follow up is essential, according to a strict regime, within centres where there are facilities for promptly available radiological surveillance, rapid feed back of markers etc. It is most important that the message coming from this conference is not misinterpreted as these patients do not need radiotherapy, because that is not what was stated.

One very important point for those who run a clinical practice is that a good marker service is essential, and the clinicians in Leeds are very fortunate if I may say so.

A further point about radiotherapy was that there is a tendency for this to adversely influence the chances of successful chemotherapy at a later date. Certainly the message came over very strongly that these patients should not be irradiated above the diaphragm.

We heard some interesting and important data about primary debulking surgery from Dr. Ozols, on behalf of Dr. Javadpour, which showed that this type of surgery had no influence on the outcome of the disease and therefore should not be performed.

We have heard evidence of return of fertility in these patients after therapy. It is also important to remember that a proportion of infertile males develop these sort of tumours, and we should be teaching this to medical students and to the doctors who run infertility clinics.

Finally, the important message from this conference is that today we can salvage the lives of the majority of these patients, who a few years ago would have been condemned to death, and a miserable death at that. However, if only men would notice these small lumps in their testes early and seek medical advice, if only all doctors learnt one cardinal lesson that there is no such thing as a benign lump in the human testis, and if only the surgeon would remove the lump through a high inguinal incision having first taken some serum for marker estimations, then perhaps the numbers of patients with these frightful metastases would decrease and we could achieve the same results with lesser treatment.

INDEX

INDEX